Numerical Methods for Engineering and Data Science

Numerical Methods for Engineering and Data Science guides students in implementing numerical methods in engineering and in assessing their limitations and accuracy, particularly using algorithms from the field of machine learning.

The textbook presents key principles building upon the fundamentals of engineering mathematics. It explores classical techniques for solving linear and nonlinear equations, computing definite integrals and differential equations. Emphasis is placed on the theoretical underpinnings, with an in-depth discussion of the sources of errors, and in the practical implementation of these using Octave. Each chapter is supplemented with examples and exercises designed to reinforce the concepts and encourage hands-on practice. The second half of the book transitions into the realm of machine learning. The authors introduce basic concepts and algorithms, such as linear regression and classification. As in the first part of this book, a special focus is on the solid understanding of errors and practical implementation of the algorithms. In particular, the concepts of bias, variance, and noise are discussed in detail and illustrated with numerous examples.

This book will be of interest to students in all areas of engineering, alongside mathematicians and scientists in industry looking to improve their knowledge of this important field.

Numerical Methods for Engineering and Data Science

Rolf Wuthrich and Carole El Ayoubi

CRC Press
Taylor & Francis Group
Boca Raton London New York

CRC Press is an imprint of the
Taylor & Francis Group, an **informa** business

First edition published 2025
by CRC Press
2385 NW Executive Center Drive, Suite 320, Boca Raton FL 33431

and by CRC Press
4 Park Square, Milton Park, Abingdon, Oxon, OX14 4RN

CRC Press is an imprint of Taylor & Francis Group, LLC

ISBN: 9781032200699 (hbk)
ISBN: 9781032200750 (pbk)
ISBN: 9781003262121 (ebk)

DOI: 10.1201/9781003262121

Typeset in Times
by Apex CoVantage, LLC

Contents

Preface xiii
Author Biographies xvii

Chapter 1 Introduction 1

1.1 Aim of Numerical Methods 1
1.2 Measuring Errors 4
1.3 Some Important Definitions Related to Numerical Methods 6
1.4 Some Important Theorems from Calculus 9
1.5 Exercises 10
1.5.1 Aim of Numerical Methods 10
1.5.2 Measuring Errors 10
Notes 11

PART I Numerical Methods for Engineering Applications 13

Chapter 2 Numerical Errors 15

2.1 Round-Off Errors 15
2.1.1 Propagation of Round-Off Errors 18
2.1.2 Machine Epsilon and Undistinguishable Numbers 20
2.2 Truncation Errors 22
2.2.1 The Truncation Error of Taylor Series 23
2.2.2 Behavior of Truncation Errors 28
2.3 Error Propagation 32
2.3.1 Propagation of an Input Error 32
2.3.2 Estimating Round-Off Errors Introduced by a Function Evaluation 37
2.4 Round-Off and Truncation Error Trade-Off 42
2.5 Exercises 46
2.5.1 Round-Off Errors 46
2.5.2 Truncation Errors 47
2.5.3 Propagation of Errors 48
2.5.4 Round-Off and Truncation Error Trade-Off 48
Notes 49

Chapter 3 Solving Algebraic Equations 50

3.1 Introduction 50
3.2. Bracketing Methods 52
3.2.1 Bisection Algorithm 55
3.2.2 False Position Algorithm 58
3.3 Open Methods 62
3.3.1 Fixed-Point Iteration Method 63
3.3.2 Newton-Raphson Method 69
3.4 Convergence Order 74
3.4.1 Convergence of the Bisection Algorithm 76
3.4.2 Convergence of the Fixed-Point Iteration Method 77
3.4.3 Convergence of the Newton-Raphson Method 80
3.4.4 Estimating the Error of Generated Approximations Based on the Knowledge of the Convergence Order 84
3.5 Error Analysis 90
3.5.1 Backward and Forward Errors 90
3.5.2 General Error Estimation Formula 94
3.6 Lab: Solving Non-Linear Equations 96
3.6.1 Roots of Legendre Polynomials 96
3.6.2 A Challenging Problem 99
3.7 Exercises 101
3.7.1 Introduction 101
3.7.2 Bracketing Methods 102
3.7.3 Open Methods 102
3.7.4 Convergence Order 103
3.7.5 Error Analysis 105
Notes 106

Chapter 4 Systems of Linear Equations 107

4.1 Introduction 107
4.2 Gauss Elimination Algorithms 108
4.2.1 Naïve Gauss Elimination 109
4.2.2 Gauss Elimination Algorithm With Partial Pivoting 115
4.3 Matrix Decompositions 119
4.3.1 LU Decomposition of a Matrix 119
4.3.2 Solving a Linear System of Equations Using LU Decomposition 123
4.3.3 PA=LU Decomposition of a Matrix 125
4.3.4 Solving Systems of Linear Equations Using PA=LU Decomposition 127

4.4 Conditioning ... 129
4.4.1 Measuring Errors for Systems of Equations ... 129
4.4.2 A System of Linear Equations With High Error Magnification Factor ... 135
4.4.3 Error Magnification Factor of a Linear System of Equations ... 136
4.4.4 Conditioning of Linear Systems of Equations ... 141
4.5 Lab: Solving Systems of Linear Equations ... 146
4.5.1 Solving a System of Linear Equations ... 146
4.5.2 Vandermond Matrix ... 147
4.6 Exercises ... 149
4.6.1 Introduction ... 149
4.6.2 Gauss Elimination Algorithms ... 149
4.6.3 Matrix Decompositions ... 150
4.6.4 Conditioning ... 150
Notes ... 151

Chapter 5 Orthogonality ... 153

5.1 Inner Spaces ... 153
5.1.1 Notations ... 153
5.1.2 Norms ... 156
5.1.3 Inner Products ... 157
5.1.4 Generalized Angle and Orthogonality ... 160
5.2 Orthogonal Projections ... 162
5.2.1 Orthogonal Projects in a General Inner Space ... 163
5.2.2 Orthogonal Projection Matrix ... 164
5.3 Gram-Schmidt Orthogonalization ... 170
5.4 QR Decomposition ... 178
5.5 Exercises ... 182
5.5.1 Inner Spaces ... 182
5.5.2 Orthogonal Projections ... 183
5.5.3 Gram-Schmidt Orthogonalization ... 184
5.5.4 QR Decomposition ... 184
Notes ... 185

Chapter 6 Linear Least Square Regression ... 186

6.1 Regression Problems ... 186
6.2 General Solution of Linear Least Square Regression ... 190
6.2.1 Definitions ... 190
6.2.2 Normal Equations ... 191
6.2.3 Least Square Regression With the Normal Equations ... 195
6.2.4 QR Decomposition for Solving the Normal Equations ... 198

6.3 Linearization of Models ... 200
6.3.1 Exponential Model ... 200
6.3.2 Power Law ... 204
6.4 Abstract Normal Equations ... 207
6.4.1 Normal Equations in an Abstract Vector Space ... 207
6.4.2 Linear Regression by Orthogonalization ... 210
6.5 Approximation of Functions ... 212
6.5.1 Polynomial Function Approximation ... 212
6.5.2 Fourier Polynomials Function Approximation ... 214
6.6 Lab: Modeling Weather Data ... 217
6.7 Exercises ... 219
6.7.1 Regression Problems ... 219
6.7.2 Normal Equations ... 220
6.7.3 Linearization of Models ... 221
6.7.4 Approximation of Functions ... 221
Notes ... 222

Chapter 7 Polynomial Interpolation ... 223

7.1 Definitions ... 223
7.2 Vandermond Interpolation ... 226
7.3 Newton Interpolation ... 229
7.4 Lagrange Interpolation ... 232
7.5 Interpolating Functions ... 235
7.6 Lab: Interpolating Surface Below the Standard Normal Distribution ... 239
7.7 Exercises ... 240
7.7.1 Definitions ... 240
7.7.2 Vandermond Interpolation ... 241
7.7.3 Newton Interpolation ... 241
7.7.4 Lagrange Interpolation ... 242
7.7.5 Interpolation Error ... 242
Notes ... 242

Chapter 8 Numerical Integration ... 243

8.1 Numerical Integration Based on Polynomial Interpolation ... 243
8.2 Numerical Integration Based on Lagrange Interpolation ... 247
8.3 Newton-Cotes Quadrature ... 249
8.3.1 Newton-Cotes Formulas ... 249
8.3.2 Truncation Errors of Newton-Cotes Formulas ... 252
8.3.3 Composite Newton-Cotes Formulas ... 255

8.3.4 Truncation Errors of Composite Newton-Cotes Formulas 259
8.3.5 Estimating Truncation Errors of Composite Quadrature Formulas 265
8.3.6 Truncation Error of the Composite Trapezoidal Method 271
8.4 Romberg Integration 274
8.4.1 Richardson Extrapolation 274
8.4.2 Romberg Quadrature Method 275
8.5 Gauss Quadrature 279
8.5.1 Legendre Polynomials 279
8.5.2 Gauss Quadrature Formulas 281
8.6 Lab: Computing a Numerical Approximation of the Error Function 285
8.7 Exercises 287
8.7.1 Numerical Integration Based on Polynomial Interpolation 287
8.7.2 Numerical Integration Based on Lagrange Interpolation 288
8.7.3 Newton-Cotes Formulas 288
8.7.4 Romberg Integration 289
8.7.5 Gauss Quadrature Formulas 289
Notes 289

Chapter 9 Initial Value Problems 291

9.1 Introduction 291
9.2 Euler's Algorithm 293
9.2.1 Basics of the Algorithm 293
9.2.2 Stability of Euler's Method 296
9.2.3 Convergence of Euler's Method 299
9.3 Analysis of Initial Value Problem Solvers 301
9.3.1 Local and Global Truncation Errors of Numerical IvP Solvers 301
9.3.2 Lipschitz Theorems 303
9.4 Runge-Kutta Methods 309
9.4.1 Second Order Runge-Kutta Methods 310
9.4.2 Fourth Order Runge-Kutta Methods 316
9.4.3 Estimating the Global Truncation Error of Numerical IvP Solvers 319
9.5 Systems and Higher Order Initial Value Problems 323
9.5.1 Systems of First Order Ordinary Differential Equations 323
9.5.2 Higher Order Ordinary Differential Equations 326

9.5.3 Estimating Errors of Numerical IvP Solvers Involving Systems of Equations 331
9.6 Adaptive Methods 335
9.6.1 Introductory Example 336
9.6.2 Adaptive Initial Value Problem Solvers 338
9.7 Exercises 343
9.7.1 Euler's Algorithm 343
9.7.2 Runge-Kutta Methods 344
9.7.3 Systems and Higher Order Initial Value Problems 345
9.7.4 Adaptive Methods 346
Notes 346

PART II Numerical Methods for Data Analysis 347

Chapter 10 Machine Learning 349

10.1 The Data, the Model, and the Performance 350
10.2 How to Learn From Data 352
10.3 Variance-Bias Trade-Off 353
Notes 359

Chapter 11 Regression Models 361

11.1 Features, Labels, and Model Weights 361
11.2 Model Training and Validation 363
11.2.1 Training a Linear Regression Model 364
11.2.2 Making Predictions 365
11.2.3 Validating the Model 366
11.3 Features Transformations 367
11.3.1 Polynomial Transformation 368
11.3.2 Pipelines 371
11.3.3 Custom Transformers 372
11.3.4 Categorical Data 375
11.3.5 Advanced Pipelines 378
11.4 Lab: Working With Real Data 382
11.4.1 Data and Preparation 382
11.4.2 Defining the Labels and Features 384
11.4.3 Model Definition and Training 385
11.4.4 Model Evaluation 386
Notes 387

Chapter 12 Model Selection 388

12.1 Types of Errors 388
12.1.1 Reference Process 388
12.1.2 Generalization Error 389
12.1.3 Training Error 391
12.1.4 Generalization and Training Errors in Function of Model Complexity 392
12.2 Estimating the Generalization Error 394
12.2.1 Training and Test Sets 394
12.2.2 Cross-Validation 397
12.2.3 Regularization 402
Notes 405

Chapter 13 Classification 406

13.1 K-Nearest Neighbors Classification 406
13.2 Classification Performance Measures 409
13.2.1 Confusion Matrix for Binary Classifiers 410
13.2.2 General Confusion Matrix 412
13.3 Logistic Regression 415
13.3.1 Binary Classifier 418
13.3.2 Multiple Classes 420
Notes 422

Chapter 14 Tree-Based Algorithms 423

14.1 Regression Trees 423
14.1.1 CART Algorithm 424
14.1.2 Regression Trees for Several Features 427
14.1.3 Model Selection 428
14.2 Classification Trees 431
14.2.1 CART Algorithm for Classification 431
14.2.2 Hyper-Parameters Fine-Tuning 433
14.2.3 Some Issues with Decision Trees 434
14.3 Random Forests 434
14.3.1 Growing Forests 434
14.3.2 Additional Properties of Random Forests 437
Notes 438

Index 439

Preface

The development of numerical methods to approximate mathematical problems has always represented a central topic in engineering and science. Indeed, most real-world problems cannot be solved analytically, and approximations are required. Numerical methods, the backbone of computational mathematics, provide the essential tools for solving mathematical problems that are otherwise intractable.

More recently, even if not entirely new, problems of a different kind have gained an increased interest. These are problems where, based on some existing data, predictions are made. The term machine learning became popular to describe methods addressing these types of problems. Machine learning, a subset of artificial intelligence, leverages algorithms and statistical models to enable computers to improve their performance on tasks through experience.

This textbook aims to present both domains. A particular focus is on a solid understanding of the concept of error. In both cases, numerical algorithms to approximate problems such as solving ordinary differential equations or computing definite integrals, and machine learning approaches aiming to predict an outcome based on existing data, errors play a key role. Any algorithm will only provide approximations. It is, therefore, essential to estimate the error associated with these approximations to decide if the provided numerical value is acceptable or not for the targeted application.

The authors also provide numerous practical examples that aim to illustrate the abstract concepts that are often introduced. If the reader encounters difficulties understanding the theoretical concepts, we encourage them to study the examples provided first to acquire a hands-on perspective of the topics discussed.

The first part of the textbook covers the fundamentals of numerical methods. We explore classical techniques for solving linear and nonlinear equations, computing definite integrals, and differential equations. Emphasis is placed both on the theoretical underpinnings, with in-depth discussion of the sources of errors, and in the practical implementation of these using Octave, an open-source software specialized for numerical calculations. Each chapter is supplemented with examples and exercises designed to reinforce the concepts and encourage hands-on practice.

The second part transitions into the realm of machine learning. We start with basic concepts and algorithms, such as linear regression and classification. As we progress, we introduce some other algorithms. As in the first part of this book, a special focus is on the solid understanding of errors and practical implementation of the algorithms. In particular, the concepts of bias, variance, and noise are discussed in detail and illustrated with numerous examples.

The aim, in both parts, is not to present a comprehensive list of algorithms but rather to present the general concepts valid for a large family of algorithms and methods. For example, in the first part of the book, the reader will learn methods that, independently of the used particular algorithm, are able to estimate the error of the computed numerical approximation. This should enable the reader to continue the studies of more specialized algorithms on their own.

We have also included numerous programming examples and code snippets to facilitate learning and experimentation. These examples are primarily presented in Octave for the first part of the book and in Python for the second part. Readers are encouraged to actively engage with the code, modify it, and observe the outcomes to gain a deeper understanding of the material.

We hope that this textbook will serve as a valuable resource for anyone interested in numerical methods and machine learning. Whether you are a student embarking on your academic journey, a researcher seeking to deepen your knowledge, or a professional looking to enhance your skills, we believe that the material presented here will provide you with the tools and inspiration to succeed.

USED SOFTWARE

Nowadays, a very rich collection of software and libraries is available to solve numerical problems. In the present book, we use Octave for the first part, and the Python library SciKit Learn in the second part. Software evolves fast, and so do the steps to follow to set them up. Instead of providing a step-by-step explanation on how to set up the codes, which will most likely be outdated quickly, we give a more generic explication as to where to find the needed information.

Octave is particularly well suited for numerical algorithms that are written in matrix forms. The software comes with an integrated development environment from which calculations can be performed either interactively or in the form of scripts. Octave uses its own language, which can be learned in a relatively short time. It can be downloaded from the official website of the Octave project at https://octave.org. Installation instructions for Windows, mac OS, and major Linux distributions are available.

The Python Library SciKit Learn is a popular, very well-documented Python library implementing a large collection of numerical methods from the field of machine learning (https://github.com/scikit-learn/scikit-learn/blob/main/COPYING). Readers not familiar with Python are encouraged to follow some of the excellent online tutorials on Python first. Among many others, a good start is available at www.w3schools.com/python. Contrary to Octave, installing and using Python is less straightforward as it will require the management of libraries. Readers unfamiliar with installing Python libraries are encouraged to use online platforms with pre-installed environments, such as Google Colab (https://colab.research.google.com). The platform can be used for free, and the needed libraries in the present book are all already pre-installed. Videos and tutorials on Colab allow you to get started quickly.

Once more familiar, the reader may wish to install a similar environment on his/her own personal computer. The project Jupyter (https://jupyter.org) is an excellent choice for this option. Even if the process of installing a running Jupyter instance is more complex than that of software such as Octave, the learning experience is excellent. The online documentation on https://docs.jupyter.org explains in detail the process to follow depending on your particular use case and operating system. Project Jupyter recently added as well a platform where it is possible to try out the system without the need to install anything on your computer (https://jupyter.org/try). You can even save your work on your machine.

Author Biographies

Carole El Ayoubi, PhD, an accomplished engineering professional, currently serves as the Director of Education at the Concordia Institute of Aerospace Design and Innovation (CIADI). In this pivotal role, she also spearheads two undergraduate programs, namely Mechanical Engineering and Aerospace Engineering. She is also a senior lecturer in the Department of Mechanical, Industrial, and Aerospace Engineering at Concordia University. Dr. El Ayoubi earned her PhD. in mechanical engineering, specializing in aerospace applications, from Concordia University in 2014. With a rich background as an aerodynamics engineer at Pratt & Whitney Canada prior to her doctoral pursuits, she brings invaluable industry experience to her academic leadership. Dr. El Ayoubi is dedicated to fostering multidisciplinary teaching methods and elevating the standards of aerospace education at Concordia University. Passionate about accessibility in higher education, Dr. El Ayoubi firmly believes in making education available to all. Her enthusiasm extends to impactful outreach activities, showcasing her dedication to inspiring the next generation of aerospace professionals.

Rolf Wuthrich, PhD, is a professor at the Department of Mechanical, Industrial and Aerospace Engineering as well as the Department of Chemical and Material Engineering at Concordia University. He earned his Master of Engineering Physics in high energy physics from the École Polytechnique Fédérale de Technologie de Lausanne (Switzerland) and his PhD in advanced manufacturing and electrochemistry from the same university in 2002. His current research focus is on advanced manufacturing and digital transformation in manufacturing, where he is leading a research laboratory on advanced manufacturing with a special focus on electrochemical technologies meeting the demand of Industry 4.0. He develops strategies involving real time data streaming, real time data processing and machine learning to enhance the performance of manufacturing processes and to reduce machining overhead. His teaching interests include numerical methods, modeling, and fundamental courses in mechanical engineering. He is heavily involved in the development of online teaching strategies.

1 Introduction

The success of engineering and scientific applications hinges, in part, on the capacity to mathematically articulate relevant issues within these fields. Typical mathematical representations include but are not limited to algebraic equations, systems of equations, defined integrals, differential equations, or optimization problems. Mathematics is a means to define the solution of such engineering and science problems precisely. Specifically, it is always possible to determine whether a claimed solution is indeed correct or not.

The challenge arises from the fact that many problems lack a closed-form solution—an explicit formula for computing the solution. However, there are numerous scenarios where such closed-form solutions are unnecessary. A numerical value suffices perfectly well in many cases. Indeed, even if a closed-form solution is available, there's often a practical need for a numerical value. Consider this scenario: a design engineer determines that one side of a rectangle to be cut from a metal sheet must measure $\sqrt{2}$ meters. This may not make much sense to the machinist who has to actually cut the metal sheet. What does $\sqrt{2}$ mean? The machinist needs a finite numerical number in order to proceed with metal cutting. The machinist can obtain this finite numerical number, but this means that the closed-form solution $\sqrt{2}$ has to be evaluated numerically.[1]

Numerical methods were developed to address two scenarios. Firstly, to approximate a numerical solution for problems where a closed-form solution remains elusive, and secondly, to numerically evaluate a known closed-form solution. The latter may appear simpler at first glance. However, as will be explored in Chapter 2, it still poses unexpected challenges. This book presents the fundamentals of numerical methods used in engineering and data science. We will investigate the application of numerical methods to various types of problems that may be encountered in those fields. The first part deals mainly with problems from algebra and calculus. The second part deals mainly with problems from statistics and optimization.

1.1 AIM OF NUMERICAL METHODS

Numerical methods are employed to tackle mathematical problems through numerical computation. While this statement may seem straightforward, it encompasses several nuanced concepts. It's crucial to unpack what it means to "solve numerically a mathematical problem". Essentially, this involves two interconnected ideas. Firstly, it entails seeking a solution to a mathematical problem in the form of a numerical value rather than a closed-form solution. Secondly, it involves employing specific methodologies to obtain this solution—methods that

DOI: 10.1201/9781003262121-1

rely on numerical calculations rather than symbolic ones. These methodologies are collectively referred to as numerical methods. It's important to note that while numerical methods aim to find a particular solution to a mathematical problem, the algorithms utilized often possess general applicability beyond specific instances.

To illustrate this abstract discussion, let us turn to an actual problem. Imagine we would like to solve numerically the following equation:

$$\sin(x) = x \tag{1.1}$$

Our aim is to find a solution $r > 0$. Equation (1.1) has no closed-form solution (except $r = 0$). As such, it is natural to try to find a numerical solution. Let us call this numerical solution x_r. In Chapter 3, several numerical methods that are capable of determining the numerical solution to the present problem will be discussed. At this stage, it is not necessary to specify which numerical method is employed nor how it actually works. Instead, let's concentrate on evaluating the consequences of assigning a numerical value to the solution of this problem.

Suppose that using a numerical method of our choice, we were able to solve Equation (1.1), and obtained the following numerical solution:

$$x_r = 1.8955 \tag{1.2}$$

Our first reflex should be to question the validity of this answer. If x_r is a solution of Equation (1.1), then one should have

$$2\sin(1.8955) = 1.8955 \tag{1.3}$$

A simple verification using a hand calculator gives, however, that

$$2\sin(1.8955) \neq 1.8955 \tag{1.4}$$

We conclude that $x_r = 1.8955$ does not satisfy the equation we want to solve and consequently 1.8955 is not the solution we are looking for. Probably, the algorithm we chose to produce this numerical solution is not a good algorithm, or, maybe, we did not apply it correctly. Let us assume that we attempted again to find the solution, but this time using another algorithm. Using this different algorithm, we obtained the following numerical solution:

$$x_r = 1.895495875 \tag{1.5}$$

Again, plugging back this numerical solution into Equation (1.1) shows that this claimed solution is still not a correct solution.

The reader may rightly argue that this behavior stems from our practice of truncating numerical answers to a finite number of digits. The challenge lies in the fact that many mathematical problems, such as Equation (1.1), do not have solutions that can be expressed by a finite number of digits. Even a simple equation such as $x^2 = 2$ does not have a solution that can be expressed by a finite number of digits (neither the number $\sqrt{2}$ nor the number $-\sqrt{2}$ can be expressed by a finite number of digits). Does this mark the limits of numerical methods? Are they only capable of yielding solutions for problems with finite-digit solutions?

If we wanted to obtain the precise solution to a mathematical problem, then, indeed, the answer would be yes. If numerical methods were strictly employed to find precise solutions, then their utility would be severely limited. Luckily, in many actual situations, we find ourselves willing to work with an approximation to the solution rather than an exact one. Recall the example of the rectangle that needs to be cut out of a metal sheet. In that particular example, we are not so much interested in the exact solution r to the problem but rather in a numerical solution x_r that would be deemed accurate enough for our application. A numerical approximation of $\sqrt{2}$ that is better than the tolerance required by the dimension of the cut will do just fine. This understanding allows us to revisit our definition of a numerical method. A numerical method aims to produce a numerical approximation of the solution of a mathematical problem, and that numerical approximation has to be deemed good enough for the intended application.

This leads to the following question: what does *good enough for the intended application* mean? The answer to that question cannot be provided entirely here. Throughout this text, the reader will progressively learn how "good enough" is measured mathematically. The reader will soon learn that "good enough" is measured via the concept of "norms". What exactly a "norm" is will be elucidated in a later chapter. For now, we can give the following definition of numerical approximation:

Definition 1.1. Numerical Approximation of an Exact Solution

To find a numerical approximation x_r of the exact, or true, solution r of a mathematical problem means to find a number x_r such that $|x_r - r|$ is smaller than a desired accuracy.

Note that the words *true solution* and *exact solution* will be used interchangeably in future.

Based on Definition 1.1, we observe that both approximations (1.2) and (1.5) of Equation (1.1) may be perfectly acceptable numerical solutions. Deciding on which numerical approximation to adopt as the solution to Equation (1.1) depends on how accurate we require this approximation to be.

An important consequence of Definition 1.1 is that two ingredients are required to produce an approximation to an exact solution. The first ingredient is an algorithm capable of producing a number x_r, a potentially acceptable approximation to the exact solution of the problem.[2] The second ingredient is a method allowing

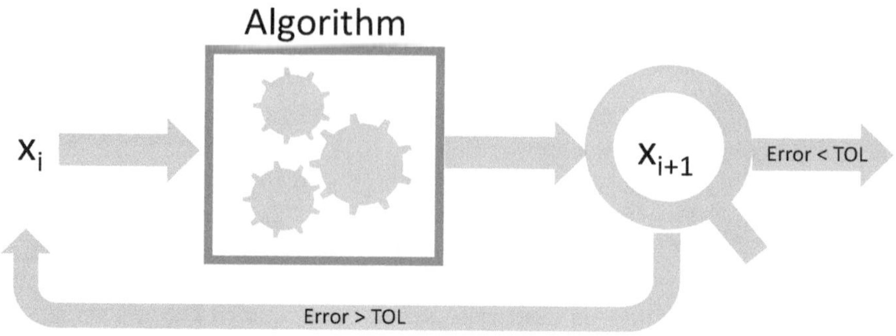

FIGURE 1.1 Schematic workflow to determine a numerical answer to a mathematical problem. An algorithm produces some approximation x_{i+1} based on some seed input x_i The numerical approximation x_{i+1} is analyzed and a decision is made depending if $| x_{i+1} - r |$ is smaller or greater than the desired accuracy TOL.

us to assess whether or not that approximation is acceptable, i.e., whether or not x_r is accurate enough.

Assessing accuracy implies checking if $|x_r - r|$ is smaller than the desired accuracy by the user, designer, engineer, sheet metal cutter, etc. This validation step is not trivial as the true solution r is not known. In general, a particular algorithm provides no indication about the accuracy of an approximation. Therefore, it is of the utmost importance to learn how to check if $|x_r - r|$ is smaller than the desired accuracy.

Figure 1.1 gives a schematic view of the application of Definition 1.1. The chosen algorithm to solve our problem produces some numerical approximation x_{i+1}. If we are able to confirm that $|x_{i+1} - r|$ is less than the desired accuracy, then we accept this approximation. If not, we reject it, and we produce another approximation. Often, but not always, algorithms can use the rejected approximation to produce a better approximation. In that case, we say that the algorithm is iterative. Or, in different words, the numerical algorithm attempts to improve an approximation x_i to produce a better approximation x_{i+1}. In other cases, parameters controlling the behavior of our algorithm have to be changed[3] prior to producing a better approximation.

The definition of a numerical solution, as outlined in Definition 1.1, will remain valid (up to a few minor modifications) in the first part of this book. In part two, which will deal with numerical methods for data analysis, we will have to revisit this definition and adapt it to the context of data science.

1.2 MEASURING ERRORS

Definition 1.1 specifies that a numerical value x_r is considered a numerical solution to a mathematical problem if the number x_r is close enough to the exact solution r of the problem. Close enough means that the quantity $|x_r - r|$ is smaller than the desired accuracy. Another, more generic way to formulate this definition would be to state that the number x_r is a numerical solution to a mathematical problem if its *error* is below the desired accuracy. The definition of the *error* will often depend on the application, i.e., on the nature of the mathematical problem that we are trying to solve and the intended use-case of its solution. However, there are some generic, commonly used definitions of errors. Among them are the absolute error and the relative error.

Definition 1.2. True Errors

The true absolute error of a numerical approximation x_r to a true solution r of a mathematical problem is the quantity

$$E(x_r) = |x_r - r| \tag{1.6}$$

The true relative error of a numerical approximation x_r to a true solution r of a mathematical problem is the quantity

$$e\,(x_r) = \left|\frac{x_r - r}{r}\right| \tag{1.7}$$

Note that sometimes we will use x_r in the denominator of Equation (1.7) instead of r.

For clarity, we added the word "true" in front of the error because we compared our numerical answer with the true mathematical solution. However, it's important to note that in practical applications, true errors are inherently unknown since we lack access to the true value of the solution to a problem. Indeed, true errors are theoretical concepts. In practice, we will have to learn to estimate these true errors. A good estimate of an error is a conservative estimation, i.e., larger than or equal to the true error.

Definition 1.3. Estimated Errors

An estimated absolute error of a numerical approximation x_r of a true solution r to a mathematical problem is a quantity E_a such that

$$E_a \geq E(x_r) = |x_r - r| \tag{1.8}$$

The estimated relative error of a numerical approximation x_r of a true solution r to a mathematical problem is a quantity e_a such that

$$\varepsilon_a \geq \varepsilon(x_r) = \left|\frac{x_r - r}{r}\right| \tag{1.9}$$

Note that sometimes we will use x_r in the denominator of Equation (1.9) instead of r.

Our task as designers, engineers, or data scientists will be to develop methods capable of estimating the errors associated with the numerical solution. The acceptance (or rejection) of a numerical solution produced by an algorithm is based on the assessment of error estimates. If a numerical solution is accepted based on its estimated error, we will be confident that the true error will also be less than the desired precision. Consequently, the numerical solution will be a good enough solution to the mathematical problem in the sense of Definition 1.1.

1.3 SOME IMPORTANT DEFINITIONS RELATED TO NUMERICAL METHODS

In this section, some important concepts and definitions of numerical methods are introduced. We start with the concept of convergence of an algorithm. A numerical algorithm aims to produce an approximation x_r of a true solution r of a mathematical problem. Not all, but several algorithms are actually producing a sequence of approximations $x_o, x_1, \ldots, x_n$. They can be seen as a mathematical series. If this series converges to the actual true solution r, we say that the algorithm is converging.

Definition 1.4. Converging Algorithm

An algorithm producing a series of approximations $x_o, x_1, \ldots, x_n$ of the true solution r of a mathematical problem is said to converge if $|x_n - r| \rightarrow 0$ as $n \rightarrow \infty$.

Let us formulate a few remarks on Definition 1.4. First, we formulated this definition for the case in which the solution to our mathematical problem is a real number. In case the solution is a different mathematical object, for example, a function, as is the case with solving ordinary differential equations, this definition needs to be revisited. In fact, the change to be made to Definition 1.4 will be minor and is essentially of a technical nature. The reader can ignore that problematic for now.

Secondly, even Definition 1.4 looks very similar to the definition of the convergence of a mathematical series, there is a fundamental difference. The series $x_o, x_1, \ldots, x_n$ produced by the algorithm may very well converge, for example

$x_n \to r'$, but r' could be different than r, the solution to the problem we wish to solve. In this case, the algorithm is not considered to be convergent, but we will say the algorithm converges to a wrong value. The wrong meaning here, not to the desired value r. The difference between r and r' is termed the bias.[4] We will often say the algorithm has a biased convergence.

Definition 1.5. Bias of an Algorithm

An algorithm producing a series of approximations $x_o, x_1, \ldots, x_n$ of the true solution r of a mathematical problem is said to have a bias if $|x_n - r'| \to 0$ as $n \to \infty$ and $r' \neq r$. The difference $|r' - r|$ is called the bias of the algorithm.

Practically, it is rather straightforward to realize if an algorithm is converging to a value or not. This can be achieved by plotting the values $x_o, x_1, \ldots, x_i$ versus i. Such a plot is called a convergence plot. It is not straightforward to assess if the value to which the series converges is indeed the true solution r. As such, care will have to be taken to develop bias-free algorithms or to understand under which conditions they are bias-free. The problem of bias will be particularly challenging in the context of data science and is at the heart of the second part of this book.

A further distinction, loosely linked to the concept of bias, is the distinction between accuracy and precision. Consider a scenario where an algorithm is generating some values $x_o, x_1, \ldots, x_n$. For example, it could be a drilling operation, and the x_i could be the diameter of the drilled hole. Mathematically, such situations are described by random variables X. Accuracy refers to how far away, on average, the outcome of the process is from the targeted value. It can be seen as a generalization of the concept of bias defined in Definition 1.5. In our example of the drilling process, how far away is the average drilled diameter from the targeted diameter? Precision refers to the extent of the spread in the outcomes of the process. In our example of the drilling process, it can be measured by the standard deviation of our drilled holes.

Definition 1.6. Accuracy and Precision

Consider a process described by a real-valued random variable X. Let r be the targeted value of the process.

1. The accuracy is the difference between r and the expected value $\mathbb{E}[X]$ of the process:

$$\text{Accuracy} = |r - \mathbb{E}[X]| \tag{1.10}$$

2. The precision is the variance $\mathbb{V}[X]$ of the process:

$$\text{Precision} = \mathbb{V}[X] = \mathbb{E}\left[\left(X - \mathbb{E}[X]\right)^2\right] \tag{1.11}$$

As a practical illustration of Definition 1.6, let us revisit our drilling process. Accuracy, in this context, implies that the average diameter of the drilled holes aligns with the target diameter. It doesn't mandate each individual hole to match a specific diameter precisely. For instance, if the drilling yields holes with varying diameters—some too small, some too large—but overall, the average diameter meets the target, we classify the process as accurate.

Precision refers to the consistency among the diameters of the drilled holes—they are very similar to each other. However, precision does not guarantee that the diameter is correct; it could still be incorrect. For instance, if all drilled holes have very similar diameters, but these diameters are consistently 10% too large, the process is deemed precise, albeit with a bias. In practice, accuracy and precision are both important. Depending on the particular situation, one may be more important than the other or both equally important. If we revisit our example of the drilling process, maybe only precision is important. In this scenario, the emphasis lies in producing identical holes, with less significance placed on the exact diameter of the hole as long as it remains consistent. This is typically the case when you design a drilling machine. It is key the drilling machine always produces the same holes. If the diameter is wrong, you can always correct for this systematic error or bias by tuning the machining parameters. In other applications, the accuracy is the most important. This is typically the case for numerical algorithms. For a numerical algorithm, it is important to be bias-free (i.e., accurate). Precision matters less as you could simply average out the produced results to get the correct value.

While the distinction between accuracy and precision may play a minor role in the initial part of this book, it will gain significant importance in the second part. There, we will discover the existence of a trade-off between accuracy and precision formulated in the bias-variance trade-off theorem.

Algorithms can further be characterized by the underlying mechanism they use to produce an approximation. Some of them are able to use an approximation x_i of the true solution r and, hopefully, improve this approximation to a new approximation x_{i+1}. Such algorithms are called iterative algorithms.

Definition 1.7. Iterative Algorithm

An algorithm that uses an approximation x_i of the true solution r of the problem it aims to solve in order to produce a new approximation x_{i+1} is called an iterative algorithm.

Note that the new approximation x_{i+1} does not necessarily have a higher accuracy than x_i. In fact, this is one of the challenges to be addressed in numerical analysis. As we will learn, even converging algorithms do not guarantee that successive iterations improve the accuracy of the result. It can very well happen that this accuracy decreases. This apparent contradiction is an effect specific to the fact one uses numbers, as opposed to symbols, in the calculations. At this stage, the reader cannot understand the exact reasons behind this effect. Our aim here is to draw the reader's attention to this problematic. The detailed reasons will become clear as different methods and concepts will be progressively outlined.

1.4 SOME IMPORTANT THEOREMS FROM CALCULUS

This section recalls some theorems from calculus that will be used several times during this book. The reader should be familiar with them. We do not give their proofs here.

Theorem 1.1. Intermediate Value Theorem

If f is a continuous function over the interval $[a, b]$ then for any number y between $f(a)$ and $f(b)$ exists a number c in $[a, b]$ such that $f(c) = y$.

Another way to formulate Theorem 1.1 is to state that a continuous function over an interval $[a, b]$ takes any value between $f(a)$ and $f(b)$.

Theorem 1.2. Mean Value Theorem

If f is a continuously differentiable function over $[a, b]$, then there exists a number c in $[a, b]$ such that $f'(c) = [f(b) - f(a)] / (b - a)$.

There also exists a mean value theorem formulated for integrals instead of derivatives.

Theorem 1.3. Mean Value Theorem for Integrals

If f is a continuous function over $[a, b]$ and g is an integrable function that does not change its sign over the interval $[a, b]$, then there exists a number c in $[a, b]$ such that

$$\int_a^b f(x) g(x) dx = f(c) \int_a^b g(x) dx \tag{1.12}$$

We wish to draw the reader's attention to the fact that the function g is not allowed to change its sign over the integration interval $[a, b]$, an element which is often overseen.

1.5 EXERCISES

1.5.1 Aim of Numerical Methods

1. Explain in simple words why errors are measured with absolute values.

2. Consider the equation $2\sin(x) = x$. Our aim is to find a numerical approximation x_r with an accuracy better than 10^{-5} in the sense of Definition 1 of chapter 1. For this exercise, consider that $r = 1.89549426703$ is the "true" solution of this equation.

   The algorithm is the following:

   Set $x_o = 1.895$. Then compute $x_1 = 2\sin(x_o)$, $x_2 = 2\sin(x_1)$, . . ., $x_{i+1} = 2\sin(x_i)$.

   For which x_i do we reach the desired precision?

1.5.2 Measuring Errors

3. We want to provide a numerical approximation with an absolute error lower than 10^{-4} of the equation $x^2 = 5$.

   Among the following proposed numerical solutions, which one(s) is/are a solution of the equation in the sense of Definition 1.1 of Chapter 1?

   (a) 2.23 (b) 2.236 (c) 2.235 (d) 2.2361 (e) 2.23607

4. Astrophysicists estimated the average distance between the Earth and the Moon with an absolute error of less than 10 km. Civil engineers measured the width of the St. Lawrence River at the height of Quebec City with an error of less than 1 cm.

   Which measurement is more accurate? Discuss your answer by considering absolute and relative errors.

5. Consider two numerical approximations x_{r1} and x_{r2} of the true solutions r_1 and r_2 of two mathematical problems. Let $E(x_{r1}) = |x_{r1} - r_1|$ and $E(x_{r2}) = |x_{r2} - r_2|$ be there true absolute errors. Define $r = r_1 + r_2$.

   Give an estimation $E_a(x_r)$ of the absolute error of the numerical approximation $x_r = x_{r1} + x_{r2}$ of r.

6. Repeat problem 5 using the absolute error approximations $E_a(x_{r1})$ and $E_a(x_{r2})$ instead of the true absolute errors $E(x_{r1})$ and $E(x_{r2})$.

7. Consider two numerical approximations x_{r1} and x_{r2} of the true solutions r_1 and r_2 of two mathematical problems. Let $e(x_{r1}) = \left|\frac{x_{r1} - r_1}{r_1}\right|$ and $e(x_{r2}) = \left|\frac{x_{r2} - r_2}{r_2}\right|$ be their true relative errors. Define $r = r_1 \cdot r_2$.

   Give an estimation $e_a(x_r)$ of the relative error of the numerical approximation $x_r = x_{r1} \cdot x_{r2}$ of r.

8. Repeat problem 7 using the relative error approximations $e_a(x_{r1})$ and $e_a(x_{r2})$ instead of the true relative errors $e(x_{r1})$ and $e(x_{r2})$.

NOTES

1 The attentive reader will have noted that, actually, even a numerical value will not be enough for the machinist. The machinist also needs to know with which precision the cut has to be performed. As we will see shortly, the same situation will arise for numerical methods.

2 Note that this algorithm could be as simple as: choose a random number.

3 Parameters that can be tuned when running a numerical algorithm are called "hyper-parameters".

4 The concept of bias plays a key role in the field of data science when one attempts to develop algorithms able to predict some outcomes based on some inputs (e.g., predict the market value of a house based on some features such as number of rooms, address, age of house, etc.). The second part of this book will be entirely dedicated to this problematic.

Part I

Numerical Methods for Engineering Applications

2 Numerical Errors

Numerical methods are used to find an approximation of the solution of a mathematical problem with an error that falls below a desired user-defined accuracy. In this chapter, we will discuss why numerical methods can only provide approximations. In engineering and science, when we develop a mathematical model to describe some reality, we must be aware of the many possible error sources: firstly, the model itself is an approximation of this reality, hence, inherently, the model cannot be fully accurate; secondly, the assumptions that are made to derive the model may be wrong; also, when developing the model, some mathematical mistakes could be introduced; the parameters needed to feed the model are usually inaccurate; finally, programming errors may also occur when coding the process to solve the model. It is worth noting that these errors are not specific to numerical methods and can occur in any field of work. In the rest of this book, we will assume that these errors can be either avoided, negligible, or quantifiable. However, there are two categories of errors that are specific to the use of numerical methods. These two error types are known as *round-off* errors and *truncation* errors. Round-off errors are due to the limited number of digits that can be stored in memory. Truncation errors appear because we solve a simplified and approximated problem in lieu of our actual problem. Both round-off and truncation errors will be defined and discussed in detail in this chapter. As we will see, round-off errors cannot be avoided. Truncation errors, for some algorithms, can be avoided. But, for most cases, both types of errors are present.

2.1 ROUND-OFF ERRORS

When executing numerical calculations in a computer, only a finite number of digits can be stored in the computer's memory. As a consequence, regardless of the number of stored digits, they constitute an approximation of the actual mathematical number that should have been used in calculations. Depending on the size of the memory allocated to store digits, more or less of them will be handled. A common case is to use *double precision* numbers, which roughly corresponds to storing 16 decimal digits. For example, if the number $\sqrt{2}$ has to be stored, only the first 16 decimal digits are actually in memory. The difference between the actual number ($\sqrt{2}$ in our case) and the stored number (the first 16 digits of $\sqrt{2}$) is called the round-off error.

Definition 2.1. Round-Off Errors

The difference between the actual mathematical number r and its representation with a finite number of digits is called the round-off error of r.

DOI: 10.1201/9781003262121-3

The difficulties begin as soon as we use a finite number of digits to represent numbers and perform calculations. Let us illustrate this with an example.

Example 2.1. Round-Off Errors

Consider the following Octave code that aims to compute the quantity $a = 0.2 + 0.2 + 0.2$:

```
>> a = 0.2+0.2+0.2
a = 0.60000
```

As expected, Octave attributes the value 0.6 to the variable a. But does it really assign 0.6 to a? Let us verify this by computing $a - 0.6$:

```
>> a-0.6
ans = 1.1102e-16
```

The answer is not equal to zero! Even if it is very close to zero, there is a difference.

Example 2.1 shows that adding three numbers can result in an erroneous answer. The reason behind this phenomenon is the accumulation of round-off errors during the additions. To understand how these round-off errors appear, a closer look at how numbers are stored in a computer is required. A computer stores numbers using base two. Technically, this consists of a sequence of bits b_i, where a bit has only two states (0 or 1). This is commonly referred to as the *binary code*. For example, the binary number $x = 10100$, consists of 5 bits, and equals to the decimal number 20, as follows:

$$x = 2^4 + 0 + 2^2 + 0 + 0 = 20$$

The IEEE754 standard specifies how decimal numbers are expressed as binary numbers in a computer using *floating-point numbers* or *floating-point representation*. A floating-point number has three parts: a sign (either positive or negative), a mantissa (a sequence of N bits b_i) and an exponent p:

$$\pm b_1 b_2 \ldots b_N \times 2^p$$

Depending on how many bits are used to store the mantissa and exponent, different families of precision are defined in the IEEE standard. Table 2.1 lists these defined families and specifies how many bits are used for each part of the floating-point number.[1] For example, the popular double precision floating-point number has a total of 64 bits (or 4 bites, as per the definition that one bite is a sequence of 8 bits). The first bit is used to encode the sign, the next 11 bits are for the exponent, and the remaining 52 are used to store the mantissa.

A double precision floating-point number can hold approximately 16 decimal digits. Indeed, typically, about 3.25 bits are needed to encode a decimal digit ($2^{3.25} \approx 10$). Since double precision has memory to store 52 bits for the mantissa, this gives about $52/3.25 = 16$ decimal digits.

Obviously, without any further rules, the decision on which part goes to the mantissa or to the exponent is open. For example, we could encode the decimal number 12.34 as 12.34×10^0, 1.234×10^1 or 0.1234×10^2 (or in any other way). The IEEE standard defines which one of the enumerated choices is actually used to store a number in a computer. The interested reader may learn more about how floating-point numbers are handled in a computer by reading in the IEEE754 standard. The technical details, as we will realize, are not important for our purpose.

Expanding further on the example of representing the number 12.34 allows us to understand the terminology "floating-point" number: by keeping the same mantissa, we can move around the location of the decimal point.

Let us go back to Example 2.1. The number 0.2, in binary code, would actually need an infinite number of digits to represent it. To see the binary representation of a decimal number with Octave, we can use the `format bit` command:

```
>> format bit
>> 0.2
ans = 0011111111001001100110011001100110011001100110011001100
1100110011001101 0
```

In fact, in binary code, 0.2 is equal to $0011111111001\overline{0011}$, a situation similar to $\frac{1}{3} = 0.\overline{3}$ for numbers represented in base ten. Consequently, it is not possible, using a finite number of digits, to represent in binary code the number 0.2 without a round-off error. As we add the number 0.2 consecutively, the round-off error accumulates and ends up becoming significant.

The takeaway here is that we should never trust a numerical calculation. We must always be aware that a numerical calculation is actually wrong (from a purely mathematical point of view) and represents only an approximation. We like to refer to this fact as the "golden rule of numerical methods":

TABLE 2.1
IEEE Standard for Different Precision of Floating-Point Number. The Table Lists the Number of Bits Used for Each Part of the Floating-Point Number

Precision	Sign	Exponent	Mantissa
Single	1	8	23
Double	1	11	52
Long double	1	15	64

Golden Rule of Numerical Methods

Any numerical calculations conducted with floating-point numbers are only an approximation and never free of round-off errors. *Never* assume a numerical calculation is error-free.

At this point, it is useful to introduce the concept of significant digits. Not all authors define this concept the same way. In the frame of this book, we adopt Definition 2.1. Note that we will use interchangeably the words *significant digits* and *significant figures*.

Definition 2.1. Significant Digits

Any digit of a number, beginning with the digit farthest to the left that is not zero and ending with the last digit farthest to the right that is either not zero or that is a zero but is considered to be exact, is called a significant digit.

To become familiar with this definition, it is useful to study Example 2.2.

Example 2.2. Significant DigitsThe number 20.1 has three significant digits, which are 2, 0, and 1.

The number 0.0201 has three significant digits, which, again, are 2, 0, and 1. This shows that leading zeros are simply ignored when identifying significant digits.

The number 2.010 has four significant digits, which are 2, 0, 1, and 0. Here, the last 0 is significant (otherwise, one would have written 2.01).

A good way to understand the concepts of significant digits is to keep in mind that numbers usually represent physical quantities associated with a physical unit. If, for example, the number 20.1 represents a distance of 20.1 meters, then changing its unit to centimeters or kilometers, for example, will not change the knowledge, nor which digits are significant or not. What is important is the actual distance.

2.1.1 Propagation of Round-Off Errors

Round-off errors enter into a calculation in a complex way, and their effect on the final result depends on how the different operations are executed. Example 2.3 gives an illustration of this fact.

Example 2.3. Propagation of Round-Off Errors

Consider the following two functions:

$$f(x) = x\left(\sqrt{x+1} - \sqrt{x}\right) \tag{2.1}$$

$$g(x) = \frac{x}{\sqrt{x+1} + \sqrt{x}} \tag{2.2}$$

It is simple to prove that $f(x) = g(x)$ for any x where f and g are defined. Let us evaluate these two functions numerically in $x = 500$. We start by defining the functions f and g in Octave:

```
>> f = @(x) x*(sqrt(x+1)-sqrt(x));
>> g = @(x) x/(sqrt(x+1)+sqrt(x));
```

Here, we used the so-called *anonymous functions* to define them in Octave. Let us now evaluate these two functions in $x = 500$:

```
>> f(500)
ans = 11.175
>> g(500)
ans = 11.175
```

They seem to agree as expected. But if we display all the digits that are stored inside our computer, we observe that actually they do not agree:

```
>> format long
>> f(500)
ans = 11.17475530074685
>> g(500)
ans = 11.17475530074720
```

The last three digits differ. This shows that at least one of the numerical function evaluations must be wrong (maybe even both are wrong). At this stage, it is impossible to know which one is correct (if any).

At this point, it should become clear to the reader that round-off errors can accumulate during calculations and result in inaccurate answers. Another phenomenon that can take place is known as *loss of significance*. This happens when, for example, two numbers are added, both of which can be free of round-off errors, but their sum requires more digits than what possibly can be stored in the computer memory. For example, let us imagine we are using a hypothetical computer that can only store four digits at a time. Let's also assume, for the sake of this example, that this hypothetical computer stores numbers using base 10. Let's consider two decimal numbers 1234 and 0.12. Both those numbers can be stored without round-off error. But their sum (which would be 1234.12) cannot be stored free of round-off error.

In numerical algorithms, loss of significance happens often when subtracting numbers that are similar.[2] For illustrative purposes, let us imagine we are using a hypothetical computer that can only store three digits in its mantissa. Again, for the sake of this example, let us assume the numbers are stored in base ten. Using this hypothetical computer, let's perform the following operation: $\sqrt{4.01}-2$. As only three decimal digits can be stored, the result becomes:

$$\sqrt{4.01}-2 \underbrace{=}_{3\,s.d} 2.00-2.00=0.00$$

In this extreme example, we end up with a numerical result that has all its significant digits wrong. We say that all significant digits were lost.

The important takeaway from the previous examples is that whenever calculations with floating-point numbers are conducted, we can almost be sure that the answer is wrong, and some (difficult to predict) additional round-off error is introduced compared to the initial round-off error. In fact, any calculations performed with floating-point numbers will result in the loss of some significant digits. Observation 2.1 summarizes this discussion.

Observation 2.1. Loosing Correct Digits

If x has m correct digits, then, in general, $f(x)$ will have less than m correct digits.

Since round-off errors are not avoidable in numerical calculations, the only solution is to learn to live with them. The consequence is that we will have to develop methods that can provide good estimates of errors (recall that a good estimate is an estimation that is larger than or equal to the actual error). Section 2.3 provides a first set of methods for estimating errors.

2.1.2 Machine Epsilon and Undistinguishable Numbers

A useful number to quantify the round-off errors is the machine epsilon ε_m. This number is defined as follows:[3]

Definition 2.2. Machine Epsilon

The machine epsilon is the smallest number ε_m such that for any number $x \leq 0.5\varepsilon_m$ we have $1+x=1$.

If calculations can be performed with infinite precision, then ε_m is equal to zero. But in numerical calculations, where only a finite number of digits are available, ε_m will not be equal to zero. Octave uses double precision floating-point numbers and stores ε_m in the built-in variable `eps`:

```
>> eps
ans = 2.220446049250313e-16
```

We can verify that, indeed, $1+x=1$ for any $x \leq 0.5\varepsilon_m$. The following lines of Octave code demonstrate this when adding 10^{-17} to the number one:

```
>> format long
>> x=1E-17;
>> 1+x
ans = 1
```

Definition 2.2, which uses the number 1 as a reference, can be generalized to an arbitrary number δ.

Theorem 2.1. Undistinguishable Numbers

If ε_m is the value of the machine epsilon and δ a real number, then for any number $x \leq 0.5 \cdot \delta \cdot \varepsilon_m$ we have $\delta + x = \delta$.

Let us illustrate this with the number $\delta = 3279$. In this case, for any number x smaller than $0.5 \cdot 3279 \cdot \varepsilon_m$ the relation $3279 + x = 3279$ holds. For example:

```
>> format long
>> x = (3279*eps)/6;
>> 3279+x
ans = 3279
```

The reader is encouraged to repeat this example with other values of x (some smaller and some larger than $0.5 \cdot 3279 \cdot \varepsilon_m$).

An alternative way to formulate Theorem 2.1 is summarized in Observation 2.2.

Observation 2.2. Undistinguishable Numbers

If ε_m is the of the machine epsilon, then the numbers x and any number in the interval $\left[x, x\left(1+0.5\varepsilon_m\right)\right]$ cannot be distinguished numerically.

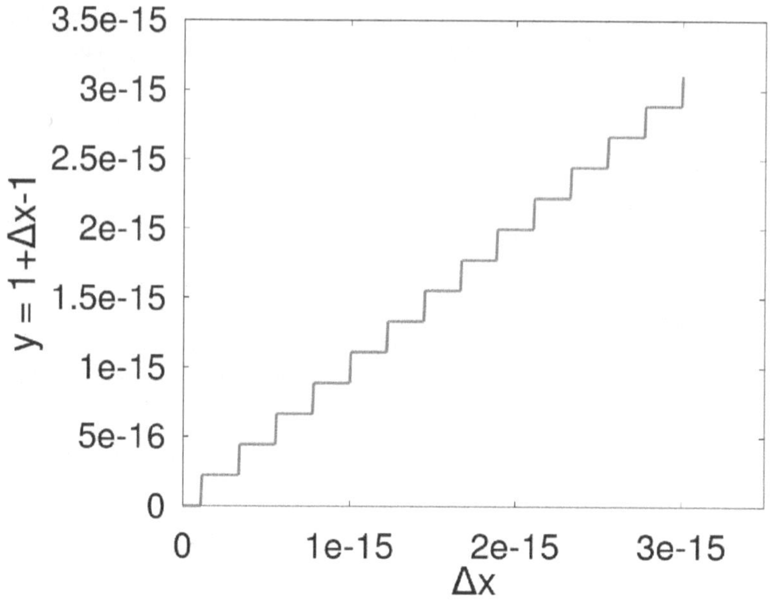

FIGURE 2.1 Numerical evaluation, using double precision floating point numbers, of the quantity $y = 1 + \Delta x - 1$.

Figure 2.1 gives a visual representation of Theorem 2.1 and Observation 2.2. The result of the numerical evaluation of $y = 1 + \Delta x - 1$ is plotted in function of Δx. If y could be evaluated with infinite accuracy, then of course $y = \Delta x$. But, when using floating-point numbers, the situation is different. Due to round-off errors, $1 + \Delta x$ evolves stepwise with Δx. The quantity $y = 1 + \Delta x - 1$ differs from Δx.

2.2 TRUNCATION ERRORS

Numerical algorithms are designed to find approximations to mathematical problems, even if a closed-form solution is unknown. Algorithms can achieve this by replacing the problem aimed to be solved with another problem that approximates the original problem. This approximated problem is chosen such that it has a known closed-form solution. The error introduced when replacing the original problem with an approximated one is called the truncation error.

Definition 2.3. Truncation Error

The error introduced by replacing a mathematical problem with another problem is called truncation error.

The word "truncation" can be misleading. First of all, it should not be confused with the concept of round-off error. The "truncation" operation in Definition 2.3 does not refer to truncating a number of digits. It refers rather to a common method used to approximate a problem: series expansion (for example, Taylor series) followed by truncating the series after a finite number of terms. Example 2.4 illustrates this idea in the case of a series expansion of the exponential function.

Example 2.4. Truncation Error

A well-known series expansion of the exponential function is

$$e^x = 1 + x + \frac{x^2}{2!} + \frac{x^3}{3!} + \cdots = \sum_{n=0}^{\infty} \frac{x^n}{n!} \tag{2.3}$$

To give an approximation of the exponential function, we can decide to include, for example, only the first three terms of the series:

$$e^x \cong 1 + x + \frac{x^2}{2!} \tag{2.4}$$

The truncation error is then the difference between the exact value of e^x and the approximation, including only the first three terms of the series (as always, errors are measured in absolute values):

$$\text{Truncation error} = \left| e^x - \left(1 + x + \frac{x^2}{2!}\right) \right| \tag{2.5}$$

Note that the truncation error does not include the round-off errors. In other words, it is assumed that the various expressions can all be evaluated with an infinite number of digits.

2.2.1 The Truncation Error of Taylor Series

A common tool used in numerical methods to generate an approximated expression of an exact one is the Taylor series expansion, as stated in Theorem 2.2.

Theorem 2.2. Taylor Series Expansion

Consider two real numbers x and h. If the function f is $k+1$ times continuously differentiable in the interval $[x, x+h]$, then there exists a number ξ in $[x, x+h]$ such that

$$f(x+h) = f(x) + f'(x)h + \frac{f''(x)}{2!}h^2 + \ldots + \frac{f^{(k)}(x)}{k!}h^k + \frac{f^{(k+1)}(\xi)}{(k+1)!}h^{k+1} \tag{2.6}$$

The term

$$R = \frac{f^{(k+1)}(\xi)}{(k+1)!} h^{k+1} \tag{2.7}$$

is called the Taylor remainder of order $k+1$ and the polynomial

$$T_k(h) = f(x) + hf'(x) + \frac{f''(x)}{2!} h^2 + \ldots + \frac{f^{(k)}(x)}{k!} h^k \tag{2.8}$$

is called the Taylor polynomial of order k or the order k approximation of the function f in $x+h$.

Taylor series expansion is a powerful tool used to generate various approximations of mathematical expressions. For example, if a function f is known in some point x, we can obtain the approximation of f in some other point $x+h$. Figure 2.2 illustrates this idea. Figure 2.2 suggests that the more terms are included in the Taylor series expansion, the more accurate the approximation becomes. This is correct when h is sufficiently small (the term: "sufficiently small" will be discussed further on).

Taylor's theorem can also generate an expression for the truncation error E_{trunc} by evaluating the remainder term of the series:

$$E_{trunc} = \left| \frac{f^{(k+1)}(\xi)}{(k+1)!} h^{k+1} \right| \tag{2.9}$$

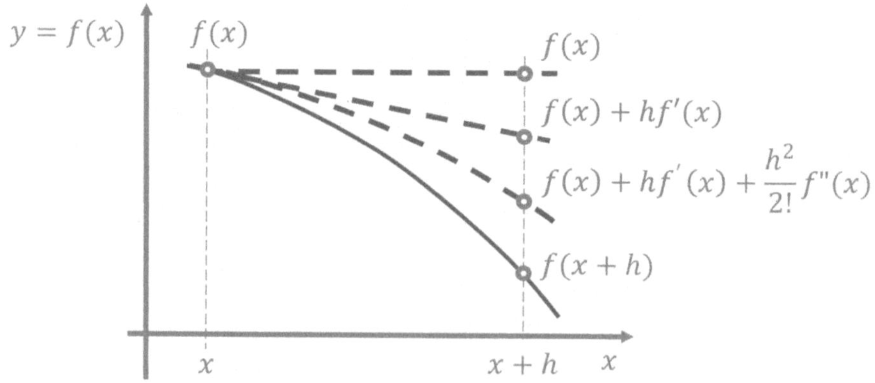

FIGURE 2.2 Graphical illustration of Taylor's theorem. In general, as the order of the expansion increases, the approximation becomes better.

Even though $\xi \in [x, x+h]$ is unknown, we can provide an upper bound to the error by considering the maximum of the Taylor reminder in $[x, x+h]$:

$$E_{trunc} \le \max_{\xi \in [x,x+h]} \left| \frac{f^{(k+1)}(\xi)}{(k+1)!} h^{k+1} \right| \tag{2.10}$$

Alternative ways of obtaining upper boundaries to the truncation error exist, but they are problem-specific. Example 2.5 illustrates such a case.

Example 2.5. Approximating the Cosine Function

The cosine function has no closed-form expression that would allow its numerical evaluation. Taylor series can be used to give numerical approximations of this function. The idea is to use a known value of the cosine function (in our case $\cos(0) = 1$) and then, with the help of the Taylor series, construct an approximation for a value x where the cosine function is not known. Using Equation (2.6), it follows that:

$$\cos(x) = \cos(0+x) = \cos(0) + x \frac{d}{dx} \cos(x) \bigg|_{x=0} + \frac{x^2}{2!} \frac{d^2}{dx^2} \cos(x) \bigg|_{x=0} + \dots \tag{2.11}$$

For an order $k = 6$ approximation, we obtain after simplification:

$$\cos(x) = 1 - \frac{x^2}{2!} + \frac{x^4}{4!} - \frac{x^6}{6!} + R \tag{2.12}$$

Based on the Taylor remainder term (2.7), an upper boundary of the truncation error is:

$$\text{Truncation error} \le \left| \frac{x^7}{7!} \right| \tag{2.13}$$

The derivative of order $k+1 = 7$ of $\cos(x)$ will always be between −1 and 1 regardless in which point it is evaluated. However, in the present example, a more precise estimation of the truncation upper boundary can be obtained. Indeed, if we are interested in small numbers x (for example, $x = 0.3$) then we can estimate the truncation error as follows:

$$\text{Truncation error} = \left| \frac{x^8}{8!} - \frac{x^{10}}{10!} + \frac{x^{12}}{12!} - \dots \right| \tag{2.14}$$

For small x, the terms in this summation decrease very rapidly, and we may approximate the truncation error by only including the first term:

$$\text{Truncation error} \cong \left|\frac{x^8}{8!}\right| \tag{2.15}$$

Note that this estimation is more precise than estimation (2.13). As an example, let us approximate $\cos(0.3)$:

$$\cos(0.3) \cong 1 - \frac{0.3^2}{2!} + \frac{0.3^4}{4!} - \frac{0.3^6}{6!} \cong 0.9553385 \tag{2.16}$$

The truncation error is:

$$\text{Truncation error} \cong \left|\frac{0.3^8}{8!}\right| \cong 7 \cdot 10^{-6} \tag{2.17}$$

Assuming Octave can compute $\cos(0.3)$ exactly (or with an error lower than our estimated truncation error), for the true error, we obtain[4]

```
>> format short E
>> x=0.3;
>> 1-x^2/2+x^4/factorial(4)+x^6/factorial(6)-cos(x)
ans = 2.0234E-06
```

This confirms that our truncation error estimation (2.17) is indeed a valid estimation as it is a conservative estimation.

It would be wrong to conclude that adding more terms to the Taylor series expansion will always result in the reduction of the truncation error. In fact, the Taylor series expansion must be conducted sufficiently close to the reference point. Sufficiently close means that the expansion must be done within the radius of convergence of the Taylor series. Definition 2.4 recalls the definition of the radius of convergence of a power series.

Definition 2.4. Radius of Convergence of a Power Series

Consider a power series such as

$$f(x) = \sum_{k=0}^{n} a_k (x-b)^k \tag{2.18}$$

The radius of convergence of this power series is the smallest positive number r such that for any x satisfying $|x-a| < r$ the series (2.18) converges as $n \to \infty$.

Example 2.6 illustrates the importance of keeping the Taylor series expansion within its radius of convergence.

Example 2.6. Approximation of the Logarithmic Function

To approximate the natural logarithmic function, we expand the function around $x=1$, where we know that $\ln(1)=0$. Using Equation (2.6), we obtain the following k order expansion:

$$\ln(1+x)=x-\frac{x^2}{2}+\frac{x^3}{3}-\frac{x^4}{4}+\ldots-(-1)^{k-1}\frac{x^k}{k}+(-1)^k\frac{\xi^{k+1}}{k+1} \quad (2.19)$$

with $\xi\in[1,1+x]$. Note that x must be larger than or equal to -1 to ensure $\ln(1+x)$ is defined. Written as a series, Equation (2.19) becomes

$$\ln(1+x)=\sum_{k=1}^{\infty}(-1)^{k-1}\frac{x^k}{k} \quad (2.20)$$

which is known to have a radius of convergence of 1 from calculus.[5] Consequently, the truncation error of an order k approximation

$$\text{Truncation error}=\left|\frac{\xi^{k+1}}{k+1}\right| \quad (2.21)$$

will only decrease with order k if $|x|<1$.

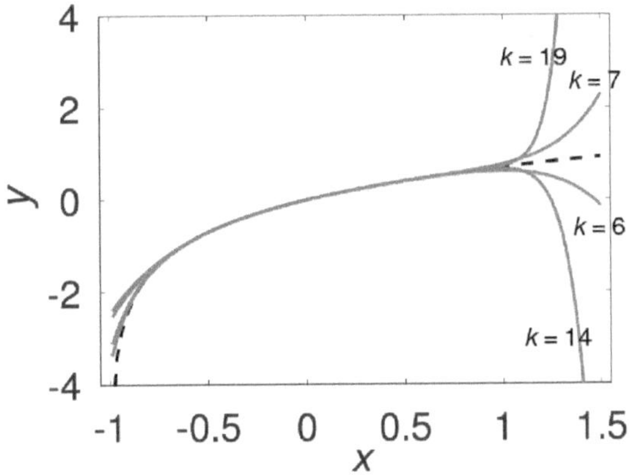

FIGURE 2.3 Taylor series expansion of $f(x)=ln(1+x)$ with increasing orders (order 6, 7, 14, and 19). Note that the Taylor series approximates the function f (dashed line) only in the interval $-1<x<1$.

Figure 2.3 illustrates this behavior. Consider x such that $-1 < x < 1$. In this case, when increasing the order of the Taylor series, better approximations of the function $f(x) = \ln(1+x)$ are obtained. The truncation error (2.21) decreases as the order k of the Taylor series increases. The situation is different for $x > 1$. In this case, increasing the order of the Taylor series results in approximations of the function $f(x) = \ln(1+x)$ that become progressively less accurate. The truncation error (2.21) increases as the order k of the Taylor series increases. The radius of convergence is consequently equal to one since, for any $|x| < 1$, the Taylor series will converge.

2.2.2 Behavior of Truncation Errors

In the present textbook, the majority of mathematical problems encountered will be approximated using the Taylor series expansion. The corresponding algorithms will make use of a parameter h. This parameter h is user-defined and necessary to accomplish the series expansion. Hence, it is essential to address how h affects the behavior of the truncation error. Equation (2.7) gives an expression for the truncation error using the remainder R of the Taylor series. However, as seen in Equation (2.9), the algorithms will not have full control over the exact value of the truncation error. The parameter h and the order k of the series expansion can be chosen, but little control is available over ξ which is only known to be a value between x and $x+h$. Luckily, to operate numerical algorithms, the exact expression of the truncation error is not needed. On the other hand, the behavior of the truncation error in the two parameters h and k is important. In fact, it is important to understand how the truncation error decreases with the parameters h and k. To quantify this decrease, it is common to use the concept of the "big O" notation. For our purposes, we define the big O notation as follows:

Definition 2.5. Big O Notation

A real function f is said to behave as $\mathcal{O}(g)$ if there exists a finite number $K \neq 0$ such that

$$\lim_{x \to a} \frac{f(x)}{g(x)} = K \tag{2.22}$$

We write $f(x) = \mathcal{O}(g(x))$ for $x \to a$ which reads as "f is of order g as x tends to a".

When a function f is of order $\mathcal{O}(g)$, this means intuitively that, as x tends to approach a, the functions f and g become increasingly similar. If we know that a function f is of order g as $x \to a$, we have more information compared to evaluating $f(a)$. Indeed, we know how $f(x)$ behaves around $x = a$.

Let us now apply this concept to the truncation error of a Taylor series expansion. We would like to quantify how this truncation error behaves as $h \to 0$. Let's consider a Taylor series expansion $T_k(h)$ of the order k of a function f around x. Equation (2.9) provides the truncation error when replacing $f(x+h)$ by $T_k(h)$ and suggests that this error is in order h^{k+1}. This is indeed the case. For h within the radius of convergence of the Taylor series expansion of $f(x+h)$ and $f^{(k+1)}(x) \neq 0$, we can write

$$\lim_{h\to 0} \frac{f(x+h)-T_k(h)}{h^{k+1}} = \lim_{h\to 0}\left\{\frac{1}{h^{k+1}}\left[\frac{f^{(k+1)}(x)}{(k+1)!}h^{k+1} + \frac{f^{(k+2)}(x)}{(k+2)!}h^{k+2} + \ldots\right]\right\} = \frac{f^{(k+1)}(x)}{(k+1)!} \tag{2.23}$$

Equation (2.23) confirms that, indeed, the truncation error of the Taylor series expansion of $f(x+h)$ is in order h^{k+1} if $f^{(k+1)}(x) \neq 0$. Using the big O notation, the Taylor series expansion in order k can, therefore, be written as:

$$f(x+h) = T_k(h) + \mathcal{O}\left(h^{k+1}\right) = f(x) + f'(x)h + \frac{f''(x)}{2!}h^2 + \ldots + \frac{f^{(k)}(x)}{k!}h^k + \mathcal{O}\left(h^{k+1}\right) \tag{2.24}$$

Let us again showcase the intuitive meaning of $\mathcal{O}\left(h^{k+1}\right)$. As $h \to 0$, the truncation error tends to become increasingly similar to h^{k+1}. In fact, from Equation (2.23), we can write

$$\left|f(x+h) - T_k(h)\right| \cong K \cdot h^{k+1} \tag{2.25}$$

with $K = \left|\frac{f^{(k+1)}(x)}{(k+1)!}\right|$. Equation (2.25) becomes more precise as h grows smaller. Example 2.7 illustrates the usage of this equation.

Example 2.7. Behavior of Taylor Series Truncation Error

To illustrate the intuitive interpretation of the big O notation, we consider the Taylor series expansion of the square root function $f(x) = \sqrt{x}$ around 1. As an example, let us consider the order 3 expansion:

$$\sqrt{1+h} = T_3(h) + \mathcal{O}(h^4) = 1 + \frac{h}{2} - \frac{h^2}{8} + \frac{h^3}{16} + \mathcal{O}(h^4) \tag{2.26}$$

The truncation error is in order 4, which means that as $h \to 0$, the truncation error becomes increasingly similar to $\left|K \cdot h^4\right|$ with the coefficient K given by:

$$K = \left|\frac{f^{(4)}(x=1)}{4!}\right| = \frac{5}{128} \tag{2.27}$$

Similarly, we can obtain approximations of $\sqrt{1+h}$ with different orders and their corresponding truncation error estimations.

Figure 2.4 illustrates how the truncation errors for various orders of approximation behave. The true truncation errors, computed by evaluating numerically $\left|\sqrt{1+h} - T_k(h)\right|$, are plotted for a range of values of h and expansion orders from $k=0$ to $k=3$. Furthermore, the estimation of the truncation errors based on $\left|K \cdot h^k\right|$ is added, too. We can observe how the truncation errors become increasingly similar to $\left|K \cdot h^k\right|$ as $h \to 0$.

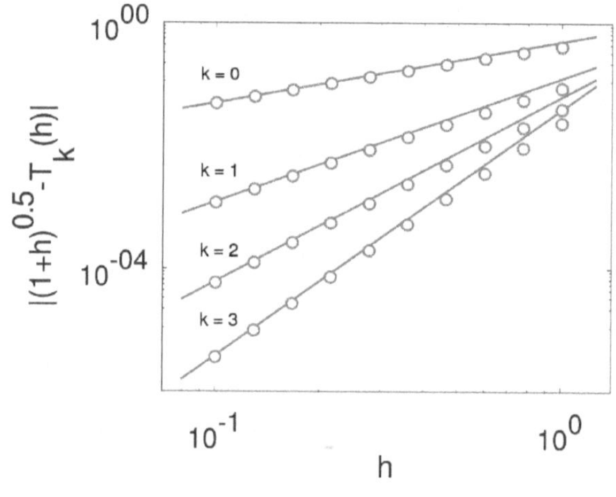

FIGURE 2.4 Truncation errors when approximating $\sqrt{1+h}$ with a Taylor series expansion of order k. The circles are the true truncation errors obtained by numerically calculating $\left|\sqrt{1+h} - T_k(h)\right|$ and the lines are the truncation error estimation from the Taylor remainder term $\left|K \cdot h^4\right|$.

Plots depicting the truncation errors in the function of some parameters controlling its behavior, the parameter h in our case, are important tools in numerical methods to understand how algorithms behave. They will also be useful in developing tools and methods to estimate errors introduced by the algorithms during their numerical calculations.

Equation (2.24) indicates that the truncation error of a Taylor series of order k (i.e., by using a Taylor polynomial $T_k(h)$ to replace the original function) the truncation error is of order $\mathcal{O}(h^{k+1})$. Let us note that if the function f is such that $f^{(k+1)}(x)=0$, then the truncation error is of order $\mathcal{O}(h^{k+2})$, which is of higher order than $\mathcal{O}(h^{k+1})$. This statement can be proved by following calculations similar to Equation (2.23). Example 2.8 gives an illustration of such a case.

Example 2.8. Truncation Error of the Cosine Taylor Series

Recall the Taylor series expansion of order $k=6$ of the cosine function around $x=0$:

$$\cos(x)=1-\frac{x^2}{2!}+\frac{x^4}{4!}-\frac{x^6}{6!}+R(x) \tag{2.28}$$

The truncation error $R(x)=|\cos x-T_6(x)|$ is of order $\mathcal{O}(x^8)$. To verify this, let's consider the expression of the remainder term R:

$$R(x)=\left|\frac{x^8}{8!}-\frac{x^{10}}{10!}+\frac{x^{12}}{12!}-\dots\right| \tag{2.29}$$

This term is indeed of order $\mathcal{O}(x^8)$ as

$$\lim_{x\to 0}\frac{R(x)}{x^8}=\lim_{x\to 0}\left\{\frac{1}{8!}-\frac{x^2}{10!}+\frac{x^4}{12!}-\dots\right\}=\frac{1}{8!} \tag{2.30}$$

In other words, as $x\to 0$, the error in replacing $\cos x$ by the Taylor series of order 6 becomes increasingly similar to $\frac{x^8}{8!}$.

The big O notation has some properties that will be used frequently.

Theorem 2.3. Properties of Big O Notation

1. The sum of two expressions of order $\mathcal{O}(h^k)$ is again of order $\mathcal{O}(h^k)$.
2. The ratio of an expression of order $\mathcal{O}(h^k)$ over an expression of order $\mathcal{O}(h^n)$ is of order $\mathcal{O}(h^{k-n})$.
3. An expression of order $\mathcal{O}(h^k)$ will eventually always become smaller than an expression of order $\mathcal{O}(h^n)$ if $k>n$ as h approaches zero.

The proof of property 1 is straightforward and based on the definition of the big O notation. Consider two expressions $f(h)$ and $g(h)$ of order $\mathcal{O}(h^k)$. Their sum is as well of order $\mathcal{O}(h^k)$ as

$$\lim_{x\to 0}\frac{f(h)+g(h)}{h^k}=\lim_{x\to 0}\frac{f(h)}{h^k}+\lim_{x\to 0}\frac{g(h)}{h^k}=K_f+K_g \tag{2.31}$$

Similarly, we can prove property 2. Let us now discuss the third property. An expression in order $\mathcal{O}(h^k)$ will become increasingly similar to $K_1 \cdot h^k$, whereas an expression in order $\mathcal{O}(h^n)$ will become similar to $K_2 \cdot h^n$. Now, regardless of how small K_2 may be, eventually, for a small enough h, we will have $K_1 \cdot h^k < K_2 \cdot h^n$ if $k > n$, which proves property 3.

Property 3 justifies notations involving the approximation sign "$\cong$". For example, when we write

$$f(x+h)\cong f(x)+hf'(x) \tag{2.32}$$

we imply that this expression is evaluated numerically using a value h sufficiently small such that higher order terms, order 2 in this case, are negligible to the first order approximation.

A further important application of property 3 is to compare the truncation errors of algorithms. An algorithm having a truncation error of a higher order than another algorithm eventually always produces approximations with a lower truncation error.

2.3 ERROR PROPAGATION

When numerical calculations are conducted, errors propagate through them. The propagation of the errors can become quite difficult to trace. Example 2.3 illustrated error propagation in the case of round-off errors. We saw that error propagation depends on how the calculations are executed. In this section, we want to delve into the details of how numerical errors propagate. Due to the complexity of how round-off errors enter into a specific calculation, we will not be able to provide a precise estimation of the effect of round-off errors. However, we will be able to discuss the portion of the error propagation that cannot be avoided, regardless of which way a specific numerical calculation is accomplished.

2.3.1 Propagation of an Input Error

Let us suppose that we would like to evaluate a function $y = f(x)$ in some value x. Furthermore, let us assume that for x, we only have an approximation x_c such that $x_c = x + \varepsilon$. Here, ε is the absolute error (for example, due to round-off errors) of x_c. The question we want to answer is the following: assuming we can evaluate with infinite precision the function f, how large will the absolute error of $y=f(x)$ be if we evaluate this function in $x_c=x+\varepsilon$? In numerical analysis, the error ε is referred to as the input error, and the error $\left|f(x+\varepsilon)-f(x)\right|$ is called the output error.

Taylor series can help us estimate the output error given an input error. We write:

$$f(x_c)=f(x+\varepsilon)=f(x)+\varepsilon f'(x)+\mathcal{O}(\varepsilon^2) \tag{2.33}$$

For the moment, we assume that $f'(x)\neq 0$. The case where $f'(x)=0$ will be discussed at the end of this section. If $f'(x)\neq 0$, we can estimate, using Equation (2.33), the output error as a first order expression in the input error ε (i.e., the input error is small enough such that the second order terms $\mathcal{O}(\varepsilon^2)$ are negligible compared to $\varepsilon f'(x)$):

$$\left|f(x_c)-f(x)\right|\cong\left|f'(x)\cdot\varepsilon\right|=\left|f'(x)\right|\cdot\left|\varepsilon\right| \tag{2.34}$$

Equation (2.34) can be understood as follows: the input error on x (the error $|\varepsilon|$ on the value x) is multiplied by the quantity $|f'(x)|$ after evaluating the function f.

From Equation (2.34), we can also obtain a relation linking the relative errors instead of the absolute errors:

$$\left|\frac{f(x_c)-f(x)}{f(x)}\right|\cong\left|\frac{f'(x)\cdot x}{f(x)}\right|\cdot\left|\frac{\varepsilon}{x}\right| \tag{2.35}$$

Equations (2.34) and (2.35) are the motivation for Definition 2.6:

Definition 2.6. Condition Number of a Function

The absolute condition number $C_f(x)$ of a function f in the point x is

$$C_f(x)=\left|f'(x)\right| \tag{2.36}$$

And the relative condition number $c_f(x)$ of a function f in the point x is

$$c_f(x)=\left|\frac{f'(x)\cdot x}{f(x)}\right| \tag{2.37}$$

In numerical analysis, conditioning is an important concept. It measures how a calculation, or more generally an algorithm, is sensitive to errors in the input. Depending on the mathematical problem of interest, conditioning has to be defined differently (depending on the input and output of the problem).[6] But the idea is

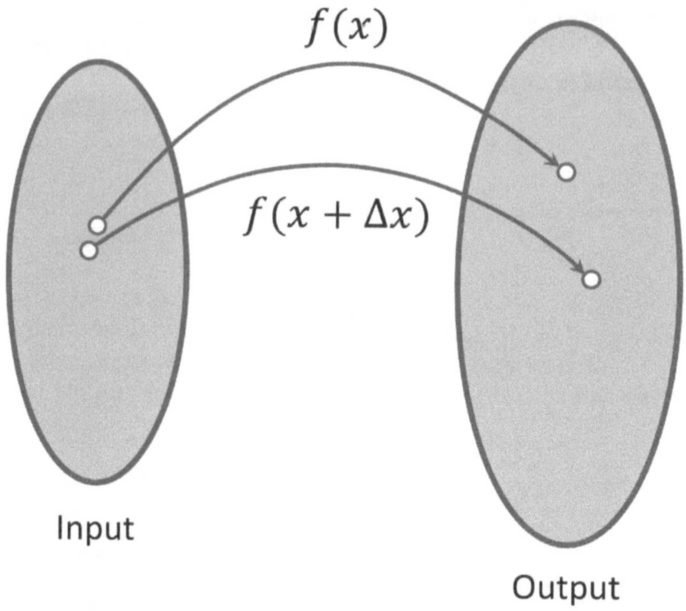

FIGURE 2.5 The concept of conditioning in numerical analysis. Conditioning measures how a small change Δx in the input is reflected in the output after the calculation is accomplished. Here $f(x)$ represents a calculation (or an algorithm) that is applied to some quantity x.

always the same: we want to estimate how a small change in the input affects the output. Figure 2.5 illustrates this idea. Some calculations, or some algorithms, f are applied to a quantity x. Conditioning measures how a small change Δx in the input is reflected in the output. If the outputs $f(x)$ and $f(x+\Delta x)$ are close to each other, the condition number will be small. Whereas if the outputs $f(x)$ and $f(x+\Delta x)$ are far from each other, the condition number will be large. In the next chapters, we will revisit this concept more thoroughly and define quantitatively what the terms "close" and "far" mean. For the moment, we restrict ourselves to situations where the inputs are numbers and the outputs are numbers too. In this case, the distance (how close or how far) two numbers x_1 and x_2 are from each other is measured with the quantity $|x_2 - x_1|$.

It is important to realize that the condition number of a function will only estimate the *unavoidable error*, which is the error due to the error in the input of the function. The actual numerical error may be larger than the estimation using the condition number. Indeed, when evaluating numerically the function, round-off errors are introduced. These are not considered by the condition number approach, which assumes that the function can be evaluated with infinite accuracy. This observation is summarized in Theorem 2.4.

Theorem 2.4. Unavoidable Output Error Estimation

When evaluating a function f, the first order estimation of the unavoidable output error Δy due to the input error $|\varepsilon|$ is:

$$\Delta y \cong C_f(x) \cdot |\varepsilon| \tag{2.38}$$

where $\Delta y = |f(x+\varepsilon) - f(x)|$ is the unavoidable output error (the error in the output assuming f can be evaluated with infinite precision) and $C_f(x)$ is the absolute condition number of f.

Similarly, we have, for the relative errors

$$\left|\frac{\Delta y}{f(x)}\right| \cong c_f(x) \cdot \left|\frac{\varepsilon}{x}\right| \tag{2.39}$$

where $c_f(x)$ is the relative condition number of f.

Example 2.9 illustrates how the concept of condition number of a function can be used to estimate the propagation of input errors through calculations.

Example 2.9. Error Propagation and Conditioning

Let us consider the function $y = f(x) = x^2$. From Definition 2.6, it follows that the absolute condition number is

$$C_f(x) = 2|x| \tag{2.40}$$

This means that if we evaluate the function f in $x+\varepsilon$, the result will have an absolute error of about $\Delta y \cong C_f(x) \cdot \varepsilon = 2|x| \cdot \varepsilon$. The initial error is multiplied by the factor $2|x|$. We can verify this by numerically evaluating the function f in $x = \sqrt{2}$. The correct answer would be $f(\sqrt{2}) = 2$. Octave finds an absolute error of $\Delta y \cong 4 \cdot 10^{-16}$:

```
>> f=@(x) x^2;
>> f(sqrt(2))-2
ans = 4.4409e-16
```

This agrees with the condition number calculated in Equation (2.40). Indeed, when evaluating numerically $\sqrt{2}$, a round-off error of about $\varepsilon \cong 0.5 \cdot \sqrt{2} \cdot \varepsilon_m$ is introduced (refer to Observation 2.2). Consequently, the expected absolute error when evaluating $f(\sqrt{2})$ will be about

$$\Delta y \cong 2\sqrt{2}\cdot 0.5\cdot\sqrt{2}\cdot\varepsilon_m \simeq 4.5\cdot 10^{-16} \tag{2.41}$$

which is similar to the error we found when evaluating $f\left(\sqrt{2}\right)-2$ numerically with Octave.

So far, we have considered situations where $f'(x)\neq 0$. If $f'(x)=0$, then according to Theorem 2.4, the first order estimation of the unavoidable error is equal to zero. As we will see, in such a case, the unavoidable error is indeed small but not equal to zero. When a more precise estimation is required, a higher-order estimation in the input error needs to be developed. For this purpose, we use the Taylor series expansion of $f(x_c)=f(x+\varepsilon)$, include higher order terms, and consider that $f'(x)=0$. We obtain:[7]

$$f(x_c)=f(x+\varepsilon)=f(x)+\frac{1}{2!}f''(x)\varepsilon^2+\mathcal{O}\left(\varepsilon^3\right) \tag{2.42}$$

The unavoidable error becomes, for small enough input errors (i.e., small enough such that the third order terms $\mathcal{O}\left(\varepsilon^3\right)$ are negligible compared to the second order term $\frac{1}{2!}f''(x)\varepsilon^2$):

$$\left|f(x_c)-f(x)\right|\cong\left|\frac{1}{2!}f''(x)\cdot\varepsilon^2\right|=\left|\frac{1}{2!}f''(x)\right|\cdot\left|\varepsilon^2\right| \tag{2.43}$$

The unavailable error is now in second order of the input error $|\varepsilon|$.

Example 2.10. Error Propagation When $f'(x)=0$

Consider the cosine function $y=\cos(x)$. Let us estimate the error when numerically computing $\cos(\pi)$ when we approximate π by 3.14159.

Let us first determine the round-off error of our numerical approximation of π. For this, we use the inbuilt variable `pi` of Octave:

```
>> format short E
>> abs(3.14159-pi)
ans = 2.6536E-06
```

The input error $|\varepsilon|$ is about $2.7\cdot 10^{-6}$.

As $\frac{d}{dx}(\cos x)=-\sin x$ and $\sin\pi=0$, we have to use a second order approximation for the unavoidable error:

$$\left|\cos(3.14159)-\cos\pi\right| \cong \left|-\frac{1}{2!}\cos\pi\cdot\varepsilon^2\right| = \frac{1}{2}\left|\varepsilon^2\right| \cong 3.5\cdot 10^{-12} \tag{2.44}$$

We can verify that this estimation is correct by calculating with Octave

```
>> abs(cos(3.14159)-cos(pi))
ans = 3.5207E-12
```

Note that the unavoidable error is indeed very small compared to the input error (as it is in the second order of the input error) but not equal to zero, as would predict a first order estimation of the unavoidable error.

2.3.2 Estimating Round-Off Errors Introduced by a Function Evaluation

As stated in Theorem 2.4, the condition number allows for the estimation of the unavoidable output error only. However, when evaluating a function, round-off errors occur as well (refer to Observation 2.1). The estimation of the round-off errors introduced during the evaluation of a function is a challenging problem. Contrary to the unavoidable error, there is no simple formula to estimate it. Further analysis is required. Here, we present one methodology that can be used.

Its principle is simple. We do not know how to compute exactly $f(x)$ (i.e., we do not know how many digits from the numerical evaluation of f are correct). However, based on Theorem 2.4, we know the error associated with computing $f(x+h)$, assuming a zero round-off error associated with the evaluation of $f(x)$. Consequently, the difference between $\left|f(x+h)-f(x)\right|$, when evaluating numerically, the different quantities and $\left|f'(x)h\right|$ equals the round-off associated with computing numerically $\left|f(x+h)-f(x)\right|$. To obtain the round-off error of $f(x)$ we need to conduct this analysis for h tending towards zero.

Let us put in action this idea. When evaluating any function f in x, the numerical answer which we will call f_{fl}, will contain an accumulated round-off error. We can write

$$f_{fl}(x) = f(x) + \varepsilon \tag{2.45}$$

where f_{fl} is the numerical evaluation of f using floating-point numbers and ε is the corresponding accumulated round-off error.

Similarly, evaluating $f(x+h)$ numerically will produce a numerical answer $f_{fl}(x+h)$ with a corresponding round-off error ε':

$$f_{fl}(x+h) = f(x+h) + \varepsilon' \tag{2.46}$$

Using Theorem 2.4, we can further write

$$f_{fl}(x+h) - f_{fl}(x) = f(x+h) - f(x) + \varepsilon' - \varepsilon \cong f'(x) \cdot h + \varepsilon' - \varepsilon \quad (2.47)$$

Rearranging (2.47) results in (recall that $|a+b| \geq |a| + |b|$):

$$\left| f_{fl}(x+h) - f_{fl}(x) \right| - \left| f'(x) \cdot h \right| \geq \left| \varepsilon' - \varepsilon \right| \quad (2.48)$$

Note that the round-off errors ε and ε' will either add up or cancel each other out, depending on the value of x and h.

By evaluating Equation (2.48) for small values of h, a conservative estimation of ε is obtained. Indeed, if h is small, it seems reasonable to assume that the round-off errors ε and ε' are similar. Their difference can be either equal to zero or to $|2\varepsilon|$, depending on whether the round-off errors add up or cancel out. Consequently, for small values of h:

$$\left| f_{fl}(x+h) - f_{fl}(x) \right| - \left| f'(x) \cdot h \right| \geq 2|\varepsilon| \quad (2.49)$$

where ε is the accumulated round-off error when numerically evaluating f in x. Hence, Equation (2.49) can be employed to obtain a conservative estimation of the round-off error.

Note that so far, our analysis has assumed that x is free of error. In practice, this is not the case. The reader is encouraged to repeat calculations from Equation (2.45) to (2.49) and verify that the input error on x will not change the final result (2.31). In fact, any effect an input error may produce will eventually be eliminated in Equation (2.47).

In practice, using Equation (2.49) requires an adequate choice of h. A lower limit is given by the effect of the round-off error on the quantity $x+h$. One needs to choose h large enough such that $x+h$ and x can be distinguished numerically. In other words, h needs to be at least equal to $0.5\varepsilon_m \cdot x$ (refer to Observation 2.2). As for the upper limit of h, there is no general rule that can be given. The value of h must be small enough such that the approximation $f(x+h) - f(x) \cong f'(x) \cdot h$ remains acceptable. Plotting $\left| f_{fl}(x+h) - f_{fl}(x) \right|$ versus h can help in judging if this is the case. Example 2.11 gives an illustration of how Equation (2.49) can be used to estimate the introduced round-off error in the case of the square root function.

Example 2.11. Round-Off Errors of the Square Root Function

This example studies how the function $y = f(x) = \sqrt{x}$ introduces round-off errors when evaluated numerically.

As a first example, let us estimate the round-off error when evaluating $\sqrt{16}$ numerically. In this case, we know that $\sqrt{16} = 4$. This allows us to verify our methodology. Octave computes

```
>> sqrt(16)-4
ans = 0
```

No digits are lost during the evaluation of $\sqrt{16}$ as all 16 decimal digits of the solution are correct. Now, let us reach the same conclusion by applying Equation (2.49), to estimate the round-off error associated with the numerical solution of $\sqrt{16}$. For this, we compute numerically

$$\Delta y_{fl} = \left| f_{fl}(16+h) - f_{fl}(16) \right| = \left| \sqrt{16+h} - \sqrt{16} \right| \tag{2.50}$$

for a range of values of h. We use, as a lower limit, $0.5 \cdot 16\varepsilon_m$ and then increase it logarithmically (by multiplying h each time by 2). Let us note that for $h = 0.5 \cdot 16\varepsilon_m$ we obtain $\Delta y_{fl} = 0$:

```
>> sqrt(16+8*eps)-sqrt(16)
ans = 0
```

This indicates the round-off errors introduced when numerically evaluating the functions must have canceled all significant digits. For larger values of h, this is no longer the case, as seen in Figure 2.6 (a). Figure 2.6 (a) shows that approximation $f(x+h) - f(x) \cong f'(x) \cdot h = \dfrac{1}{2\sqrt{x}} h$ is reasonable over the range of h values.

Figure 2.6 (b) depicts the difference between Δy_{fl} and $f'(x) \cdot h = \dfrac{1}{2\sqrt{x}} h$. This is, according to Equation (2.49), an estimation of twice the round-off error introduced when evaluating numerically $\sqrt{16}$. The figure reveals that the round-off error associated with the numerical computation of $\sqrt{16}$ is about $2 \cdot 10^{-16}$. This is smaller than $0.5 \cdot \sqrt{16}\varepsilon_m \cong 4.5 \cdot 10^{-16}$, which means that all the computed decimal digits are correct.

The round-off error introduced when numerically evaluating the square root function will depend on the value of x where this evaluation takes place. Let us repeat the same exercise, but this time for $\sqrt{401}$. In this case, Octave computes

```
>> format long
>> sqrt(401)
ans = 20.02498439450079
```

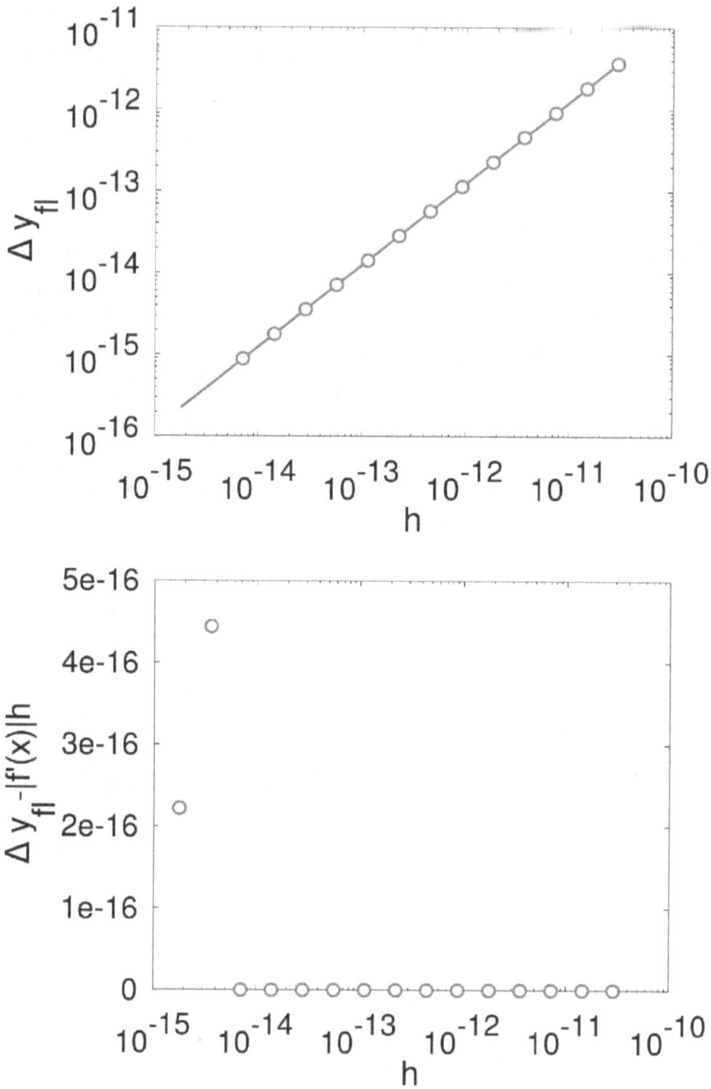

FIGURE 2.6 (a) Plot of $\Delta y_{fl} = \left|\sqrt{16+h} - \sqrt{16}\right|$ versus h. The straight line is y= $f'(16)h = 1/2\sqrt{16} \cdot h$. (b) Plot of $\Delta y_{fl} - f'(16)h$ versus h.

Let us try to estimate how many of the computed digits are correct. For this, we need to study the behavior of the numerical evaluation of

$$\Delta y_{fl} = \left|f_{fl}(401+h) - f_{fl}(401)\right| = \left|\sqrt{401+h} - \sqrt{401}\right| \tag{2.51}$$

as a function of h. Again, we choose a range of h with a lower bound of $0.5 \cdot 401\varepsilon_m$.

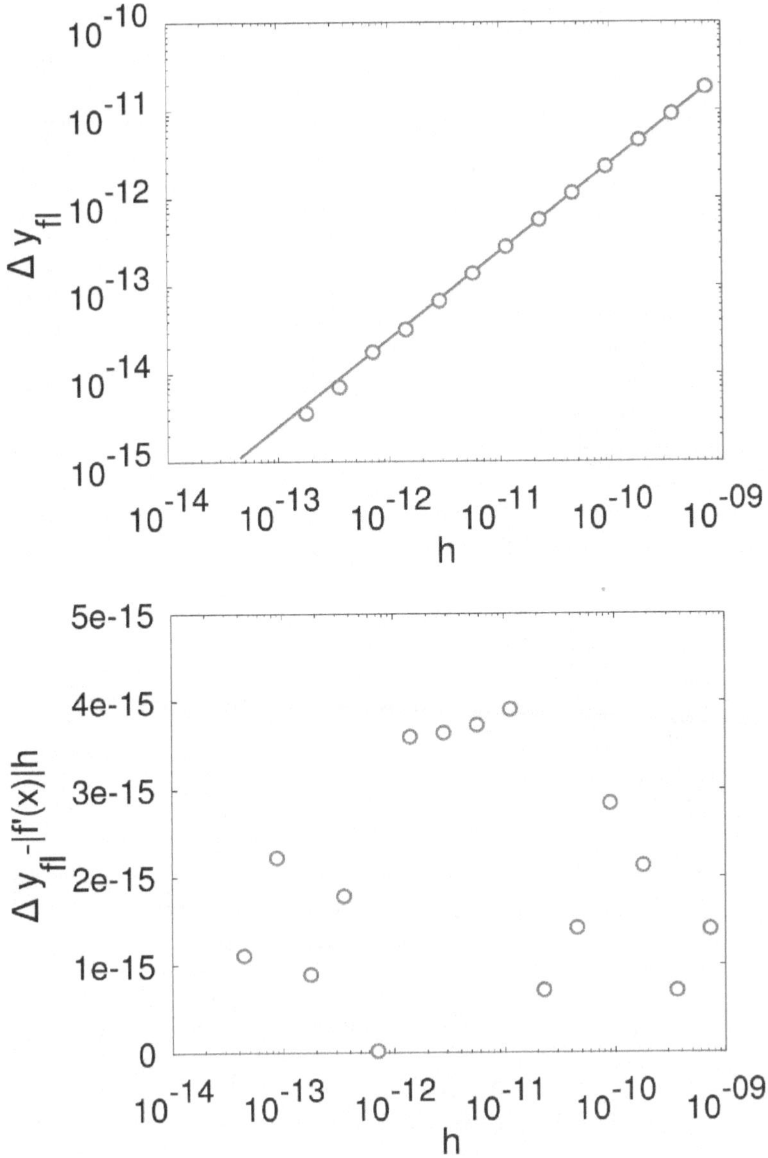

FIGURE 2.7 (a) Plot of $\Delta y_{fl} = \left|\sqrt{401+h} - \sqrt{401}\right|$ versus h. The straight line is y=$f'(401)h = 1/2\sqrt{401}\cdot h$. (b) Plot of $\Delta y_{fl} - f'(401)h$ versus h.

Figure 2.7 (a) confirms that the approximation $f(x+h) - f(x) \cong f'(x)\cdot h = \frac{1}{2\sqrt{x}}h$ is reasonable for all h values considered. From Figure 2.7 (b), we can estimate the round-off error when evaluating $\sqrt{401}$. We find a round-off error of about $2 \cdot 10^{-15}$. This indicates that the last digit (maybe even the last two, depending on how the round-off error is added or subtracted) of the numerical solution of $\sqrt{401}$ is wrong.

2.4 ROUND-OFF AND TRUNCATION ERROR TRADE-OFF

As discussed so far, when performing numerical calculations, two sources of errors are present. The round-off errors cannot be avoided when using floating-point numbers. The truncation errors occur when some form of mathematical approximation is employed. Both sources contribute to the total error. For obvious reasons, it is desirable to minimize their effect.

Reducing the impact of the round-off error can solely be accomplished by increasing the number of digits used in calculations. Reformulating the mathematical problem may mitigate the round-off error impact but not as efficiently as employing a larger number of digits during computations.

The truncation error, on the other hand, can be reduced via two strategies derived from the Taylor series expansion. The first strategy consists of employing a higher order expansion, whereas the second is to conduct the expansion as locally as possible, i.e., by reducing the magnitude of h in Equation (2.6). However, both strategies have their limitations. As the truncation error is reduced by using a higher order expansion, the higher order terms themselves become of increasingly small magnitude. This may lead to an increase in the round-off error, thereby negating the impact of a higher order expansion.

Example 2.12. Round-Off and Truncation Error Trade-Off in Taylor Series

Let us estimate Euler's number with as many correct digits as possible using a Taylor series expansion $T_k(x)$ of the exponential function around zero given, as shown by Equation (2.3).

The exact value of Euler's number to 16 decimal digits is $e = 2.718281828459045$. Figure 2.8 depicts the true error when approximating the 16-digit Euler's number with the kth order Taylor series expansion $T_k(x=1)$ of the exponential function. As expected, the error decreases with increasing order k. But, when the 17th order is reached, the error no longer decreases. This is due to the additional terms in the Taylor series expansion that have become small enough such that the numbers $T_k(1)$ and $T_{k+1}(1)$ are now undistinguishable numerically:

```
>> format long
>> Tk = @(k) sum(1./factorial(linspace(0,k,k+1)));
>> Tk(17)
```

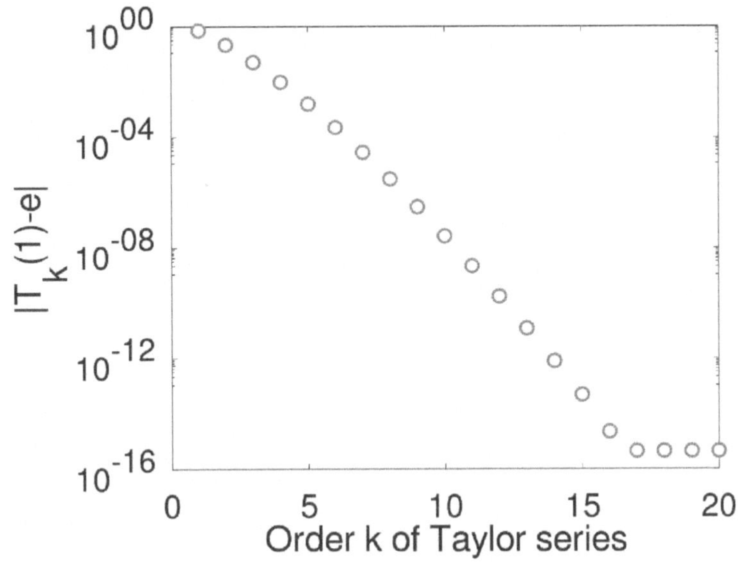

FIGURE 2.8 True error when approximating Euler's number e with a Taylor series expansion of order k. The true value to 16 digits was used to represent Euler's number e.

```
ans = 2.718281828459046
>> Tk(17)+1/factorial(18)
ans = 2.718281828459046
```

The numerical evaluation of the Taylor series $T_k(1)$ cannot approximate e to 16 decimal digits despite using high order expansions. For example, if we use an expansion of the order $k = 19$, the truncation error is about

$$\text{Truncation error} \cong \frac{1}{20!} \cong 4 \cdot 10^{-19} \tag{2.52}$$

However, the total error is about $\left|T_{19}(1) - e\right| \cong 4 \cdot 10^{-16}$:

```
>> abs(Tk(19)-e)
ans = 4.440892098500626e-16
```

which is larger than the truncation error.[8] Consequently, the round-off error must have become significantly larger than the truncation error and limits the accuracy of the result.

The next example illustrates the issues that algorithms face if one attempts to reduce the truncation error by conducting the expansion as locally as possible.

Example 2.13. Round-Off and Truncation Error Trade-Off

A simple algorithm to estimate numerically the first derivative of a function $f(x)$ can be developed based on the first order Taylor series expansion of f:

$$f(x+h) = f(x) + hf'(x) + \mathcal{O}(h^2) \tag{2.53}$$

After rearrangement and using Theorem 2.3, it follows:

$$f'(x) = \frac{f(x+h) - f(x)}{h} + \mathcal{O}(h) \tag{2.54}$$

Let us apply Equation (2.54) to the exponential function $f(x) = e^x$. We want to estimate the first derivative in $x = 0$. The numerical approximation becomes

$$\left.\frac{d}{dx}e^x\right|_{x=0} \cong \frac{e^h - 1}{h} \tag{2.55}$$

whereas the exact value is $f'(0) = 1$. Figure 2.9 displays the true absolute error in the function of h of the numerical estimation (2.55) computed as

$$\text{True error} = \left|\frac{e^h - 1}{h} - 1\right| \tag{2.56}$$

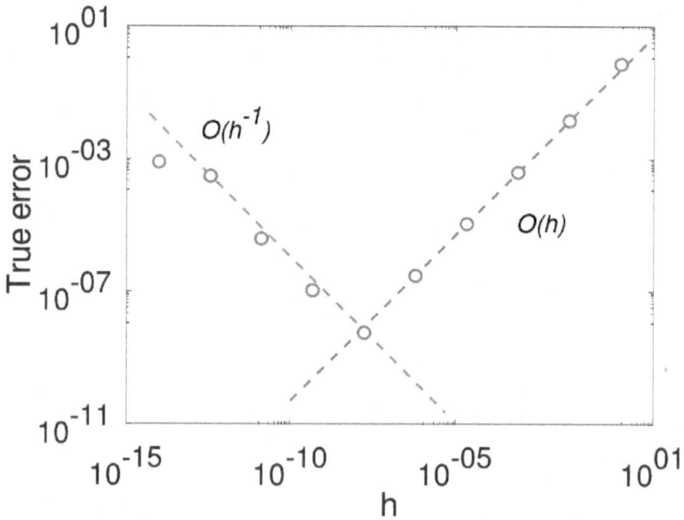

FIGURE 2.9 True error when numerically approximating the first derivative of the exponential function $f(x) = e^x$ in $x = 0$ with Equation (2.54) in function of h. The error has two contributions: the truncation error of order $\mathcal{O}(h)$, and the round-off error, of order $\mathcal{O}(h^{-1})$.

As expected, as h decreases, so does the error. At first, the total error consists essentially of the truncation error. This error decreases with an order of h as predicted by Equation (2.54). However, at a certain point where h reaches values lower than 10^{-8}, the true error starts increasing again. This occurs because the round-off errors are no longer negligible and contribute significantly to the total error.

It is possible to obtain a theoretical expression for the total error by careful analysis of the approximation of the first derivative:

$$\begin{aligned}\text{True error} &= \left| f'(x) - \frac{f_{fl}(x+h) - f_{fl}(x)}{h} \right| \\ &= \left| f'(x) - \frac{f(x+h) + \varepsilon_1 - f(x) - \varepsilon_2}{h} \right|\end{aligned} \tag{2.57}$$

where f_{fl} stands for the floating-point number representation of the result of the function evaluation, and ε_1 and ε_2 are the round-off errors. Rearranging and simplifying these expressions leads to

$$\begin{aligned}\text{True error} &= \left| f'(x) - \frac{f(x+h) - f(x)}{h} + \frac{\varepsilon_1 - \varepsilon_2}{h} \right| \\ &\le \left| f'(x) - \frac{f(x+h) - f(x)}{h} \right| + \left| \frac{\varepsilon_1 - \varepsilon_2}{h} \right|\end{aligned} \tag{2.58}$$

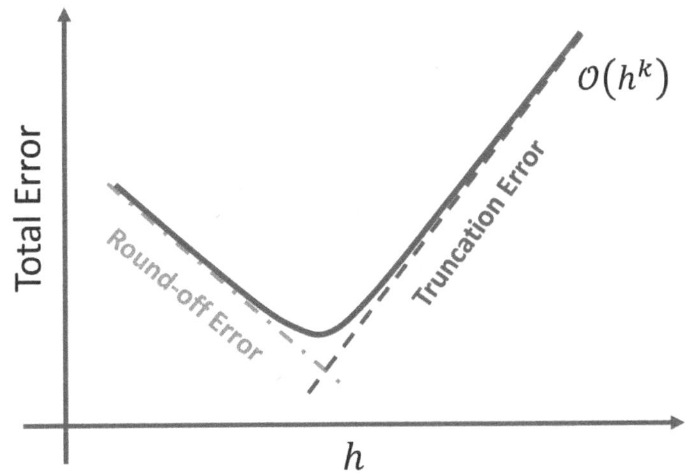

FIGURE 2.10 Round-off and truncation error trade-off. Total error, round-off, and truncation errors are shown. In this case, the approximation is based on a series expansion of order $\mathcal{O}(h^k)$.

The true error is the sum of two contributions. The first is the truncation error, which, in our case, is of the order $\mathcal{O}(h)$. The second is the accumulated round-off error, which, in this case, is of order $\mathcal{O}(h^{-1})$.

The two aforementioned examples illustrate the important concept of round-off and truncation error trade-off. The total error consists of two contributors: the round-off and the truncation error. While the truncation error may be reduced by employing higher order expansions or conducting more local expansions, the round-off errors may increase. This showcases the limitations of numerical methods and demonstrates how the total error cannot be decreased infinitely.

Figure 2.10 illustrates the trade-off between truncation and round-off errors. In this case, the approximation is based on a series expansion of order $\mathcal{O}(h^k)$. As h decreases, the total error decreases according to $\mathcal{O}(h^k)$ because the round-off error is still several order of magnitudes smaller than the truncation error. However, there comes a point where the truncation error becomes of similar magnitude and eventually significantly smaller than the truncation error. Round-off errors will limit the accuracy of the result. Usually, round-off errors will increase as h decreases. Figure 2.10 is proof that running an algorithm with as many local approximations (small values of h) as possible will not lead to accurate results. On the contrary, for too small values of h, the results may become less accurate than for large values of h.

2.5 EXERCISES

2.5.1 ROUND-OFF ERRORS

1. In Octave, compute 4.9–4.845. Compare it with the number 0.055. What do you observe? Can you explain this observation?

2. Using Octave, find the binary representation of the decimal number 0.1. What do you observe?

3. In this exercise, we want to illustrate how numerical calculations can depend on the way they are conducted.

   From elementary trigonometry, we know that $\sin^2 x + \cos^2 x = 1$. Using Octave, evaluate the following two expressions: $f(x) = \sin^2 x + \cos^2 x - 1$ and $g(x) = 1 - \sin^2 x - \cos^2 x$ for $x = 0.1$.
   What do you observe?

4. For the following numbers a find numbers $x \neq 0$ such that the equation $a + x = x$ holds when evaluated numerically with floating numbers:

   a. $a = 291$
   b. $a = 8924$

c. $a = 32$
d. $a = 48098$

5. Reproduce Figure 2.1 by computing with Octave $y = 1 + \Delta x - 1$ for values of Δx from 0 to $3 \cdot 10^{-15}$.

   Repeat the exercise by computing with Octave $y = 1 - 1 + \Delta x$ instead of $y = 1 + \Delta x - 1$. Explain why the results are radically different.

2.5.2 Truncation Errors

6. Provide a numerical approximation of Euler's number using the series expansion of the exponential function. Repeat these calculations while including three to 10 terms and estimate for each case the truncation error (use exp(1) from Octave as the "true" value for Euler's number).

7. Provide a numerical estimation and an upper boundary of the estimated error of the following quantities using Taylor series expansions of order 3. For each case, choose an appropriate point (i.e., a point in which you know the exact value) around which to conduct the Taylor series expansion.

   a. $\sqrt{4.03}$
   b. $\cos(3.1)$
   c. $\sin(0.05)$

8. Compute the Taylor series expansion of the sine function around $x = 0$ for various orders k. What is the order of the truncation error? Construct a graph showing the truncation error in the function of h used to perform the Taylor series expansion. Add the estimated truncation error based on the remainder term.

9. Consider the function $f(h) = \sin(\pi + h)$. Construct the Taylor series expansion $T_k(h)$ of order k of the function f around $h = \pi$ for various values of k and determine the value n in the equation $\sin(\pi + h) = T_k(h) + \mathcal{O}(h^n)$. Construct a graph similar to Figure 2.4 of Chapter 2 for some values of n. Observer how $|f(h) - T_k(h)|$ becomes increasingly similar to $|K \cdot h^n|$ as $h \to 0$.

10. Consider the following two functions with their order in h: $f(h)$ of order $\mathcal{O}(h^k)$ and $g(h)$ of order $\mathcal{O}(h^n)$ where $k > n > 1$. Determine the order of the following expressions and discuss all the possible cases if required:

    a. $3f(h) + 5g(h)$
    b. $4f(h)g(h)$

c. $12f(h)g(h)-5g(h)$
d. $3f(h)\cdot h^3-5g(h)$
e. $[f(h)]^3+345$
f. $f(h)/g(h)+4g(h)$

2.5.3 Propagation of Errors

11. Using Octave, compute $\cos\pi$ and $\cos(\pi/2)$. Use the inbuilt variable pi for the numerical representation of the number π. Compare the numerical values with their exact values. Can you explain the errors following the approaches of Examples 2.9 and 2.10 of Chapter 2?

12. Repeat Example 2.9 with the function $f(x)=x^3$ when evaluated in $x=\sqrt[3]{2}$.

13. Consider the function $f(x)=(x-1)^n$. We want to evaluate this function in $f(1+\varepsilon)$ where ε is some positive small number.

    Determine the error $\Delta y=|f(1+\varepsilon)-f(1)|$ for $n=1$, $n=2$ and $n=3$ using a Taylor series expansion following the approach used in Section 2.3.1. Are the errors Δy becoming smaller or larger with increasing n?

14. Consider the two functions $f(x)=x\left(\sqrt{x+1}-\sqrt{x}\right)$ and $g(x)=\dfrac{x}{\sqrt{x+1}+\sqrt{x}}$ from Example 2.3. After proving that $f=g$ for all values of x where both functions are defined, answer the following questions:

    a. Using the methodology explained in Section 2.3.2, estimate the round-off error of $f(500)$ when using double precision floating numbers.
    b. Using the methodology explained in Section 2.3.2, estimate the round-off error of $g(500)$ when using double precision floating numbers. Verify that $g(500)$ computes all decimal digits correctly.
    c. Using the function g, which is known according to (b) to calculate all digital decimals correctly close to $x=500$ when using double precision floating numbers, compute the true round-off error of $f(500+h)$ for the values of h used in question (a) and compare them with the estimated round-off errors from question (a).

2.5.4 Round-Off and Truncation Error Trade-Off

15. Consider again Example 2.13 in which the first derivative of a function $f(x)$ is approximated by

$$f'(x)\cong\frac{f(x+h)-f(x)}{h}$$

Prove, by continuing the calculations presented in the example, that the true error can be written as

$$\text{True error} \cong \left|\frac{f'(x)}{2!}h\right| + \left|\frac{f'(x)\varepsilon_m}{h}\right|$$

where ε_m is the machine epsilon.
For the particular case of $f(x) = e^x$ and $x = 1$, what is the minimal error one can achieve when using double precision floating-point numbers? Does the result agree with Figure 2.9 of Example 2.13?

NOTES

1 The IEEE 475 standard defines how the exponent (and its sign) are encoded in bits. For the purpose of our presentation, we do not have to know how this is done technically. We can simply consider the exponent to be encoded as a decimal number p with sign. The interested reader is referred to the IEEE 475 standard.
2 As we will see in this book, there are situations in numerical algorithms where similar numbers are subtracted from each other's, leading to calculations very sensitive to round-off errors.
3 The factor 0.5 in front of ε_m comes from the rounding convention Octave is using.
4 The command format short E is used to present the answers in a more human friendly way.
5 To be precise, series (2.20) converges as well for $x = 1$, i.e. for $-1 < x \leq 1$
6 In Chapter 3, which discusses the numerical solution of algebraic equations, the mathematical expression of the conditioning number will be the inverse of the one Definition 2.6. The reason is that inputs and outputs are exactly reversed compared to the situation discussed here.
7 If $f''(x) = 0$, or higher orders of derivatives are zero, too, an even higher order expansion is required. The principle will remain the same as outlined in the text.
8 Note that the error $4 \cdot 10^{-16}$ is about the to-be-expected round-off error when representing e by a flowing point number. Indeed, $0.5 \cdot e \cdot \varepsilon_m \cong 3 \cdot 10^{-16}$.

3 Solving Algebraic Equations

In this chapter, we will look into the process of solving an algebraic equation or finding the solution of an equation. Solving an equation consists of finding a number r that, when we plug it back into the equation in question, satisfies it. In accordance with our previously presented nomenclature, we refer to such number r as a true solution to the equation. However, from the discussions we had in Chapters 1 and 2, we know that it is possible to find numbers x_r that are different than any true solution r but still satisfy the equation when evaluated numerically. It is also possible to find a true solution r that will not satisfy the equation when evaluated numerically.

To adequately address these issues, we must formulate the definition of a numerical solution of an equation using Definition 1.1 from Chapter 1. In the present chapter, we will learn how this definition can be applied in practice. To that effect, we will need two ingredients: a numerical algorithm and a verification method. The numerical algorithm must generate approximations of the solution to the equation in question. Two families of algorithms will be presented here: bracketing methods (which generate an interval in which the real root is located) and open methods (which generate a series of approximation x_i converging toward r). The second ingredient, the verification method, must verify if the generated approximations are acceptable for our purpose (i.e. if the error of our approximation is below a targeted accuracy). We will learn here that this verification step can become quite challenging depending on the nature of the equation to be solved.

Besides studying how equations are solved numerically, we will also compare the speed of the various algorithms investigated. This fuzzy statement "speed of an algorithm" will be formulated precisely and mathematically with the help of the concept of order of convergence. It will allow us to compare algorithms and answer the question of how much calculation effort is required to reach a certain approximation accuracy. But, maybe, and even more importantly, it will allow us to detect potential issues occurring during the numerical solution process.

3.1 INTRODUCTION

From Definition 1.1 of Chapter 1, the following general approach for solving an equation numerically can be derived:

1. Specify to which accuracy the equation has to be solved.

DOI: 10.1201/9781003262121-4

2. Find an approximation x_r of the solution of the equation.
3. Verify if this approximation is such that $|x_r - r|$ is smaller than the desired accuracy.

The first step, the specification of the target accuracy, is derived from the purpose of solving the equation. From the point of view of numerical methods, it is an external input. The second step is achieved by applying a numerical algorithm to generate the approximation. Finally, the third step is the key step. During the coming sections, we will describe methods that allow us to verify, without the knowledge of the true solution r, if our approximation is such that $|x_r - r|$ is smaller than the desired accuracy. If the approximation isn't acceptable (the error is larger than the target accuracy), another approximation needs to be generated.

There is actually a further problematic to be addressed when numerically solving equations. Can the desired accuracy with a given algorithm be reached or not? The answer to this question not only depends on the selected algorithm but also on the number of digits used in the calculations. From Theorem 2.1 of Chapter 2, we know that the number r, when represented as a floating-point number, has a typical round-off error of $0.5 \cdot r \cdot \varepsilon_m$ where ε_m is the machine epsilon. Consequently, even an ideal algorithm is not expected to find an approximation to r with an accuracy better than $0.5 \cdot r \cdot \varepsilon_m$. As we will see later, algorithms cannot always reach that accuracy, an effect we encountered already in Example 2.13 of chapter 2. One question we will address in this chapter is how we will estimate, with a given number of digits used in the calculations, the best possible approximation (i.e., what is the best accuracy). As we will learn, the answer to this question depends on the number of digits used in the calculations, i.e., the value of the machine epsilon ε_m, but also on the equation we need to solve. It is important to realize that a given algorithm may be able to reach a certain accuracy for a certain equation but may not be able to reach that same accuracy for another equation.

To end this general discussion on solving numerical equations, let us recall some general terminology on equations. There are different ways to formulate an equation. In this chapter, we will use two common approaches: root fining problems and fixed-point finding problems. We will also restrict ourselves to finding solutions of equations that are real numbers. The reader interested in finding solutions that are complex numbers is referred to other textbooks. The root-finding problems refer actually to the roots of a function and are outlined in Definition 3.1:

Definition 3.1. Root of a Function

A real number r is a root of a real function $y = f(x)$, if $f(r) = 0$.

In the reminder of the text, we will often use equivalently the terms roots of a function and roots of an equation. Similarly, a fixed-point finding problem is defined by Definition 3.2.

Definition 3.2. Fixed-Point

A real number r is a fixed-point of a real function $y = g(x)$, if $g(r) = r$.

There are certainly many more ways to arrange an equation. Interestingly, we will learn that the convergence of a numerical algorithm can depend on the way the equation is formulated.

3.2. BRACKETING METHODS

Bracketing methods aim to find an interval $[a, b]$ that contains the true solution r to an equation we are solving numerically. The equation-solving-problem is formulated as a root-finding problem of the form $f(x) = 0$. Bracketing methods proceed iteratively. They start with an initial interval $[a_0, b_0]$, which contains r, and then progressively produce a series of intervals $[a_1, b_1], [a_2, b_2], \ldots, [a_i, b_i]$, still containing r but of decreasing size, meaning $[a_{i+1}, b_{i+1}] \subset [a_i, b_i]$ as i inceases. Hence, the true solution r is being determined with increasing accuracy.

The core of bracketing methods is Theorem 3.1, which is used to decide if a given interval $[a, b]$ contains or not a root of a function f.

Theorem 3.1. Root of a Continious Real-Valued Function

Let f be a real-valued continuous function on the interval $[a, b]$ such that $f(a)f(b) < 0$, then there exists at least one number r in the interval $[a, b]$ such that $f(r) = 0$.

Let us add some comments on Theorem 3.1. First, it is important to realize that Theorem 3.1 is an existence theorem. This means that if $f(a)f(b) < 0$, then there exists a root in $[a, b]$. But if there is a root in the interval, it doesn't imply that $f(a)f(b) < 0$. For example, consider $f(x) = (x-1)(x+1)$. There is a root in $r = 1$, but $f(-2)f(2) > 0$ (Figure 3.1a).

There can as well be situations where there exists a root, but it is impossible to give an interval $[a, b]$ such that $f(a)f(b) < 0$. Consider $f(x) = x^2$. Even though there is a root in $r = 0$, it is impossible to find any interval $[a, b]$ such that $f(a)f(b) < 0$ (Figure 3.1b).

Furthermore, Theorem 3.1 doesn't state anything about the number of roots within the interval $[a, b]$ when $f(a)f(b) < 0$. There could be one root, as in Figure 3.1a, where there is a unique root within the interval $[-2, 0]$. But, there could also be more than one root, as in Figure 3.1c, where there are three roots in the interval $[-1.5, 1.5]$.

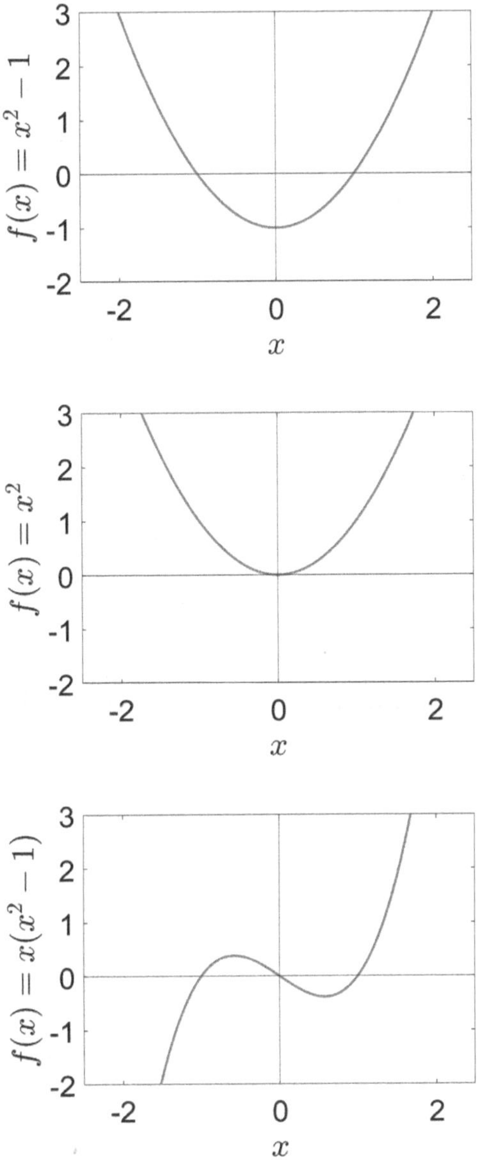

FIGURE 3.1 Illustrations for Theorem 3.1. (a) Even that $f(-2)\,f(2) > 0$, there are roots in the interval $[-2, 2]$. However, as $f(-2)\,f(0) < 0$ there is at least one root in the interval $[-2, 0]$. (b) Even $f(x) = x^2$ has a root in $r = 0$, it is impossible to find any interval $[a, b]$ such that $f(a)\,f(b) < 0$. (c) There can be more than one root in an interval where $f(a)\,f(b) < 0$: in the interval $[-1.5, 1.5]$ there are three roots. (d) Theorem 3.1 requires the function to be continuous. The none-continuous function $f(x) = 1/x$ has no root in the interval $[-1, +1]$ but $f(-1)f(1) < 0$.

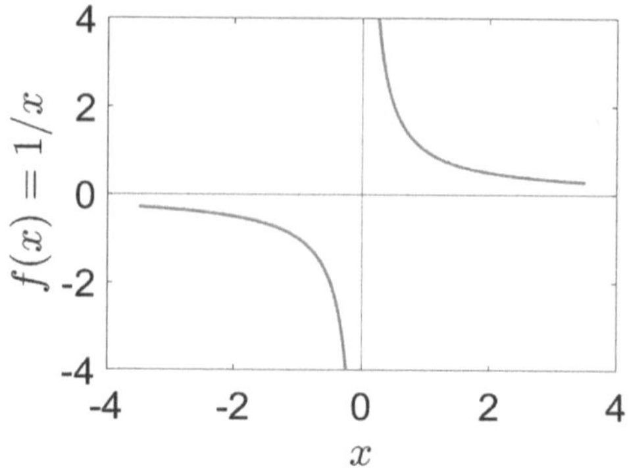

FIGURE 3.1 Continued.

Finally, let us stress the importance of the function f to be a continuous function. Consider the non-continuous function $f(x) = 1/x$. There is no root in the interval $[-1, +1]$. But, $f(-1)f(1) < 0$ (Figure 3.1d) shows that applying the criteria $f(a)f(b) < 0$ to a none-continuous function can lead to wrong conclusions.

As mentioned, Theorem 3.1 is the core of bracketing methods. It allows us to be confident that a given interval contains at least one root. Consequently, once we have such an interval we can give an approximation of the root and even an estimation of the error. Indeed, if an interval $[a, b]$ contains the root, then the mid-point of this interval is an approximation x_r of the root:

$$x_r = \frac{a+b}{2} \tag{3.1}$$

and as the root must be within this interval, the absolute error $|x_r - r|$ can not be larger than half of the size of the interval:

$$|x_r - r| \leq \left|\frac{b-a}{2}\right| \tag{3.2}$$

The major advantage of bracketing methods is their ability to provide an error estimation of the produced approximation. This is a unique characteristic of bracketing methods that is lacking in other families of algorithms.

We are now in the position of outlining a bracketing algorithm to estimate the root of a function.

Algorithm 3.1. Bracketing Method

Input: A real-valued continuous function f and a target accuracy TOL

Output: An approximation x_r of a root of the function f with an absolute error $|x_r - r|$ lower than the target accuracy TOL

Algorithm:

1. Find an interval $[a_0, b_0]$ such that $f(a_0)f(b_0) < 0$
2. As long as $\left|\frac{b_i - a_i}{2}\right| > TOL$

   find an interval $[a_{i+1}, b_{i+1}]$ such that $[a_{i+1}, b_{i+1}] \subset [a_i, b_i]$ and $f(a_{i+1})f(b_{i+1}) < 0$
3. Compute the approximation of the root as

$$x_r = \frac{a_i + b_i}{2}$$

   and the upper boundary of the absolute error as

$$|x_r - r| \le \left|\frac{b_i - a_i}{2}\right|$$

Note that the algorithm does not specify how to find the initial interval $[a_0, b_0]$ needed to start the bracketing method. It has to be found by the user of the algorithm. This can be done, for example, by plotting the function f, by an incremental search method or trial and error.

At this stage, nothing is said about how to compute a new interval $[a_{i+1}, b_{i+1}]$ from the interval $[a_i, b_i]$ in step number 2. The procedure to find $[a_{i+1}, b_{i+1}]$ depends on which method or algorithm we decide to use. There are several existing methods, but next, we will introduce two popular examples and show how this calculation is done in each of these two methods.

3.2.1 Bisection Algorithm

The bisection algorithm implements the bracketing Algorithm 3.1 by cutting the interval $[a_i, b_i]$ into two equal-sized intervals in order to produce the next interval $[a_{i+1}, b_{i+1}]$. After cutting the interval $[a_i, b_i]$, the bisection algorithm decides in which of the two new sub-intervals the root is located using Theorem 3.1.

Algorithm 3.2. Bisection Method

Input: A real-valued continuous function f and a target accuracy TOL
Output: An approximation x_r of a root of the function f with an absolute error $|x_r - r|$ lower than the target accuracy TOL

Algorithm:

1. Find an interval $[a_0, b_0]$ such that $f(a_0)f(b_0) < 0$
2. As long as $\left|\frac{b_i - a_i}{2}\right| > TOL$

Compute the cut point $x_c = \dfrac{a_i + b_i}{2}$

If $f(a_i)f(x_c) < 0$ then $[a_{i+1}, b_{i+1}] = [a_i, x_c]$ else $[a_{i+1}, b_{i+1}] = [x_c, b_i]$

3. Compute the approximation of the root as

$$x_r = \frac{a_i + b_i}{2}$$

and the upper boundary of the absolute error as

$$|x_r - r| \le \left|\frac{b_i - a_i}{2}\right|$$

Let us illustrate this algorithm with an example.

Example 3.1. Bisection Method

In order to illustrate the application of the bisection method, a root-solving problem with a known true solution is chosen. Whenever investigating a new numerical algorithm, it is recommended to try it out on some problems with known true solutions before applying it to more realistic problems where the true solution is not known. In this example, we consider the following function:

$$f(x) = x(x-1)(x+3) \tag{3.3}$$

This function has three distinct roots. We aim to approximate the root $r = 1$ with the bisection algorithm. As an initial guess for the starting interval, we choose $[a_0, b_0] = [0.5, 2]$. According to Theorem 3.1, this interval contains indeed at least one root:

$$f(0.5)f(2) = -8.75 < 0 \tag{3.4}$$

In a real problem, where the true root is not known, the starting interval has to be determined, for example, using a plot of the function f.

The bisection algorithm starts by cutting the interval $[a_0, b_0]$ into the two sub-intervals $[0.5, 1.25]$ and $[1.25, 2]$. Since

$$f(0.5)f(1.25) = -1.1621 < 0 \tag{3.5}$$

the interval $[0.5, 1.25]$ contains at least one root and will be used for the next iteration. This interval is again cut into two sub-intervals $[0.5, 0.875]$ and $[0.875, 1.25]$. As

$$f(0.5)f(0.875) = 0.37085 > 0 \tag{3.6}$$

the root is not located within the interval $[0.5, 0.875]$ and consequently must be in the interval $[0.875, 1.25]$. One can carry out in the same way using as many iterations as desired. In each iteration, the approximation of the root and the estimated error can be computed using Equations (3.1) and (3.2). Table 3.1 summarizes the results for the first 10 iterations. If, for example, an accuracy of 0.02 is targeted, at least six iterations would have to be carried out, and the root would be approximated as $x_r = 1.00$ with an estimated error of 0.01.

TABLE 3.1
Bisection Algorithm in order to Approximate a Root of $f(x) = x(x-1)(x+3)$

I	a_i	b_i	x_r	Error estimation
0	0.50000000	2.00000000	1.25000000	0.75000000
1	0.50000000	1.25000000	0.87500000	0.37500000
2	0.87500000	1.25000000	1.06250000	0.18750000
3	0.87500000	1.06250000	0.96875000	0.09375000
4	0.96875000	1.06250000	1.01562500	0.04687500
5	0.96875000	1.01562500	0.99218750	0.02343750
6	0.99218750	1.01562500	1.00390625	0.01171875
7	0.99218750	1.00390625	0.99804688	0.00585938
8	0.99804688	1.00390625	1.00097656	0.00292969
9	0.99804688	1.00097656	0.99951172	0.00146484
10	0.99951172	1.00097656	1.00024414	0.00073242

It is instructive to verify that for each iteration, the estimated errors are indeed larger than the true error.

By observing the behavior of the estimated error produced in Example 3.1, we notice that this error estimation is decreased by two in each iteration. This is normal as the bisection algorithm will always cut the original interval into two sub-intervals in order to produce the interval of the next iteration. For the initial guess, we can write for the initial error estimation

$$|x_0 - r| \le \left|\frac{b_0 - a_0}{2}\right| \tag{3.7}$$

Proceeding with the first iteration leads to

$$|x_1 - r| \le \left|\frac{b_1 - a_1}{2}\right| = \left|\frac{b_0 - a_0}{2^2}\right| \tag{3.8}$$

Or, in general, at iteration number i, the estimated error will become

$$|x_i - r| \leq \left|\frac{b_i - a_i}{2}\right| = \left|\frac{b_{i-1} - a_{i-1}}{2^2}\right| = \ldots = \left|\frac{b_0 - a_0}{2^{i+1}}\right| \tag{3.9}$$

If an accuracy TOL is targeted, it follows from Equation (3.10) that the minimal number of iterations n required to reach $|x_n - r| \leq TOL$ is given by

$$n \geq \frac{\log(b_0 - a_0) - \log(TOL)}{\log(2)} - 1 \tag{3.10}$$

Example 3.2. Number of Iterations Required in Bisection Method

Consider again Function (3.3) used in Example 3.1. If we start with the initial interval $[a_0, b_0] = [0.5, 2]$, how many iterations are required to reach an accuracy below 10^{-4} using the bisection method?

According to Equation (3.10), we find:

$$n \geq \frac{\log(2 - 0.5) - \log(10^{-4})}{\log(2)} - 1 = 12.9 \tag{3.11}$$

Consequently, at least 13 iterations are required.

3.2.2 False Position Algorithm

Another example of implementing the bracketing algorithm is the false-position method. It is among the oldest known algorithms and descriptions similar to this method were found in Egyptian papyri and Babylonian clay tablets from the 18th century B.C.[1] In western culture, it became later known under the Latin name *Regula Falsi* (introduced in the 16th century in Europe), which translates loosely to "the rule of falseness".

The false position method cuts the interval $[a_i, b_i]$ into two sub-intervals in order to produce the next interval $[a_{i+1}, b_{i+1}]$. But, contrary to the bisection method, it does not produce two equally-sized sub-intervals but uses instead a cut point constructed by computing the intersection of a line joining the points $(a, f(a))$ and $(b, f(b))$ with the x-axis (Figure 3.2).

The cut point x_c is computed according to (the reader is invited to derive this equation based on geometrical considerations):

$$x_c = \frac{a_i f(b_i) - b_i f(a_i)}{f(b_i) - f(a_i)} \tag{3.12}$$

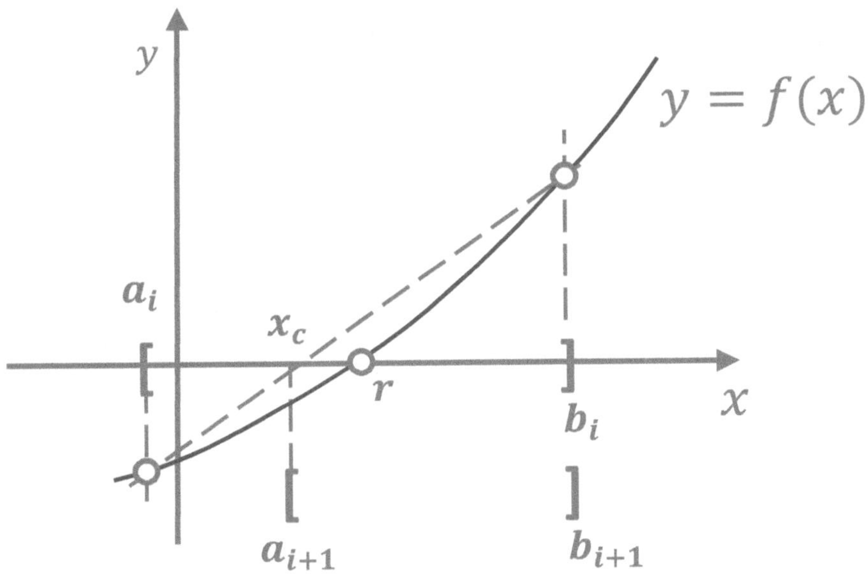

FIGURE 3.2 Graphical illustration of the false position method.

Once the False position algorithm produces the two sub-intervals from the original interval $[a_i, b_i]$, the algorithm then uses Theorem 3.1 to decide in which of the sub-intervals $[a_i, x_c]$ or $[x_c, b_i]$ the root is located.

Algorithm 3.3. False Position Method

Input: A real-valued continuous function f and a target accuracy TOL
Output: An approximation x_r of a root of the function f with an absolute error $|x_r - r|$ lower than the target accuracy TOL

Algorithm:

1. Find an interval $[a_0, b_0]$ such that $f(a_0)f(b_0) < 0$

2. As long as $\left|\frac{b_i - a_i}{2}\right| > TOL$

   Compute the cut point $x_c = \dfrac{a_i f(b_i) - b_i f(a_i)}{f(b_i) - f(a_i)}$

   If $f(a_i)f(x_c) < 0$ then $[a_{i+1}, b_{i+1}] = [a_i, x_c]$ else $[a_{i+1}, b_{i+1}] = [x_c, b_i]$

3. Compute the approximation of the root as

$$x_r = \frac{a_i + b_i}{2}$$

and the upper boundary of the absolute error as

$$|x_r - r| \leq \left|\frac{b_i - a_i}{2}\right|$$

Contrary to the bisection algorithm, the size of the new interval $[a_{i+1}, b_{i+1}]$ is not determined in advance. Consequently, the number of iterations to be carried out in order to reach a given accuracy can not be predicted using Equation (3.11). Let us illustrate this algorithm with an example.

Example 3.3. False Position Method

In order to illustrate the application of the false position method, a root-solving problem for the same function (3.3) as in Example 3.1 is chosen. Here again, we aim to approximate the root $r = 1$. As an initial guess for the starting interval, we choose $[a_0, b_0] = [0.5, 2]$ which we know to contain at least one root of f according to Theorem 3.1.

The false position algorithm starts by calculating a cut point cutting x_c of the interval $[a_0, b_0]$ according to Equation (3.12):

$$x_c = \frac{0.5f(2.0) - 2f(0.5)}{f(2.0) - f(0.5)} = 0.62069 \tag{3.13}$$

As

$$f(0.62069)f(2.0) = -8.5243 < 0 \tag{3.14}$$

the interval $[0.62069, 1.25]$ contains at least one root and will be used for the next iteration. This interval is again cut into two sub-intervals with a cut point x_c computed with Equation (3.12):

$$x_c = \frac{0.62069f(2.0) - 2f(0.62069)}{f(2.0) - f(0.62069)} = 0.72903 \tag{3.15}$$

As

$$f(0.72903)f(2.0) = -7.3665 < 0 \tag{3.16}$$

the root is located within the interval $[0.72903, 0.875]$. One can carry out in the same way for as many iterations as desired. In each iteration, the approximation of the root and the estimated error can be computed using Equations (3.1) and (3.2).

TABLE 3.2
False Position Algorithm in order to Approximate a Root of $f(x) = x(x-1)(x+3)$

i	a_i	b_i	x_r	Error estimation
0	0.50000000	2.00000000	1.25000000	0.75000000
1	0.62068966	2.00000000	1.31034483	0.68965517
2	0.72903128	2.00000000	1.36451564	0.63548436
3	0.81623347	2.00000000	1.40811674	0.59188326
4	0.88032599	2.00000000	1.44016300	0.55983700
5	0.92430065	2.00000000	1.46215033	0.53784967
6	0.95304778	2.00000000	1.47652389	0.52347611
7	0.97124541	2.00000000	1.48562270	0.51437730
8	0.98252996	2.00000000	1.49126498	0.50873502
9	0.98943811	2.00000000	1.49471906	0.50528094
10	0.99363375	2.00000000	1.49681687	0.50318313

Table 3.2 summarizes the results for the first 10 iterations. According to these calculations, the algorithm seems to converge to a number close to 1.5 and an estimated error of 0.5. This is an example of an algorithm converging with a bias and not converging to the desired value.

Example 3.3 shows that the false position algorithm in the form we have presented is not always bias-free. In the case of Example 3.2, it is possible to give an intuitive explanation for this effect. In fact, the algorithm converges to an interval $[1, 2]$. According to Equation (3.2), this produces an approximation $x_r = 1.5$ of the root with indeed a bias of 0.5. This comes from the shape of the function f. The algorithm keeps the end point 2.0 of the interval fixed while progressively moving the cut point x_c toward the root $r = 1$ (Figure 3.3). This example illustrates a major drawback of the false position algorithm. Contrary to the bisection algorithm, the false potion algorithm does not have to converge to an interval $[a_i, b_i]$ such that $|b_i - a_i| \to 0$ as $i \to \infty$. Consequently, the root cannot be located as accurately as desired.

Obviously, this effect comes from the way the approximation x_r of the root is computed in each iteration. If one would use the cut point x_c as an approximation to x_r (instead of the mid-point used in the bisection algorithm), then the algorithm would converge to the correct root. However, without the knowledge of the true value of the root, we cannot draw this conclusion. Changing the way

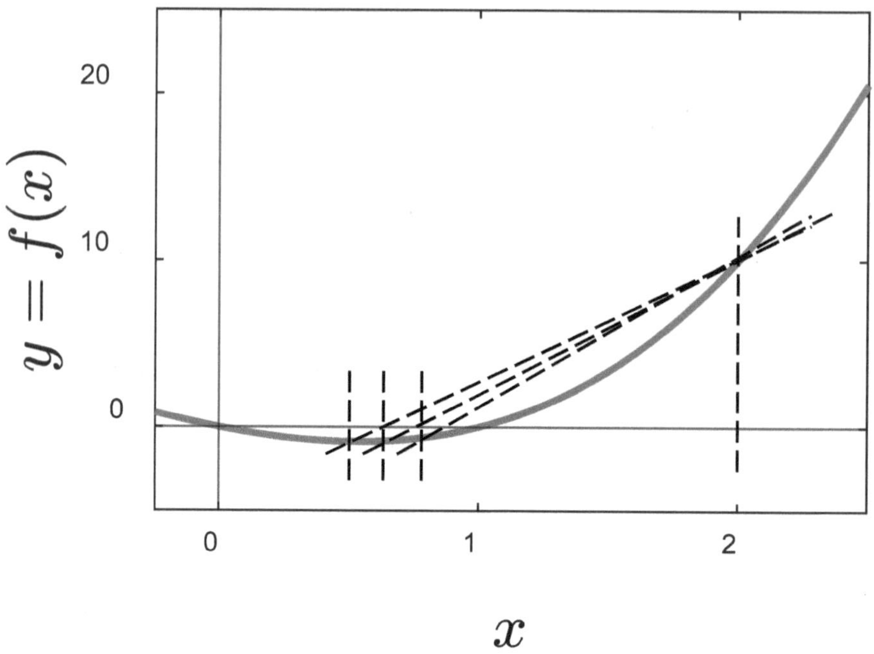

FIGURE 3.3 False position algorithm applied to the function $f(x) = x(x-1)(x-3)$. First, three iterations are shown. Note how the left point of the bracketing interval is moving slowly towards x=1 while the right point remains fixed at x=2.

of estimating the root will not change the way the successive intervals are computed. The algorithm will still converge to the interval $[1, 2]$. The only certitude we will have is that the root is located within $[1, 2]$, and the best approximation we can give is $x_r = 1.5 \pm 0.5$. Had we available a different way to estimate the error associated with the approximation x_r, then we could potentially know that $x_r = x_c$, the cut points of the intervals would be a better approximation.[2] However, we will postpone this discussion to later, as we first need to learn other methods of error estimation.

3.3 OPEN METHODS

Open methods no longer aim to find an interval containing a root but rather attempt to directly approximate r by producing an approximation x_r. Open methods proceed iteratively by starting with an initial guess x_o of the root. In each iteration, they use the approximation x_i in order to produce a hopefully better approximation x_{i+1}. As no bracketing interval is available, a new methodology to estimate the error will have to be developed. Two error estimation techniques are presented in Sections 3.4 and 3.5. In this section, we focus on presenting the algorithms of two popular open methods: the fixed-point iteration and the Newton-Raphson method. Special attention is given to the fact that the mere knowledge of an iteration scheme is insufficient to solve equations correctly. Finally, by studying various examples, we will learn that, unlike bracketing methods, open methods do not always converge. We will have to find conditions under which a given algorithm converges.

In summary, to apply open methods, the following aspects have to be mastered:

1. The mechanism of the algorithm (known as the iteration scheme).
2. The conditions under which the algorithm converges to the desired solution r of the equation.
3. A method to estimate the error of a produced approximation x_r.

The present section focuses mainly on the first aspect.

3.3.1 Fixed-Point Iteration Method

The fixed-point iteration method aims to find a fixed-point r of a function $y = g(x)$:

$$g(r) = r \tag{3.17}$$

The principle is straightforward. Starting with an initial guess x_o, the algorithm produces successive, hopefully improved approximations x_i using the following equation:

$$x_{i+1} = g(x_i) \tag{3.18}$$

The popularity of this algorithm comes, among others, from the simplicity of implementing it, as demonstrated in Example 3.4.

Example 3.4. Fixed-Point Iteration—Approximation of the Square Root of Two

Consider the following function g from which a fixed-point has to be determined:

$$g(x) = \frac{1}{2}\left(x + \frac{2}{x}\right) \tag{3.19}$$

It is straightforward to prove that finding the fixed-point of g is the same as solving

$$x^2 = 2 \tag{3.20}$$

In other words, the fixed-point algorithm will approximate the square root of two. This is a famous example from the history of mathematics. The description of this algorithm was found on a Babylonian clay tablet (tablet number YBC 7289) dating somewhere between 1800 and 1600 B.C.

To apply the fixed-point algorithm, an initial guess x_o is required. We choose $x_o = 0.5$. We will discuss shortly how such an initial guess can be chosen. Besides the initial guess, we only need the function g. The following Octave code defines the function and performs the first iteration:

```
>> g = @(x) 0.5*(x+2/x);
>> xo = 0.5;
>> x1 = g(xo)
x1 = 2.2500
```

It is straightforward to perform a given number of iterations (10 in the example that will be further on) with the help of a for-loop and afterward construct a convergence plot (the vector `iterNumber` is an auxiliary vector that allows plotting the results in function of the iteration number and can be omitted):

```
>> x = [0.5]; iterNumber[0];
>> for i = 1:10
x = [x g(x(i))];
iterNumber = [iterNumber i];
endfor
>> plot(iterNumber, x, 'o-')
```

The results of the iterations are stored in a vector x, which grows as the algorithm progresses. Figure 3.4a displays the iterations x_i in function of i. The produced values x_i indeed converge to a fixed value.

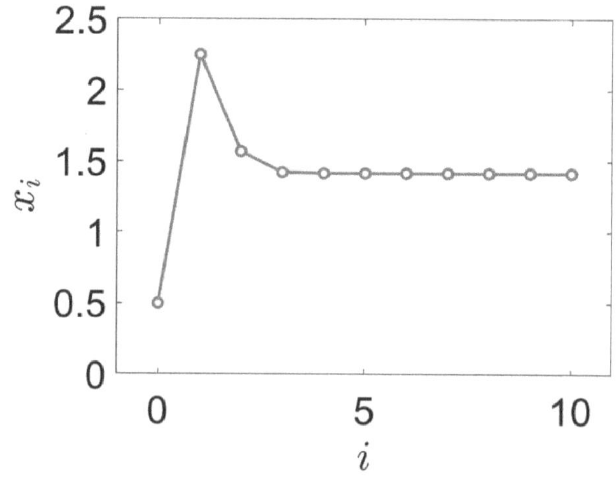

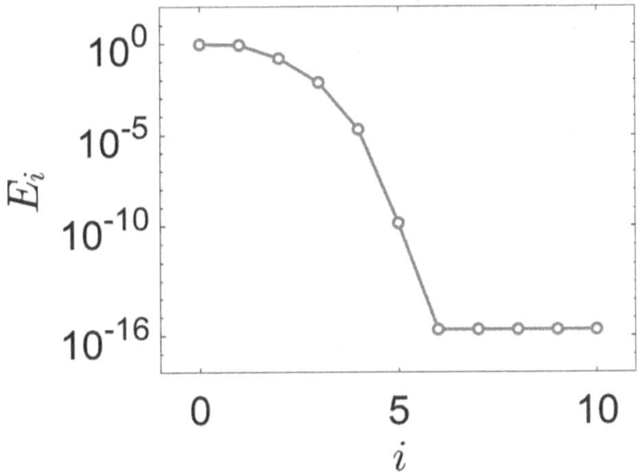

FIGURE 3.4 (a) Convergence plot of the fixed-point iterations method to approximate the square root of two. (b) True error in function of performed iterations.

In the example, we already know that the true solution is $r = \sqrt{2}$, and the true error $E_i = \left|x_i - \sqrt{2}\right|$ can be computed:

```
>> err = abs(x-sqrt(2));
>> semilogy(iterNumber, err, 'o-')
```

As illustrated in Figure 3.4b, the true error drops until iteration number six and remains constant afterward. This is because the algorithm has reached the limit imposed by round-off errors. The true error E_6 of iteration number six is (note that iteration number six is stored in position number seven of the vector `err`, as Octave starts numbering the indexes of vectors with 1):

```
>> err(7)
ans = 2.2204e-16
```

In fact, due to round-off errors, the two numbers $g(x_6)$ and $g(x_6 + E_6)$ can no longer be distinguished:

```
>> g(x(7)) - g(x(7)+err(7))
ans = 0
```

This explains why the algorithm can no longer generate better approximations. Note that the true error in iteration six is very close to the round-off error when storing the number $\sqrt{2}$ using double-precision floating-point numbers. Indeed, the round-off error is estimated by (see Chapter 2, Section 2.1.2):

```
>> 0.5*sqrt(2)*eps
ans = 1.5701e-16
```

The previous observations allow us to conclude that the fixed-point iteration algorithm, when applied to the present example, can be deemed as a very good algorithm because it converges to the true solution bias-free and with nearly the smallest possible error, which is associated with the round-off error of the true solution.

As demonstrated by Example 3.4, the fixed-point iteration method can be a very good algorithm in the sense that it converges towards the true solution, and this with an error reaching nearly the smallest possible error, only limited by the round-off error resulting from representing the true solution with a floating-point number. Let us, however, insist on the fact that, for the moment, we are unable to provide an error estimation of the produced approximations.

Let us now turn to the question as to why this algorithm is converging. To give a first intuitive understanding, it is helpful to consider a graphical representation

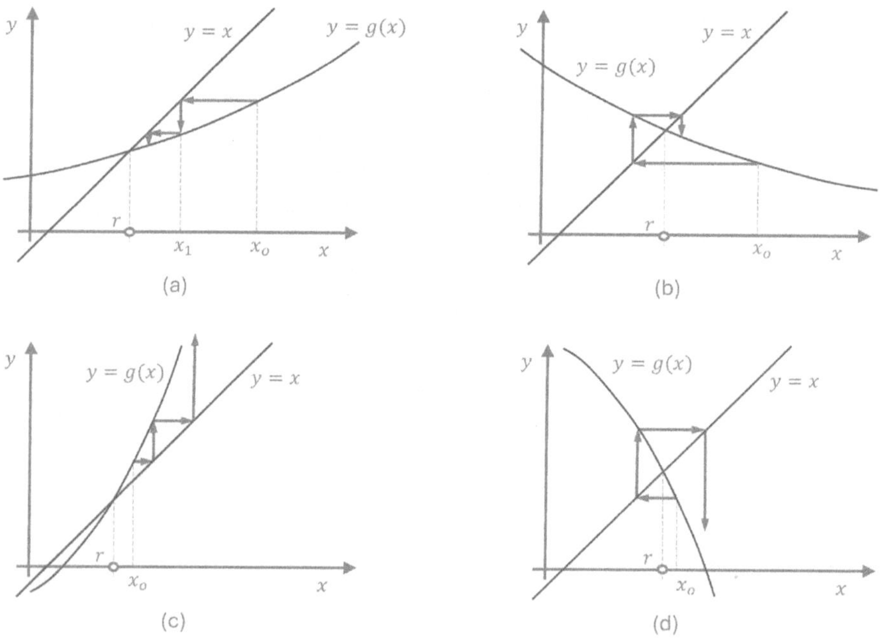

FIGURE 3.5 Four possible situations for the fixed-point iteration algorithm. The algorithm can either converge (cases a and b) or diverge (cases c and d). In cases a and c, the algorithm progresses in a staircase manner, whereas in situations b and d, a spiral-type evolution is observed. The value of the first derivative of g in the fixed-point determine which situation the algorithm will be in.

of the algorithm known as cobweb diagrams. Consider Figure 3.5a. The function $y = g(x)$ together with the function $y = x$ are displayed. The intersection of these two curves is the fixed-point $x = r$, the point that the fixed-point method aims to reach. Let us now choose an initial guess x_o. According to the fixed-point method, the first approximation x_r of r is given by $x_1 = g(x_o)$. The value x_1 can be reported graphically onto the x-axis from the value $g(x_o)$ found on the y-axis with the help of the line $y = x$. The process can be repeated in a second iteration using the evaluation of g at $x_1$to produce x_2. This scheme is repeated again and again to result in a staircase-type evolution, as depicted in Figure 3.5-a, where the successive values of x_i converge towards $x = r$. A symmetric situation is illustrated in Figure 3.5c, where this time, the successive values of x_i diverge away from $x = r$. Figure 3.5b and Figure 3.5d show two other possible cases that exist where the evolution is now in the form of a spiral, either converging towards or diverging from the fixed-point.

As can be observed in these figures, it is the value $g'(r)$ of the first derivative of g at the fixed-point that determines how the algorithm will operate. For $|g'(r)| < 1$ the algorithm will converge (in a staircase manner if $g'(r)$ is positive and in a spiral type if it is negative). The diverging case if obtained for $|g'(r)| > 1$ (again in a staircase manner if $g'(r)$ is positive and in a spiral type if negative).

Note that, besides giving an intuitive understanding of the mechanism of the fixed-point iteration method, cobweb diagrams are very helpful in selecting an initial guess x_o to start the algorithm and deciding if the algorithm will converge or not.

Theorem 3.2 formalizes the observations of Figure 3.5. We postpone the proof to Section 3.4, where we will discuss in more detail the convergence of open methods.

Theorem 3.2. Convergence of the Fixed-Point Iteration Method

If g is a continuously differentiable function with a fixed-point $r = g(r)$ and $|g'(r)| < 1$, then the fixed-point iteration algorithm $x_{i+1} = g(x_i)$ will converge when x_o is chosen sufficiently close to r.

"Sufficiently close" means that there exists a positive number δ such that for any x_o satisfying the inequality $|x_o - r| < \delta$ the algorithm will converge. This is called local convergence and is quite common for numerical algorithms.

Definition 3.3. Local Convergence

A method iteratively producing a succession of numbers x_i is called locally convergent if there exists a positive number δ such that for any x_o satisfying the inequality $|x_o - r| < \delta$ the method will converge to r.

The practical implementation of the fixed-point iteration algorithm is very simple, however, there are two major challenges that limit its applicability. First of all, the method is only locally convergent. This means one has to choose carefully the initial guess x_o. As neither the true solution r nor the radius of convergence is known, selecting an appropriate initial guess can be challenging. Secondly, as stipulated by Theorem 3.2, local convergence can be achieved only under very restrictive conditions. Furthermore, since the true solution r is not known, assessing the condition $|g'(r)| < 1$ is not feasible in practice.

In any case, plotting, when possible, a cobweb diagram can help determine whether the fixed-point method will converge or not. Unfortunately, these diagrams are of little use when aiming to construct a function g such that $|g'(r)| < 1$. Example 3.5 illustrates this aspect.

Example 3.5. Nonconverging Fixed-Point Iteration Method

Consider the problem of approximating numerically the square root of two using the fixed-point iteration method. The equation one aims to solve is $x^2 = 2$. This is not a fixed-point finding problem. One has to reformulate this equation to the form $g(x) = x$ first. This is, of course, not unique. A possible formulation is obtained by adding x to both sides of the equation $x^2 = 2$ and re-arranging:

$$x^2 + x - 2 = x \tag{3.21}$$

A few iterations for the fixed-point method reveal that our choice of $g(x) = x^2 + x - 2$ will not result in convergence despite the fact that x_o is chosen close to $r = \sqrt{2}$:

```
>> g = @(x) x^2+x-2;
>> x=[1.5]; for i=1:5; x=[x g(x(i))]; endfor;
>> x
x =
  1.5000 1.7500 2.8125 8.7227 82.8074 6937.8709
```

The results clearly show that the algorithm is diverging. The reader can verify that, indeed, $|g'(r)| > 1$.

Another simple way to reformulate the equation $x^2 = 2$ into a fixed-point finding problem is to multiply both sides by x and rearranging afterwards:

$$x^3 - x = x \tag{3.22}$$

Again, the fixed-point iteration method diverges:

```
>> g = @(x) x^3-x;
>> x=[1.5]; for i=1:5; x=[x g(x(i))]; endfor;
```

```
>> x
x =
 1.5000e+00 1.8750e+00 4.7168e+00 1.0022e+02
1.0066e+06 1.0200e+18
```

In fact, quite a bit of imagination and trial and error is necessary prior to identifying an appropriate reformulation of the original equation $x^2 = 2$ that will lead to a converging algorithm.

3.3.2 Newton-Raphson Method

The major drawback of the fixed-point iteration method is that local convergence can be achieved only under very restrictive conditions. This means that the application of the fixed-point method is not straightforward in practice. In this section, we will introduce another open method, the Newton[3]-Raphson[4] algorithm, which does not suffer from this drawback. The algorithm numerically solves equations written as a root-finding problem $f(x) = 0$. The method can be derived using the Taylor series expansion of the function f:

$$f(x_i + h) = f(x_i) + hf'(x_i) + \dots \tag{3.23}$$

Defining $h = x_{i+1} - x_i$ and keeping only terms up to the first order approximation, Equation (3.23) is written as:

$$f(x_{i+1}) = f(x_i) + (x_{i+1} - x_i) f'(x_i) \tag{3.24}$$

Solving for x_{i+1} leads to the Newton-Raphson iteration scheme:

$$x_{i+1} = x_i - \frac{f(x_i)}{f'(x_i)} \tag{3.25}$$

Besides the abstract derivation based on the Taylor series, which will turn out to be very useful for analyzing the behavior of the algorithm in Section 3.4, there is a visual interpretation of the algorithm shown on Figure 3.6. In order to determine x_{i+1}, the algorithm computes the intersection between the tangent to the function f in $x = x_i$ and the x-axis. The Newton-Raphson iteration scheme (3.25) can be derived using simple geometrical considerations based on this construction.

In order to apply the Newton-Raphson algorithm, an initial guess x_o of the root has to be provided. Such a guess has to be found by the user of the algorithm. Plotting the function f is helpful when selecting an initial guess. Note as well that once $f'(x_i)$ is calculated and the iteration scheme is simplified as much as possible, the practical implementation of the Newton-Raphson algorithm is identical to the fixed-point iteration method. Let us illustrate this algorithm with an example.

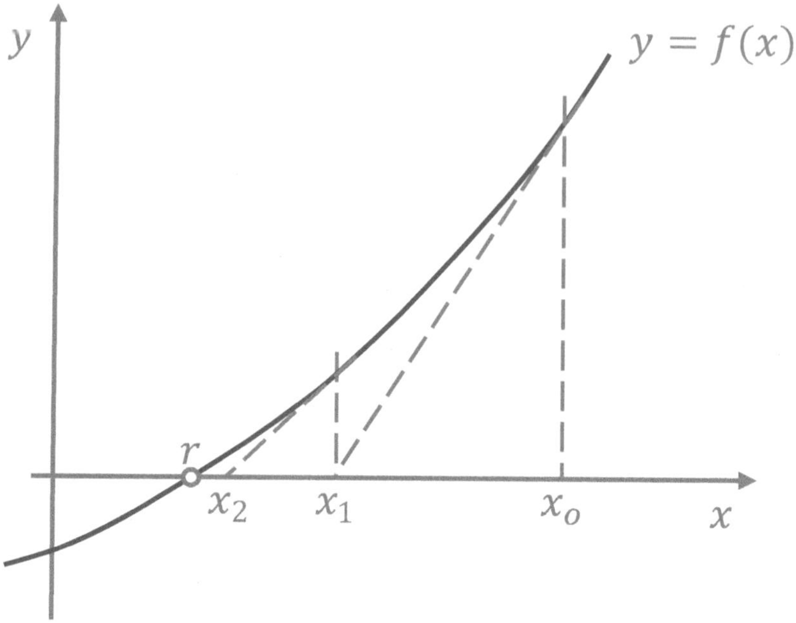

FIGURE 3.6 Graphical representation of the Newton-Raphson root-finding algorithm. The true root is r and x_{i+1} is computed from x_i by constructing the intersection between the tangent to the function f in $x = x_i$ and the x-axis.

Example 3.6. Newton-Raphson Method

In order to illustrate the application of the Newton-Raphson method, a root-solving problem for the same function (3.3) as in Example 3.1 is chosen. Again, we aim to approximate the root $r = 1$.

For the first step, let us write down explicitly the iteration scheme of the algorithm for our particular function $f(x) = x(x-1)(x+3)$. After calculating $f'(x)$ and simplifying the expression, one finds:

$$x_{i+1} = x_i - \frac{f(x_i)}{f'(x_i)} = g(x_i) = \frac{2x_i^2(x_i+1)}{3x_i^2 + 4x_i - 3} \quad (3.26)$$

With the help of the simplified expression $g(x_i)$, the implementation of the Newton-Raphson algorithm is straightforward (we chose as an initial guess $x_o = 1.5$):

```
>> g = @(x) 2*x^2*(x+1)/(3*x^2+4*x-3);
>> x=[1.5]; for i = 1:10; x = [x g(x(i))]; endfor
```

As, in this example, we know the true solution $r = 1$, we can compute the true error $E_i = |x_i - 1|$ in order to study how our algorithm behaves:

```
>> err = abs(x-1);
```

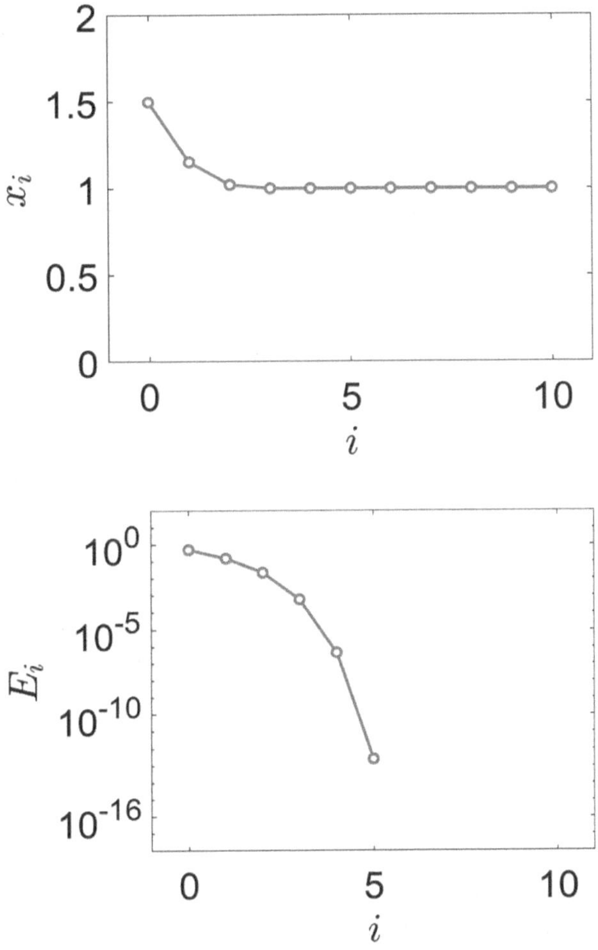

FIGURE 3.7 Newton-Raphson algorithm applied to the root-finding problem $x(x-1)(x+3)=0$. (a) Convergence plot (b) True error of the generated approximations by the algorithm.

Figure 3.7 displays the convergence plot and the true error of the generated approximations x_i. Note how, after only a few iterations, the error drops to very low values. In fact, the true error drops at iteration 6, in this example, to exactly 0 (note that the result of iteration number six is stored in position seven of the vector `err`):

```
>> format long
>> err(7)
ans = 0
```

As the true error at iteration six is exactly equal to zero, it can no longer be displayed on the semi-log plot from Figure 3.7b.

Example 3.6 illustrates an important characteristic of the Newton-Raphson algorithm: it reaches a high accuracy after only a few iterations. At this stage, without knowing the true solution, we are unable to quantify the error associated with the approximations. A methodology for estimating the errors of approximations generated by open methods needs to be developed. A further question that arises is under which conditions the Newton-Raphson algorithm converges. It turns out that contrary to the fixed-point iteration method, the Newton-Raphson algorithm is always locally convergent under very mild conditions:

Theorem 3.3. Local Convergence of the Newton-Raphson Algorithm

If f is a twice continuously differentiable function with a root $f(r)=0$, then the Newton-Raphson algorithm is locally convergent.

A partial proof, for the case where $f'(r)\neq 0$, can be conducted using Theorem 3.2. Indeed, if $f'(r)\neq 0$ the Newton-Raphson algorithm can be seen as a fixed-point iteration method with g defined as

$$g(x)=x-\frac{f(x)}{f'(x)} \tag{3.27}$$

The condition of local convergence $|g'(r)|<1$ leads directly to the conclusion of Theorem 3.3, keeping in mind that $f(r)=0$. The full proof, including the case $f'(r)=0$ will be given in the next section.

Note that Theorem 3.3 guarantees that, under mild conditions, the Newton-Raphson algorithm is always locally convergent. It does, however, not state that the algorithm is always convergent. Example 3.7 gives a counter-example.

Example 3.7. Diverging Newton-Raphson Method

Consider the root-finding problem for the function

$$f(x)=2-\frac{1}{x} \tag{3.28}$$

The true root is $r = 0.5$. Newton-Raphson algorithm writes for this example:

$$x_{i+1} = x_i - \frac{f(x_i)}{f'(x_i)} = g(x_i) = x_i - \left(2 - \frac{1}{x_i}\right) x_i^2 \tag{3.29}$$

For $x_o = 0.7$ the algorithm converges to the root.

```
>> g = @(x) x-(2-1/x)*x^2
>> x=[0.7]; for i = 1:5; x = [x g(x(i))]; endfor; x
x =
 0.70000 0.42000 0.48720 0.49967 0.50000 0.50000
```

But for $x_o = 3$, the algorithm diverges:

```
>> x=[3]; for i = 1:4; x = [x g(x(i))]; endfor; x
x =
 3.0000e+00 -1.2000e+01 -3.1200e+02 -1.9531e+05
-7.6294e+10
```

The reason is that Theorem 3.3 guarantees only local convergence. If the initial guess is too far away from the root, the algorithm may not converge.

Let us enumerate some common cases where the Newton-Raphson algorithm fails.

1. The function is not twice continuously differentiable in the root
   An example is the root-finding problem $f(x) = \sqrt[3]{x} = 0$. Here, the derivative in the root $r = 0$ is not defined, and the algorithm, which writes in this case $x_{i+1} = -2x_i$ will diverge.
2. For some functions, depending on the initial guess x_o, the algorithm may enter into an infinite loop. An example is the root-finding problem $f(x) = x^3 - 2x + 2 = 0$ with $x_o = 0$. In this case, the algorithm will enter an infinite loop, producing the sequence 0, 1, 0, 1, 0,... These are rather exceptional cases that will not be encountered in real examples. Should it still happen, a change of the initial guess will address the problem right away.
3. If, during the execution of the algorithm $f'(x_i)$, it evaluates to zero, then the algorithm will fail. As for the previous case, these are rather exceptional cases that will not be encountered in real examples. However, what can happen in practical applications is that $f'(x_i)$ becomes a small number. This is the case if $f'(r) = 0$. Indeed, as the iterations progress, x_i will become closer to r and, as such, $f'(x_i)$ will become closer to zero. The algorithm will not exactly fail, but as will be discussed in the next section, it still behaves differently compared to cases where $f'(r) \neq 0$.

However, in general, the Newton-Raphson algorithm is an excellent choice due to its guaranteed local convergence and, as will be discussed in the next section, its superior speed in converging to the root.

3.4 CONVERGENCE ORDER

So far, a few selected examples of root-finding algorithms have been presented. In this section, we want to take a step back and discuss some common properties of root-finding algorithms. Of particular interest is the question of how fast an algorithm converges to the desired solution. As will be shown, this knowledge will help estimate the error of the generated approximations. Using the knowledge of how fast an algorithm converges to the true solution to compute the estimation of the error of a generated approximation is a very common approach in numerical methods. We will use it for other types of algorithms in the coming chapters. Let us start with an example to better understand the problematic.

Example 3.8. Convergence Speed of Equation Solving Algorithms

In the example, three equation solving algorithms are compared for a same equation. The test equation used is:

$$\sin x = \cos x \tag{3.30}$$

Equation (3.30) has a known true root in $r = \frac{\pi}{4}$. The aim is to approximate this root with the bisection, the fixed-point, and the Newton-Raphson algorithms. In order to make a fair comparison, the same initial guess is provided for the three algorithms. For the fixed-point and Newton-Raphson algorithms, the initial guess $x_o = 0.5$ is used. For the bisection algorithm, the initial interval $[a_o, b_o] = [0, 1]$. This way, all three algorithms will start with an identical initial error of $\left|\frac{\pi}{4} - 0.5\right|$.

To apply the various algorithms, Equation (3.30) has to be written in the form the algorithm requires. For the bisection and Newton-Raphson algorithms, the equivalent root-solving problem $f(x) = \sin x - \cos x$ is used, and for the fixed-point method, the equivalent fixed-point finding problem $g(x) = \sin x - \cos x + x = x$ is used.

Figure 3.8 displays, in a semi-log plot, the true error of the produced approximations of the three algorithms in function of the number of iterations. Several interesting observations can be drawn. First of all, one notes that all algorithms are such that the error of their generated approximations decreases, i.e., they all converge. Furthermore, we observe that the bisection and the fixed-point methods (the circles and the squares) produce

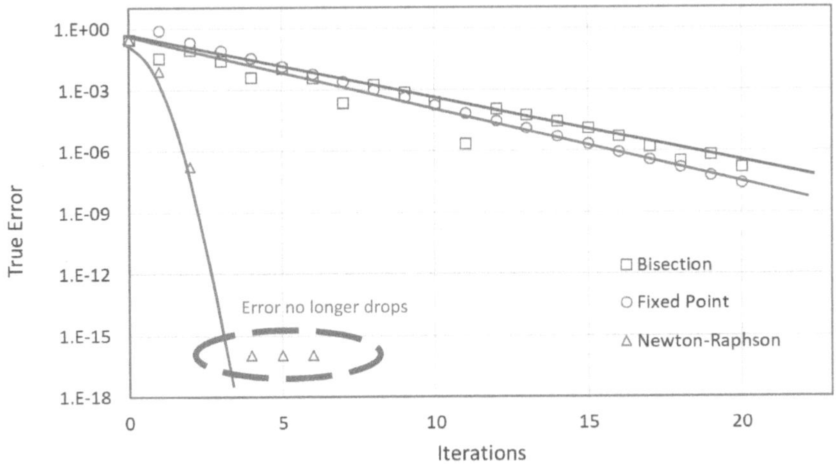

FIGURE 3.8 True absolute error $\left|\frac{\pi}{4} - x_i\right|$ in function of the number of iterations for the bisection (○), fixed-point (□), and Newton-Raphson (△) algorithm.

approximations with very similar true errors. From that point of view, both algorithms are equivalent. The fixed-point algorithm produces slightly better approximations, but the difference is insignificant. Since the bisection algorithm allows us to estimate the error associated with three approximations, this algorithm is preferred over the fixed-point iteration algorithm.

On the other hand, the Newton-Raphson algorithm (the triangles) produces significantly better approximations, and this is after only a very few iterations. One can say that the Newton-Raphson algorithm is faster than the other two. Note as well that, starting from iteration number four and forward, the error remains constant as it reaches the limit imposed by round-off errors. The reader is invited to verify that the computed approximations x_{i+1} become numerically indistinguishable from x_i for $i \geq 4$.

Example 3.8 illustrates that different algorithms can converge at different speeds. Practically, this means that after the same number of iterations, the error of the generated approximations will differ. Definition 3.4 allows comparing quantitatively the different algorithms in terms of their respective speed of convergence.

Definition 3.4. Convergence Order

Let $\{x_i\}$ be a numerical series converging to the number r. Define $E_i = |x_i - r|$. If there exist two constants $\lambda \geq 0$ and $\alpha \geq 1$ such that

$$\lim_{i\to\infty}\frac{E_{i+1}}{E_i^\alpha}=\lambda \tag{3.31}$$

exists, then the series $\{x_i\}$ is said to converge to r with order α and asymptotic error constant λ.

In the context of numerical algorithms, the series consists of the generated approximations by the algorithm, and the quantity $E_i = |x_i - r|$ is the corresponding true absolute error. Finally, it is common to use the terminology linear convergence when $\alpha = 1$ and quadratic convergence when $\alpha = 2$.

Writing Definition 3.4 in a slightly less rigorous way as

$$E_{i+1} \cong \lambda E_i^\alpha \tag{3.32}$$

gives an intuitive interpretation. When an algorithm converges with order α and asymptotic error constant λ, it means that at each iteration, the error is reduced approximatively according to (3.32). At a large enough i, E_i eventually becomes less than unity. Note that for linear convergence, λ must be lower than unity, otherwise, the series would not converge.

The usefulness of Definition 3.4 comes from the fact that the order of convergence α is essentially a fixed number for a given algorithm (i.e., it does not depend on the specific equation one aims to solve). The asymptotic error constant λ often depends on the specific equation one solves but is of less importance for characterizing the decrease of the error (for $\alpha > 1$). Indeed, from Equation (3.32), it is clear that (except when $\alpha = 1$), the parameter α controls the decrease of the error, whereas the asymptotic error constant[5] affects that decrease only in a minor way. In what follows, we will discuss the values of the convergence order and asymptotic error constant of the three algorithms presented thus far.

3.4.1 Convergence of the Bisection Algorithm

Theorem 3.4. Convergence Order of the Bisection Algorithm

If f is a continuous function over $[a, b]$ with a root within the interval $[a, b]$, the bisection algorithm converges linearly to the root with an asymptotic error constant equal to $\frac{1}{2}$.

The proof follows directly from the fact that the bisection algorithm cuts into half the root bracketing interval in each step:

$$E_{i+1} = |x_{i+1} - r| \le \frac{|x_i - r|}{2} = \frac{1}{2}E_i \tag{3.33}$$

3.4.2 Convergence of the Fixed-Point Iteration Method

As expected from Example 3.8, the fixed-point iteration method is linearly convergent. However, for the special case where $g'(r)=0$, the algorithm converges faster.

Theorem 3.5. Convergence Order of the Fixed-Point Iteration Method

Let g be a continuously differentiable function over $[a,b]$ such that $|g'(x)|=\lambda<1$ for all x in $[a,b]$. If the function g has a fixed-point r within $[a,b]$ then:

1. For $g'(r)\neq 0$, the fixed-point iteration method converges linearly to the fixed-point with asymptotic error constant $\lambda=|g'(x)|$ for any initial guess x_o within $[a,b]$.
2. For $g'(r)=0$, the fixed-point iteration method converges at least quadratically to the fixed-point for any initial guess x_o within $[a,b]$.

We start by proving the local convergence (for both cases) with the help of the mean value theorem. Indeed, with the help of the mean value theorem, one can write for the first iteration $x_1=g(x_o)$:

$$|x_1-r|=|g(x_o)-r|\leq|g'(c)||x_o-r|\leq\lambda|x_o-r| \tag{3.34}$$

Proceeding the same way for the next iterations, it follows

$$|x_i-r|=\leq\lambda^i|x_o-r| \tag{3.35}$$

As $\lambda<1$, this proves that for $i\to\infty$ the error $|x_i-r|\to 0$ and, consequently, the fixed-point iteration algorithm does indeed converge.

To prove the first case, we use again the mean value theorem to write

$$\frac{|x_{i+1}-r|}{|x_i-r|}=\frac{|g(x_i)-r|}{|x_i-r|}=|g'(c_i)| \tag{3.36}$$

where the number c_i is between x_i and r. As $|x_i-r|\to 0$ one has that $c_i\to r$ as $i\to\infty$ which proves the first case of our theorem.[6]

To prove the second case, we need to establish that the algorithm does not only converge linearly but even quadratically. Let's consider the Taylor series expansion of g around the fixed-point, taking into account that $g'(r)=0$:

$$g(x) = g(r) + g'(r)(x-r) + \frac{1}{2}g''(c)(x-r)^2 = g(r) + \frac{1}{2}g''(c)(x-r)^2 \tag{3.37}$$

with c between x and r. Using this result, it follows

$$\frac{|x_{i+1} - r|}{|x_i - r|^2} = \frac{|g(x_i) - r|}{|x_i - r|^2} = \left|\frac{1}{2}g''(c_i)\right| \tag{3.38}$$

where the number c_i is between x_i and r. As $|x_i - r| \to 0$, one has that $c_i \to r$ as $i \to \infty$, which proves the second case of our theorem. It shows as well that if $g''(c_i) \neq 0$, the asymptotic error constant will be equal to $\left|\frac{1}{2}g''(r)\right|$. If $g''(r) = 0$, the convergence is at least cubic, and one can repeat the same process by expanding the Taylor series (3.37) to one higher order.

Let us illustrate Theorem 3.5 with an example. Since we still do not know a way to estimate the error associated with the fixed-step method, we will restrict our example to a case where the true solution r is known.

Example 3.9. Convergence Order of the Fixed-Point Iteration Method

In this example, we revisit the fixed-point iteration method used to approximate the square root of two presented in Example 3.4. From Example 3.4, we know already that the algorithm is locally convergent.

Let us first investigate if the convergence is linear. For this, we compute the ratio between the true errors E_{i+1} and E_i for an increasing number of iterations and observe if this ratio converges to a fixed number:

```
>> g = @(x) 0.5*(x+2/x);
>> x = [0.5]; for i = 1:10; x = [x g(x(i))]; endfor
>> err = abs(x-sqrt(2));
>> ratio = []; for i = 1:10; ratio = [ratio
err(i+1)/err(i)]; endfor
```

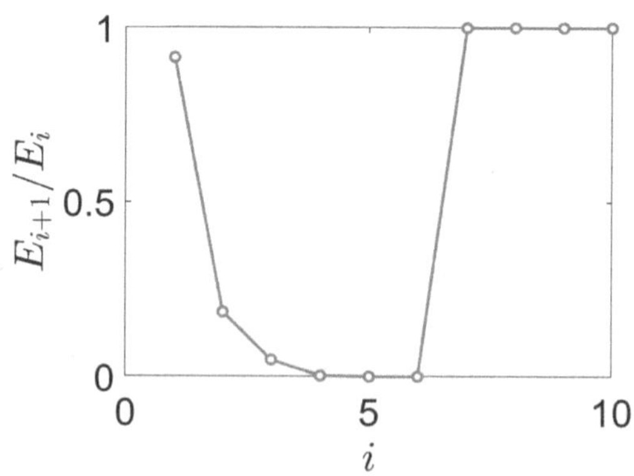

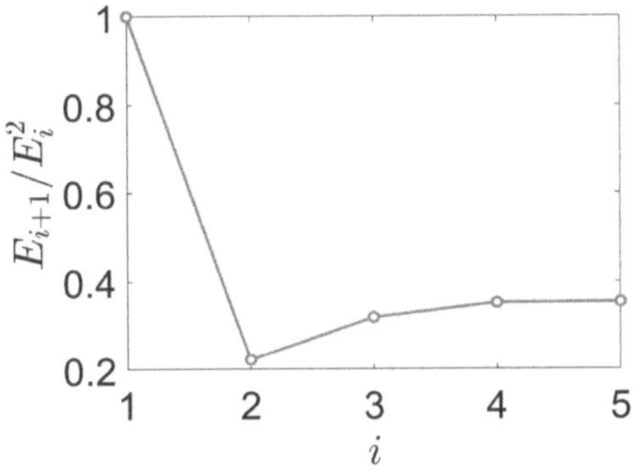

FIGURE 3.9 (a) Ratio between the true errors E_{i+1} and E_i in function of the iteration number i. (b) Ratio between the true errors E_{i+1} and E_i^2 in function of the iteration number.

Plotting the ratio between the true errors E_{i+1} and E_i in function of the iteration number shows that it tends towards zero and then, at iteration number 6, to one (Figure 3.9a). The jump towards one comes from the fact the algorithm hits the limit imposed by the round-off errors at iteration number six (see Example 3.4) and, consequently, the numerically computed error becomes constant from iteration six onwards. As the ratio between the true errors E_{i+1} and E_i tends toward zero, the asymptotic error constant is zero, too, and we are in the second case of Theorem 3.5. This is indeed the case: one can verify that $g'(r) = 0$. The convergence is consequently at least quadratic.

To investigate if the algorithm converges quadratically, one computes the ratio between the true errors E_{i+1} and E_i^2 in function of the iteration number:

```
ratio = [];
for i = 1:5;
  ratio = [ratio err(i+1)/err(i)^2];
endfor;
```

Starting from iteration number six, the round-off errors take over; only the first five iterations are considered and shown in Figure 3.9b. The ratio converges indeed to a constant value, showing that the algorithm is quadratically convergent. This is indeed the case. The reader is invited, with the help of Equation (3.38), to show that the asymptotic error constant is equal to $\lambda = \frac{1}{2\sqrt{2}}$. The numerical values calculated agree well with this theoretical value.

That algorithms behave differently when some of their derivatives vanish is a situation we will encounter in several other places. This isn't surprising as several algorithms are based on some form of Taylor series expansion. To simplify the discussion for such cases, it is useful to introduce the following definition:

Definition 3.5. Multiplicity of a Root

A root r of a m-times differentiable function f is of multiplicity m_r, if all its derivates up to order $m_r - 1$ are zero, and its derivative of order m_r is non zero: $f(r)=0,\ f'(r)=0,\ f''(r)=0,\ \ldots,\ f^{(m_r-1)}(r)=0$ and $f^{(m_r)}(r)\neq 0$.

It is common to call a root where $m_r = 1$ a single root and any root where $m_r > 1$ a multiple root. Note that this definition is a generalization of the concept of root multiplicity known from polynomials. Several properties well known from polynomials with multiple roots hold for this definition. For example, if r is a root of multiplicity m_r of the function f, then one can factor out this root as:

$$f(x)=(x-r)^{m_r}\cdot p(x) \tag{3.39}$$

with the function $p(x)$ such that $p(r)\neq 0$. Indeed, if $p(r)$ is equal to zero, then the root must be of multiplicity $m_r + 1$ (see exercises). In Section 3.5, some additional properties of multiple roots are discussed, and it is shown how multiple roots present serious challenges for root-finding algorithms.

3.4.3 Convergence of the Newton-Raphson Method

The convergence of the Newton-Raphson algorithm is quadratic.

Theorem 3.6. Convergence Order of the Newton-Raphson Method

Let f be a twice continuously differentiable function. If the function f has a root r then

1. For $f'(r)\neq 0$, the Newton-Raphson method converges locally, at least quadratically, to the root with asymptotic error constant $\lambda = \left|\frac{f''(r)}{2f'(r)}\right|$.
2. For $f'(r)=0$, the Newton-Raphson method converges locally linearly to the root with asymptotic error constant $\lambda = \left|\frac{m_r - 1}{m_r}\right|$ where m_r is the multi1plicity of the root r.

As already discussed for the proof of Theorem 3.3, the Newton-Raphson method can be seen as a fixed-point iteration algorithm where $g(x)$ is defined by Equation (3.27). If $f'(r) \neq 0$ one can calculate that $g'(r) = 0$. Indeed, from Equation (3.27) it follows:

$$g'(x) = 1 - \frac{[f'(x)]^2 - f(x)f''(x)}{[f'(x)]^2} = \frac{f(x)f''(x)}{[f'(x)]^2} \tag{3.40}$$

As $f(r) = 0$, it follows that $g'(r) = 0$. From Theorem 3.5, it follows that the Newton-Raphson method has at least quadratic convergence. Furthermore, the asymptotic error constant λ is equal to $\left|\frac{1}{2}g''(r)\right|$, which for Newton-Raphson's method results in $\left|\frac{f''(r)}{2f'(r)}\right|$.

From the expression of $g''(r)$, we can further understand under which condition Newton-Raphson's method has a higher convergence order than quadratic. Revisiting the discussion in the proof of Theorem 3.5 allows the conclusion that Newton-Raphson's method is quadratic if $g''(r) \neq 0$, i.e., if $f''(r) \neq 0$. The order of convergence becomes cubic if $g''(r) = 0$, i.e., when $f''(r) = 0$. If higher order derivatives are equal to zero too, the order of convergence becomes even higher. Such cases are rather exceptional and, in most practical cases, the order of convergence will be quadratic.[7]

In case $f'(r) = 0$ this approach fails as $g'(r)$ is no longer defined. As $f'(r) = 0$ the root r is a multiple root. If m_r is its multiplicity, one can factor f as $f(x) = (x-r)^{m_r} p(x)$ with p such that $p(r) \neq 0$. By factoring out the multiple root in f, the expression (3.40) of g' simplifies and becomes defined in $x = r$. Indeed, as $f'(x) = m_r (x-r)^{m_r - 1} p(x) + (x-r)^{m_r} p'(x)$, Newton's algorithm writes:

$$x_{i+1} = x_i - \frac{f(x_i)}{f'(x_i)} = x_i - \frac{(x_i - r)p(x_i)}{m_r \cdot p(x_i) + (x_i - r)p'(x_i)} \tag{3.41}$$

Computing the first derivative in $x = r$ of the fixed-point function g leads to (exercise[8]):

$$|g'(r)| = \left| \frac{d}{dx}\left[x - \frac{(x-r)p(x)}{m_r \cdot p(x) + (x-r)p'(x)} \right]_{x=r} \right| = \left| \frac{m_r - 1}{m_r} \right| < 1 \tag{3.42}$$

This shows that the Newton-Raphson iteration scheme (3.41) is locally linearly convergent with asymptotic error constant $\lambda = \left|\frac{m_r - 1}{m_r}\right|$.

Example 3.10 illustrates Theorem 3.6. Here again, since we still do not know of a way to estimate the error we will use an example with a known true solution.

Example 3.10. Convergence Order of the Newton-Raphson Method

Consider the function $f(x) = x^2 - 1$ and the root $r = -1$. As $f'(r = -1) = -2 \neq 0$, the Newton-Raphson algorithm is at least quadratically convergent. As further $f''(r = 0) = 2 \neq 0$, the convergence will be exactly quadratic. Running the algorithm for a few iterations shows that the algorithm converges indeed quickly to the root $r = -1$:

```
>> f = @(x) x^2-1;
>> fd = @(x) 2*x;
>> x = [-1.5]; for i = 1:5; x = [x x(i)-f(x(i))/
fd(x(i))]; endfor
>> x
x =
 -1.5000 -1.0833 -1.0032 -1.0000 -1.0000 -1.0000
```

A semi-log plot depicting the true error of the first three generated approximations in function of the iteration number confirms this behavior (Figure 3.10a).

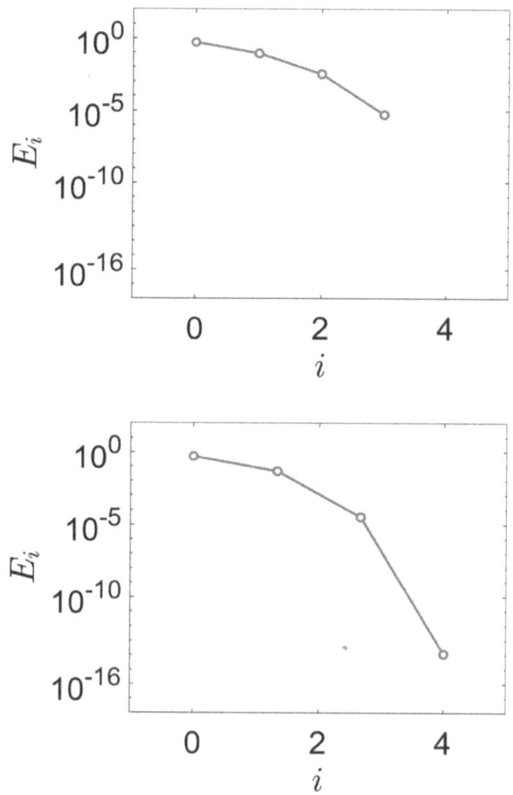

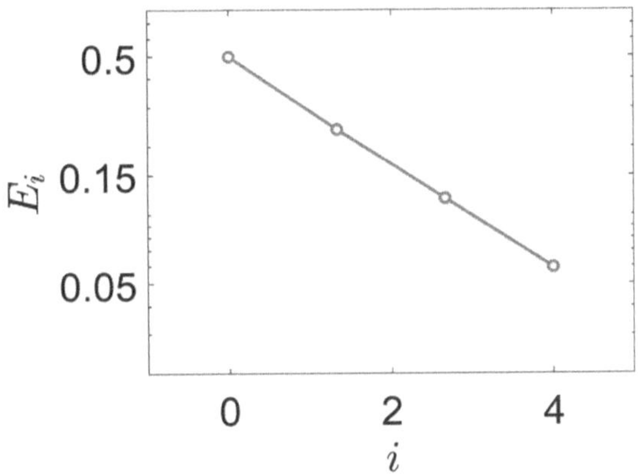

FIGURE 3.10 Errors of the first three iterations of Newton-Raphson approximations of roots from various functions a) $f_1(x) = x^2 - 1$, root $r = -1$ b) $f_2(x) = \sin x$, root $r = 0$ and b) $f_3(x) = x \sin x$, root $r = 0$. The roots are a single for f_1 and f_2, and a double for f_3. In the first case, the Newton-Raphson algorithm converges quadratically; in the second case, the convergence order is cubic as $f''(r) = 0$ whereas in the third case, a linear convergence is observed.

Consider now the function $f(x) = \sin x$ and the root $r = 0$. As $f'(r = 0) = \cos 0 \neq 0$, the Newton-Raphson algorithm is at least quadratically convergent. In fact, as $f''(r = 0) = -\sin 0 = 0$ the convergence will even be of third order. Running the algorithm for a few iterations shows that the algorithm converges indeed very quickly to the root $r = 0$:

```
>> f = @(x) sin(x);
>> fd = @(x) cos(x);
>> x = [0.5];
>> for i=1:5; x=[x x(i)-f(x(i))/fd(x(i))]; endfor;
>> x
x =
 0.50000 -0.04630 0.00003 -0.00000 0.00000 0.00000
```

A semi-log plot depicting the true error of the generated approximations in function of the iteration number confirms this behavior (Figure 3.10b).

Finally, consider $f(x) = x \sin x$. This function still has the same root $r = 0$, but this time, the root is a double root as $f'(r) = \sin r + r \cos r = 0$ and $f''(r) \neq 0$. According to Theorem 3.6, the convergence is linear. Running a few iterations shows that the convergence is indeed linear:

```
>> f = @(x) x*sin(x);
>> fd = @(x) sin(x)+x*cos(x);
>> x = [0.5];
```

```
>> for i=1:5; x=[x x(i)-f(x(i))/fd(x(i))]; endfor;
>> x
x =
 0.500000 0.238937 0.118316 0.059020 0.029493
0.014744
```

The convergence is indeed significantly slower. Plotting the true error of the generated approximations in function of the iteration number confirms the convergence is linear (Figure 3.10c).

3.4.4 Estimating the Error of Generated Approximations Based on the Knowledge of the Convergence Order

A major motivation to learn about the convergence order of algorithms is that the knowledge of how an algorithm converges allows giving an estimation of the errors of the generated approximations without the knowledge of the true solution. This approach is not specific to equation-solving algorithms. On the contrary, it is a general methodology that will be applied to algorithms solving other mathematical problems, too.

Consider an algorithm generating a series of approximations $\{x_i\}$ of a solution r of an equation. One can write:

$$|r - x_{i-1}| \leq |r - x_i| + |x_i - x_{i-1}| \tag{3.43}$$

Assume further that the algorithm convergences to r with order α with an asymptotic error constant λ. For large enough i, Equation (3.32) holds, and (3.43) can be written as

$$|r - x_{i-1}| \leq \lambda |r - x_{i-1}|^{\alpha} + |x_i - x_{i-1}| \tag{3.44}$$

Rearranging and using the notation $E_i = |r - x_i|$ for the true errors, one obtains

$$E_{i-1} - \lambda E_{i-1}^{\alpha} \leq |x_i - x_{i-1}| \tag{3.45}$$

Our aim is to calculate the error E_i from Equation (3.45). If we succeed, we have an error estimation in function of only known quantities.

For linearly converging algorithms, i.e. $\alpha = 1$, this task is straightforward. Multiplying both sides of Equation (3.45) by λ and using once more Equation (3.32) to write $E_i = \lambda E_{i-1}$, it follows:

$$E_i - \lambda E_i \leq \lambda |x_i - x_{i-1}| \tag{3.46}$$

Solving for E_i gives the desired estimation:

$$E_i \leq \frac{\lambda}{1-\lambda}\left|x_i - x_{i-1}\right| \tag{3.47}$$

In case $\alpha \geq 2$ we use a further approximation. In this case, we can say that $E_{i-1} - \lambda E_{i-1}^{\alpha} \cong E_{i-1}$ as, for large enough i, $E_{i-1} >> \lambda E_{i-1}^{\alpha}$ Using this approximation in Equation (3.45):

$$E_i \leq \left|x_i - x_{i-1}\right| \tag{3.48}$$

As both quantities are positive, one can take both sides to the power α. Subsequently multiplying both sides by λ and using Equation (3.32) to write $E_i = \lambda E_i^{\alpha}$, leads to the desired error estimation:

$$E_i \leq \lambda\left|x_i - x_{i-1}\right|^{\alpha} \tag{3.49}$$

Theorem 3.7 summarizes these results.

Theorem 3.7. Estimation of the Absolute Error Based on Convergence Order

The error $E_i = \left|r - x_i\right|$ of the generated approximation x_i of the solution r of an equation by an algorithm converging to r with order α and asymptotic error constant λ, is bounded by

$$E_i \leq \frac{\lambda}{1-\lambda}\left|x_i - x_{i-1}\right| \tag{3.50}$$

for a linearly converging algorithm and by

$$E_i \leq \lambda\left|x_i - x_{i-1}\right|^{\alpha} \tag{3.51}$$

for algorithms converging quadratically, or faster, to the root.

As can be observed from Theorem 3.7, for high order algorithms (such as the Newton-Raphson algorithm), the order of convergence α is the parameter that has to be known best in order to give a good estimation of the error E_i. Having some error on λ is much less problematic. Luckily, the order of convergence α for a given algorithm is a defined number independent of the particular equation one solves. In fact, only the asymptotic error constant needs to be estimated as it often depends on the unknown solution r (see Theorem 3.6). But, as shown in Example 3.11, this task can be addressed in simple ways.

Example 3.11. Estimating Errors Based on Convergence Order

In this example, the root $r = 3$ of function (3.3) from Example 3.1 is approximated. But contrary to what we did previously, we are going to assume we don't know the true root r and, consequently, we have to estimate the error without this knowledge. We will afterward use the true root to verify our error estimations to gain confidence in using the results of Theorem 3.7.

We use the Newton-Raphson algorithm. According to Theorem 3.6, the convergence order of the algorithm is two or one, depending on whether the root is a single one or a multiple root. Plotting our function f shows right away that the root is a single root as clearly $f'(r) \neq 0$. This can be observed on the plot even without precise knowledge of r. Consequently, the algorithm is of quadratic convergence. The only unknown remaining is the asymptotic error constant λ, which requires the knowledge of r. Instead of r, we can, however, use the approximation x_i. At the start of the algorithm, this approximation may be poor, and, consequently, the error estimation, according to Equation (3.50), will be of poor quality too. But as we progress with the iterations and the algorithm converges toward r, the difference between x_i and r becomes smaller and will no longer have any significant effect on the evaluation of Equation (3.50).

We start by calculating a few iterations of the Newton-Raphson algorithm:

```
>> f = @(x) x^3+2*x^2-3*x;
>> fd = @(x)3*x^2+4*x-3;
>> x = [0.75];
>> for i=1:5; x=[x x(i)-f(x(i))/fd(x(i))]; endfor;
>> x
x =
 0.75000 1.16667 1.02576 1.00079 1.00000 1.00000
```

As expected, the algorithm converges rapidly to a constant value. To give error estimates E_i, we use Equation (3.50) together with Theorem 3.6 to compute the asymptotic error constant λ. As mentioned, we use x_i instead of r to compute λ:

```
>> lambda = @(x) abs((6*x+4)/(2*fd(x)));
>> E = [];
>> for i = 2:6
E = [E lambda(x(i))*(x(i)-x(i-1))^2];
endfor
>> E
E =
 1.6606e-01 2.3664e-02 7.7869e-04 7.7316e-07
7.4769e-13
```

Note that we can not give an estimation of the error for the initial guess x_o. Further, as Octave starts numbering vectors with the index 1, the for-loop to compute the error estimations runs from 2 to 6. Comparing with the true error, this shows that Equation (3.50) appears to be an excellent estimation:

```
>> abs(x-1)
ans =
 2.5000e-01 1.6667e-01 2.5765e-02 7.8724e-04
7.7340e-07 7.4762e-13
```

As desired, the estimation is a conservative estimation of the error.

Example 3.11 presented a case where a single root is approximated. In the case of multiple roots, the problem becomes more involved. The Newton-Raphson algorithm now converges linearly, and the asymptotic error constant λ has to be known in order to perform an accurate error estimation. The calculation of λ requires the knowledge of the multiplicity of the root (see Theorem 3.6). Luckily, the Newton-Raphson algorithm can be used to estimate the multiplicity of a root, as demonstrated in te following Example 3.12.

Example 3.12. Estimating the Multiplicity of a Root

Consider the function f which has a root in $r = 0$:

$$f(x) = \sin x - x \tag{3.52}$$

As $f'(r=0) = f''(r=0) = 0$ and $f^{(3)}(r=0) = -1 \neq 0$, Definition 3.5 implies that the root $r = 0$ is a triple root. Let us show how one can find this result using the algorithm of Newton-Raphson without the knowledge of the true value r.

To estimate the multiplicity of a root, we start by applying the algorithm of Newton-Raphson:

```
>> f = @(x) sin(x)-x;
>> fd = @(x)cos(x)-1;
>> x = [0.25];
>> for i=1:5; x=[x x(i)-f(x(i))/fd(x(i))]; endfor;
>> x
x =
 0.250000 0.166493 0.110944 0.073947 0.049294
0.032861
```

Considering the first few iterations, it is obvious that the algorithm is converging rather slowly, indicating we are dealing with a multiple root. The convergence plot of Figure 3.11a confirms this conclusion.

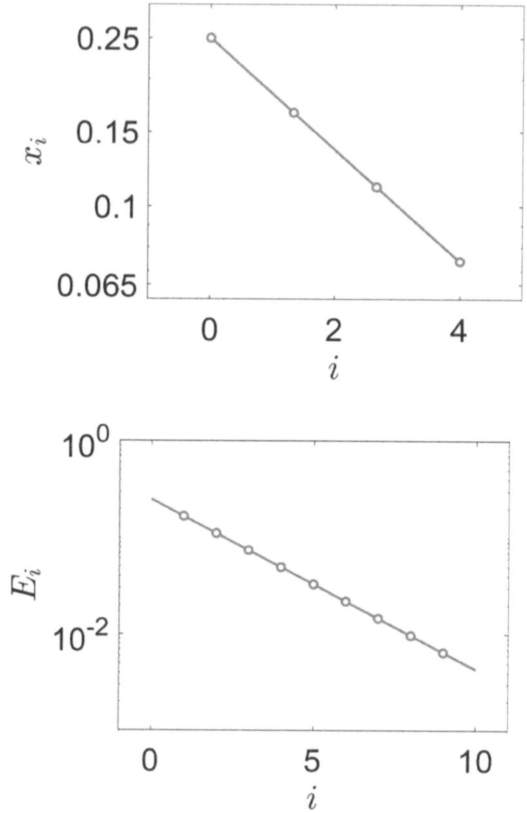

FIGURE 3.11 (a) Convergence plot of the Newton-Raphson algorithm for the root-finding problem $f(x) = sin\ x - x = 0$ (b) Estimation of the errors of the approximations of the root of f computed by the Newton-Raphson algorithm (points ○) together with the true errors (continuous line).

For multiple roots, the Newton-Raphson algorithm converges linearly with an asymptotic error constant $\lambda = \frac{m_r - 1}{m_r}$, where m_r is the root multiplicity. To estimate the asymptotic error constant for a linearly converging algorithm, we compute the ratio between E_i and E_{i-1} and find the value towards which the algorithm converges. Using Equation (3.50) from Theorem 3.7, we can write:

$$\lambda \cong \frac{E_i}{E_{i-1}} \cong \frac{|x_i - x_{i-1}|}{|x_{i-1} - x_{i-2}|} \tag{3.53}$$

The knowledge of λ is not required as it cancels out when calculating the ratio. Let us put this in practice. We first compute a few more iterations of the Newton-Raphson algorithm and then compute the ratio according to (3.53):

```
>> x = [0.25];
>> for i=1:10; x=[x x(i)-f(x(i))/fd(x(i))]; endfor;
>> ratio = [];
for i = 3:10
ratio = [ratio abs(x(i)-x(i-1))/abs(x(i-1)-
x(i-2))];
endfor
>> ratio
ratio =
 0.66520 0.66602 0.66638 0.66654 0.66661 0.66664
0.66666 0.66666
```

Note that the index used in the "for-loop" is not the same index used for the number of iterations of the algorithm (they are shifted by one). But this is not important as we are not interested in attributing a specific index number to the ratio between E_i and E_{i-1}, but are only interested in investigating to which value the ratio converges.

The Octave calculations indicate that the ratio converges towards 0.66666. Keeping in mind that the multiplicity m_r of a root is an integer number, it is clear that in this case, m_r must be equal to 3 such that $\lambda = \frac{3-1}{3} = \frac{2}{3}$.

Once the multiplicity m_r of the root is determined, it is possible to estimate the error of the approximations produced by the Newton-Raphson algorithm with the help of Theorem 3.7. Once more, it is important to be careful with the index numbering (keep in mind Octave starts numbering vector indexes with 1, which means that `x(1)` contains x_o considering how we computed the vector x). Furthermore, it is not possible to provide an estimation of the error for x_o:

```
>> mr = 3; lambda = abs((mr-1)/mr);
>> err = [];
for i = 2:10
 err = [err lambda/(1-lambda)*abs(x(i-1)-x(i))];
endfor
```

The vector `err` contains the error estimations, with `err(i)` being the estimation of the error for x_i, for $i = 1,..9$. Figure 3.11b plots the error estimations together with the true errors $E_i = |x_i - r| = |x_i|$ computed based on the true value $r = 0$. Note that the agreement can be deemed excellent.

3.5 ERROR ANALYSIS

In Section 3.4.4, based on Theorem 3.7, a method to estimate the errors of the approximations produced by an algorithm was presented. The computed error approximations are accurate. However, they rely on the approximations generated by the algorithm. This means that if there is any mistake in computing these approximations, for example, a calculation error or a programming bug, the error approximations will be wrong, too. Such a problem will be hard to detect. It is, therefore, desirable to devise an alternate way for estimating the errors, not relying on the output of the algorithm but rather allowing a direct verification, independent of the used algorithm.

In fact, one can wonder why, once an approximation x_r is available, one does not simply plug back this approximation in the original equation and verify if it is satisfied or not. From Chapter 2, we know that such an approach is flawed when using numerical calculations. For example, even if x_r would be the exact solution, it could be that when evaluated numerically, the equation is not satisfied due to the round-off errors introduced in numerical calculations. In this section, this approach is explored further, and a method to implement it properly is explained.

3.5.1 Backward and Forward Errors

If r is the true root of a function f, then $f(r)=0$. If, on the other hand, $x \neq r$ (and as well different to any other root of f), then $f(x) \neq 0$. This motivates the following definition.

Definition 3.6. Backward Error

The backward error of an approximation x_r of the root r of a function f is the quantity $|f(x_r)|$.

The terminology "backward" error comes from the fact that we are plugging back into the original problem and the presumed solution. Although Definition 3.6 is formulated for the particular case of a root-finding problem, the concept is more general. The idea will be the same for broader numerical problems: we plug the presumed numerical solution back into the problem and compare it with the value we *should* obtain (in the case of a root-finding problem, we should obtain zero). The difference, in absolute value, between what we *should* obtain and what we *do* obtain is the backward error. Therefore, for root-finding problems, the backward error is, indeed, $|f(x_r)-0|=|f(x_r)|$.

The backward error can always be computed. However, as Example 3.13 shows, in itself, the backward error does not allow us to judge how accurate the approximation x_r of the root r is.

Example 3.13. Backward Error

Consider the function $f(x)=(x+1)(x-2)$, which has two roots in $x=-1$ and $x=2$. Let us compare the backward error of the two approximations $x_1=-2$ and $x_2=1$ of the root $r=-1$. The backward error of x_1 is

$$|f(x_1)|=4 \tag{3.54}$$

And the backward error of x_2 evaluates to

$$|f(x_2)|=2 \tag{3.55}$$

In other words, $|f(x_2)|<|f(x_1)|$. But the approximation x_2 of the root $r=-1$ is less accurate than the approximation x_1. In fact, as $|x_2-r|=2>|x_1-r|=1$, the approximation x_1 is actually the better approximation in terms of accuracy, the reverse of what would suggest comparing naively the backward errors.

As a second example, consider the function $f(x)=\frac{1}{3}x^3-x^2+x-\frac{1}{3}$. This function has a root in $r=1$. However, when evaluated with floating-point numbers, one obtains a backward error different than zero:

```
>> f = @(x) 1/3*x^3-x^2+x-1/3;
>> abs(f(1))
ans = 5.5511e-17
```

This is an example where, even though the function is evaluated in the true root, the backward error does not evaluate numerically to zero.

Finally, consider the function $f(x)=x^3-x^2+\frac{1}{3}x-\frac{1}{27}$. This function has a root in $r=1/3$. However, for numbers $x_r\neq 1/3$ the backward error can still be evaluated numerically to exactly zero:

```
>> format long
>> f = @(x) x^3-x^2+1/3*x-1/27;
>> f(0.333334)
ans = 0
```

The reason for the unexpected behaviors is the round-off error. The reader is invited to plot the functions to understand intuitively the origin of the observed behaviors.

Example 3.13 demonstrates that the backward error alone cannot definitively assess the accuracy of an approximation. The more relevant error we need to consider is the error $E = |x_r - r|$ referred to as the forward error.

Definition 3.7. Forward Error

The forward error of an approximation x_r of the root r of a function f is the quantity $|x_r - r|$.

Contrary to the backward error, the forward error cannot be computed when the true root r is unknown. However, we could attempt to determine a relationship between the backward and forward errors. Such a relationship would help estimate the forward error based on our knowledge of the backward error. Definition 3.8 gives a formal description of this relationship.

Definition 3.8. Condition Number of a Root-Finding Problem

For a root-finding problem of a function f with root r, the absolute condition number M of an approximation x_r of the root r is defined as

$$M = \left| \frac{x_r - r}{f(x_r)} \right| \tag{3.56}$$

which is the ratio between the forward and backward errors. The condition number is often referred to as the error magnification factor.

The concept of conditioning has already been introduced in Chapter 2 when discussing the propagation of errors. Conditioning was shown to be a measure of how errors in the input can propagate to errors in the output of any numerical problem. This corresponds exactly to Definition 3.8. For root-finding problems, the input is the value $f(x)$. An error in the input is the difference between $f(x_r)$ and $f(r)$, which is the backward error. The output is the value of the root. Consequently, the output error is the difference between the root approximation and the root, which is measured by the forward error.

As Equation (3.56) suggests, the condition number depends on the particular function f for which it is defined. That is indeed the case. As depicted in Figure 3.12, a given backward error does not always result in a same forward error. For a stiff function around the root r, a small backward error will result in a very small forward error. On the other hand, for a rather flat function around the root r, a small backward error does not mean that the forward error is small as

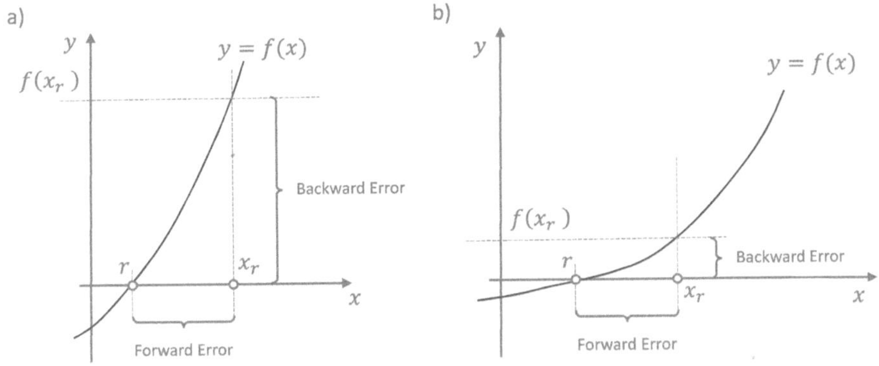

FIGURE 3.12 Two examples of how backward and forward errors relate in the context of root-finding problems. (a) For rather stiff functions around the root r, a large backward error results in a rather small forward error. (b) In the case of rather flat functions around the root, a small backward error can still result in a large forward error.

well. In fact, in Figure 3.12b, we observe a very large forward error corresponding to a much smaller backward error. This was the case as well for the function $f(x) = x^3 - x^2 + \frac{1}{3}x - \frac{1}{27}$ investigated in Example 3.13.

Figure 3.12 can be used to derive a simple estimation for the condition number M using the first derivative of f in x_r:

$$\frac{1}{M} = \left|\frac{f(x_r)}{x_r - r}\right| \cong f'(x_r) \tag{3.57}$$

Rearranging this equation provides an estimation of the forward error in function of the backward error:

$$|x_r - r| \cong \left|\frac{f(x_r)}{f'(x_r)}\right| \tag{3.58}$$

Example 3.14 illustrates how Equation (3.58) can be used to estimate the error of an approximation of a root.

Example 3.14. Estimation of the Error of the Approximation of a Root Based on the Backward Error

For this example, we use the root-finding problem discussed in Example 3.11. In Example 3.11, we generated some approximations of the root $r = 1$ of the function $f(x) = x(x-1)(x+3)$. Let us now use Equation (3.58) to estimate their errors. In order to simplify our implementation on Octave, we define the functions f and f' slightly differently than in Example 3.11:

```
>> f = @(x) x.^3+2*x.^2-3*x;
>> fd = @(x)3*x.^2+4*x-3;
```

Note the dot in front of the power operator. The difference between the Octave operator ^ (the regular power operator) and the operator `.^` (the element-by-element power operator) appears when they are applied to vectors. When the element-by-element operator is applied to a vector, each element of the vector is raised to the power individually. In this way, one can, for example, compute `f(x)` of a vector, and the function will be applied to each element of the vector.

Using these Octave implementations of the functions f and f', we compute the approximations and errors as (note the usage of the element-by-element division operator):

```
>> x = [0.75];
>> for i=1:5; x=[x x(i)-f(x(i))/fd(x(i))]; endfor;
>> err = abs(f(x)./fd(x))
err =
 4.1667e-01 1.4090e-01 2.4978e-02 7.8647e-04
7.7340e-07 7.4762e-13
```

These estimations are in excellent agreement with the true errors:

```
>> abs(x-1)
ans =
 2.5000e-01 1.6667e-01 2.5765e-02 7.8724e-04
7.7340e-07 7.4762e-13
```

Also note that the error estimation can be computed for every approximation, including for the initial guess x_o. In fact, it is not at all important where the approximation comes from. Equation (3.58) estimates the error of an approximation x_r totally independently of the method that was used to generate it.

3.5.2 General Error Estimation Formula

An important remark about Equation (3.58) is that for flat functions around the root, the forward error estimation becomes hazardous as $f'(x_r)$ becomes very small. In fact, for a multiple root, where $f'(r)=0$, the estimation of the forward error from the backward error cannot be performed using this equation. To develop a more accurate estimation, which is also valid for multiple roots, we start with the Taylor series expansion of the function f around its root r. Defining $x_r = r + \Delta r$ one can write:

$$f(r+\Delta r) = f(r) + \Delta r \cdot f'(r) + \frac{(\Delta r)^2}{2!} f''(r) + \ldots \tag{3.59}$$

If the root is of multiplicity m_r, the Taylor series expansion becomes

$$f(r+\Delta r) = \frac{(\Delta r)^{m_r}}{m_r!} f^{(m_r)}(r) + O\left(\Delta r^{m_r+1}\right) \tag{3.60}$$

Solving for $|\Delta r| = |x_r - r|$ results in Theorem 3.8.

Theorem 3.8. Error Estimation Based on Backward Error

Consider a m_r times differentiable function f with a root r of multiplicity m_r. If x_r is an approximation of the root r, then the forward error $|r - x_r|$ is given by

$$|r - x_r| = \left| \frac{m_r! \cdot f(r)}{f^{(m_r)}(r)} \right|^{1/m_r} + O\left(\Delta r^{m_r+1}\right) \tag{3.61}$$

In practice, as the true root is not known, Equation (3.61) will have to be replaced by

$$|r - x_r| \cong \left| \frac{m_r! \cdot f(x_r)}{f^{(m_r)}(x_r)} \right|^{1/m_r} \tag{3.62}$$

where we approximated r by x_r. Note that Equation (3.58) is a particular case of Equation (3.62) when r is a single root. Example 3.15 illustrates the usage of Theorem 3.8 and Equation (3.62).

Example 3.15. Estimating the Error of an Approximation of a Multiple Root

Consider the root-finding problem of the function $f(x) = \sin x - x$ of Example 3.12. In Example 3.12, we already showed how the multiplicity of a root can be estimated without the knowledge of the true root. In the present case, the root has a multiplicity of $m_r = 3$. Using Equation (3.62), we can estimate the errors of the approximations x_i of r generated in Example 3.12 (note that here $f^{(3)}(x) = \cos x$):

```
>> mr = 3;
>> err = abs((factorial(mr)*f(x)./cos(x))).^(1/mr)
err =
 0.252383 0.167190 0.111149 0.074008 0.049312
0.032866
```

These estimations are in excellent agreement with the true errors:

```
>> abs(x-0)
ans =
 0.250000 0.166493 0.110944 0.073947 0.049294
0.032861
```

The agreement becomes better the closer the estimation is to the true root. This originates in the approximation $r \cong x_r$ used in Equation (3.61).

Compared to the error estimation based on the convergence order of an algorithm, the present method has the advantage that it is an independent estimation of the algorithm. Even if an algorithm is applied incorrectly (a programming bug, for example), the error estimation is still correct. From that point of view, it is a safer method and a more general approach to estimating errors of root-finding problems than the method described in 3.4.4. It can, in fact, be applied even if the estimation x_r is provided by a totally unknown algorithm or by a simple guess (like for the initial guess to start an algorithm). In practice, it is recommended, whenever possible, to use both methods and verify that their results agree.

3.6 LAB: SOLVING NON-LINEAR EQUATIONS

In all the examples used so far to illustrate the various concepts, problems with known true solutions were used. In this lab, a problem with unknown true roots is solved in order to show how the learned methods are applied to real-world situations.

3.6.1 Roots of Legendre Polynomials

Legendre polynomials are a special class of polynomials with important applications in several fields of mathematics. In Chapters 5 and 8, some of them will be discussed. An example of a Legendre Polynomial is the following one:

$$P_5(x) = \frac{1}{8}\left(231x^5 - 315x^3 + 105x\right) \tag{3.63}$$

The roots of the Legendre polynomial are of particular interest. They play, for example, a key role in numerical algorithms to approximate definite integrals, as we will discuss in Chapter 8. The present goal is to find these roots numerically, and this will be done with an error lower than 10^{-8}.

The first important step consists of plotting $P_5(x)$.

```
>> p5 = @(x) 1/8*(231*x.^5-315*x.^3+105*x);
>> x = linspace(-1.05, 1.05, 100);
>> plot(x, p5(x))
```

Several aspects can be observed in the figure. First, one notes that the polynomial possesses five single roots within the interval $[-1, 1]$. Second, the plot is helpful to get a rough approximation of the roots, which can be used as an initial

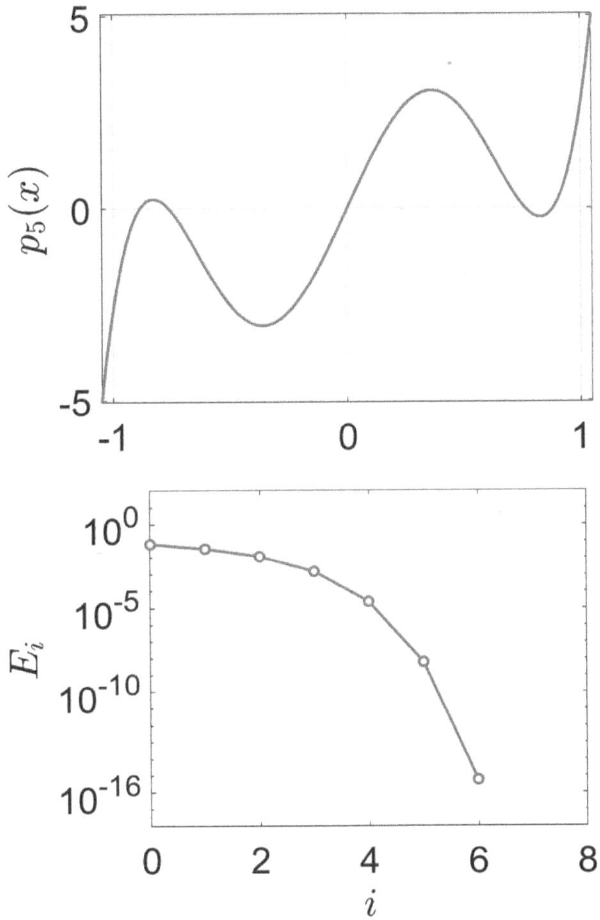

FIGURE 3.13 (a) Legendre polynomial of order five $p_5(x)$. (b) Estimated errors of Newton-Raphson approximations of the first root of the Legendre polynomial of order five $p_5(x)$.

guess for algorithms. Two roots are close to -1; another is $r = 0$, as inspection of Equation (3.63) confirms, and the last two are close to 1.

To approximate the other roots, we use Newton-Raphson's algorithm. As the roots are single roots, the algorithm will be quadratically convergent. We further need $p_5'(x)$, the first derivative of the Legendre polynomial:

```
>> p5d = @(x) 1/8*(5*231*x.^4-3*315*x.^2+105);
```

We can now conduct a few iterations with Newton-Raphson's algorithm:

```
>> x = [-1];
>> for i = 1:10
>> x = [x x(end)-p5(x(end))/p5d(x(end))];
>> endfor
```

As the roots are single roots, Equation (3.58) can be used to estimate the errors of the approximations produced by the algorithm:

```
>> err = abs(p5(x)./p5d(x));
```

The results, approximations, and corresponding estimated errors can be displayed in a table form:

```
>> format long e
>> [x', err']
ans =
 -1.000000000000000e+00     6.666666666666667e-02
 -9.333333333333333e-01     3.474205231036714e-02
 -8.985912810229661e-01     1.192242524315103e-02
 -8.866688557798151e-01     1.513465341376912e-03
 -8.851553904384383e-01     2.396182683266599e-05
 -8.851314286116057e-01     5.973757870865331e-09
 -8.851314226378478e-01     5.794175483535177e-16
 -8.851314226378473e-01                         0
 -8.851314226378473e-01                         0
 -8.851314226378473e-01                         0
 -8.851314226378473e-01                         0
```

The estimated errors of the approximations are plotted on Figure 3.13b. The errors become lower than the desired target of 10^{-8} at iteration 5 and beyond. Starting from iteration number seven, the algorithm hits the limit imposed by the round-off errors.

From these calculations, the first root of $p_5(x)$ is approximated as $x_r = -8.851314286116057 \cdot 10^{-1}$. We are confident that for this approximation, the error is lower than 10^{-8}. As we did conduct more iterations, we can also use the better approximation of $x_r = -8.851314226378473 \cdot 10^{-1}$, which has an error close to the machine epsilon. It represents, in fact, the best approximation we can generate with floating-point numbers.

In a similar way, the other roots of $p_5(x)$ can be approximated. Adequate initial guesses to start the Newton-Raphson algorithm can be determined from Figure 3.13a.

3.6.2 A Challenging Problem

We aim to find the positive root of the following equation:

$$x^3 + 2x^2 - 2.75x + 0.75 = 0 \tag{3.64}$$

A plot generated with

```
>> f = @(x) x.^3+2*x.^2-2.75*x+0.75;
>> x = linspace(-5, 5, 50);
>> plot(x, f(x))
```

shows that the positive root, between $x = 0$ and $x = 1$, is likely a multiple root (Figure 3.14a). As a cubic polynomial has three roots, the root we look for is probably a double root. We can confirm this by following the method described in Example 3.12. We first compute a few iterations with Newton-Raphson's method:

```
>> fd = @(x) 3*x.^2 + 4*x -2.75;
>> x = [1.5];
>> for i = 1:5;
x = [x x(i)-f(x(i))/fd(x(i))];
endfor;
```

and then study the evolution of the ratio (3.53):

```
>> ratio = [];
>> for i = 3:15;
ratio = [ratio abs(x(i)-x(i-1))/abs(x(i-1)-x(i-2))];
endfor
>> ratio(8:end)
ans =
 0.5011 0.5005 0.5003 0.5001 0.5001 0.5000
```

As expected, it converges to 0.5, which means that the root multiplicity is, indeed, $m_r = 2$. As the root has a multiplicity higher than 1, the numerical root-finding process is challenging. To estimate the error of the approximations, Equation (3.62) has to be used:

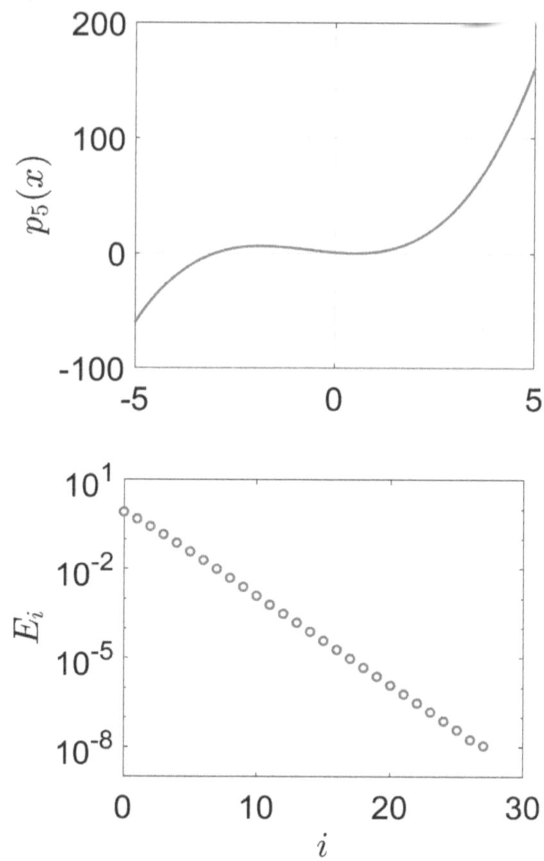

FIGURE 3.14 (a) Plot of $f(x) = x^3 + 2x^2 - 2.75x + 0.75$ (b) Estimated errors of Newton-Raphson approximations of the first root of the Legendre polynomial of order five $p_5(x)$

```
>> fdd = @(x) 6*x + 4;
>> err = sqrt(abs(2*f(x)./fdd(x)));
>> err(12:end)
ans =
  2.4852e-02 1.7574e-02 1.2427e-02 8.7876e-03 6.2138e-03
```

The error decreases slowly, as the Newton-Raphson method is of first order. Figure 3.14b illustrates the estimated error for the first 30 iterations. Starting from iteration $i = 28$, the algorithm reaches the limit imposed by the round-off errors. Further iterations no longer improve the approximations, and the error is wrongly estimated as zero. The lowest error the algorithm can reach is 10^{-8}. The corresponding approximation is $x_{27} = 0.500000010368919$.

In summary, the best approximation for the positive root of Equation (3.64) is

$$x_r = 0.50000001 \pm 0.00000001 \qquad (3.65)$$

This is rather a poor approximation, considering we use floating point numbers (about 16 significant digits used in the calculations). But, due to the nature of the problem, a better approximation cannot be given. Indeed, the backward error of x_{27} becomes in the order of magnitude of the machine epsilon, and it can no longer be distinguished from 0. To increase accuracy, more digits would have to be used in the calculations.

Switching the algorithm does not allow us to obtain a better result. To illustrate this, let us find the `roots` of Equation (3.64) using the Octave function `roots`. This algorithm, specialized to find the roots of polynomials, uses a different approach than Newton-Raphson's algorithm. The interested reader can find more information in the Octave documentation. For our purpose, it is enough to know that this algorithm is very different from Newton-Raphson. Despite this fact, it finds approximations of the root with a similar error. To use the `roots` Octave function, first, the polynomial is defined as a column vector containing its coefficients, starting with the highest degree:

```
>> p = [1 2 -2.75 0.75];
```

The roots are then approximated as

```
>> format long
>> r = roots(p)
r =
  -2.999999999999997
   0.499999991516732
   0.500000008483268
```

The error of, for example, the last approximation is about

```
>> sqrt(abs(2*f(r(3))./fdd(r(3))))
ans = 9.755099837897146e-09
```

Which is about 10^{-8}, for Newton-Raphson's method.

3.7 EXERCISES

3.7.1 INTRODUCTION

1. Reformulate each of the following equations as a root-finding problem and as a fixed-point problem:

    a. $4x^2 - 2x + 5 = x^3 + 3x$

b. $\sin(x^2) - 5x = \ln(4x)$

c. $4\cos(2x) + 5\sin(x) = 4$

3.7.2 Bracketing Methods

2. Specify for the functions f if bracketing methods can be used to find their roots or not. If the bracketing methods cannot be used, explain why.

   a. $f(x) = x^2 - 6x + 9$

   b. $f(x) = \sin(x) + 0.5$

   c. $f(x) = \cos(x) + 1$

   d. $f(x) = \dfrac{x-4}{x-3}$

3. For the functions from Exercise 2 for which bracketing methods can be used, find approximations of root(s) of the function f with an accuracy better than 10^{-3} using the bisection algorithm.

4. Using the bisection method, find the number x, with an absolute error less than 10^{-3} that makes the determinant of the matrix

$$A = \begin{bmatrix} x & 1 & 2 & 3 \\ 4 & x & 5 & 6 \\ 7 & 8 & x & 9 \\ 10 & 11 & 12 & x \end{bmatrix}$$

   equal to 10000.
   Hint: use the Octave command `det` *to compute the determinant of a matrix.*

3.7.3 Open Methods

3.7.3.1 Fixed-Point Method

5. We want to apply the fixed-point method to solve the equation $x = g(x)$. Choose among the following examples of functions g the ones that will lead to a converging fixed-point algorithm when using the initial guess x_0 specified:

   a. $g(x) = e^{x-2} + x^3$ and $x_0 = 0.5$

   b. $g(x)\dfrac{x^4 - \cos x}{3 - 5x^3}$ and $x_0 = 0.8$

c. $g(x) = \ln x + x^3 \cos x$ and $x_0 = 5$

d. $g(x) = e^x - 4x^2$ and $x_0 = -1$

e. $g(x) = x^2 - 3x \cos x$ and $x_0 = 1$

6. Find approximations of root(s) of the function f of Exercise 2. For this, the root-finding problems must first be reformulated as fixed-point problems with a function $g(x)$ such that the fixed-point algorithm will converge. In each case, conduct four iterations and choose an initial value such that the algorithm will converge.

3.7.3.2 Newton-Raphson Method

7. Derive the Newton-Raphson algorithm based on its graphical interpretation shown in Figure 3.6.

8. Find approximations of root(s) of the function f of Exercise 2 using Newton's-Raphson's method. In each case, conduct four iterations and choose an initial value such that the algorithm will converge.

9. Consider the following function

$$f(x) = \left(1 - \frac{3}{4x}\right)^{\frac{1}{3}}$$

which has a root in $r = 3/4$.

a. Explain why Newton-Raphson's method will likely fail
b. Compute the first 50 iterations of Newton-Raphson's method using different initial guess x_o. What do you observe?
c. Explain why Newton-Raphson's method produces such chaotic behavior (you may want to plot the function f to help in the analysis).

3.7.4 Convergence Order

10. Consider the following fixed-point iteration scheme:

$$x_{i+1} = w \cdot x_i (1 - w) \frac{A}{x_i^2}$$

with w a number between 0 and 1.

a. Assuming the algorithm will converge, to which value will it converge?

b. How should we choose the value of w such that the algorithm will converge as fast as possible?

11. The equation $x^5 - 2x^4 + 2x^2 - x = 0$ has a solution in $r = -1$. You are using Newton's method to estimate this solution. If the absolute error in iteration 4 is about 0.02, how large do you expect to be the absolute error in iteration 5?

12. Using theorem 3.7, prove by recursion that for a linearly converging algorithm, the absolute error $E_i = |r - x_i|$ of iteration number i can be estimated as

$$E_i \leq \frac{\lambda^i}{1-\lambda}|x_1 - x_o|$$

13. For the converging cases of exercise 5 and using the result from exercise 12, estimate the number of iterations required by the fixed-point iteration method to reach an absolute error smaller than an absolute error of 10^{-5}. (Estimate the asymptotic error constant the best way you can).

    Verify your prediction by conducting the required number of iterations and estimating the error of the computed approximation using Theorem 3.7.

14. Prove that if a root r of a function f is of multiplicity m_r, then one can write

$$f(x) = (x - r)^{m_r} \cdot p(x)$$

    with $p(x)$ such that $p(r) \neq 0$.

15. Recall that Newton-Raphson's method can be seen as a fixed-point iteration method with g(x) defined as

$$g(x) = x - \frac{f(x)}{f'(x)}$$

    By computing the second derivative of $g(x)$ and using Theorem 3.5, prove that the asymptotic error constant λ of Newton-Raphson's algorithm is equal to when $f'(r) \neq 0$:

$$\lambda = \left|\frac{f''(r)}{2f'(r)}\right|$$

3.7.5 Error Analysis

16. Consider the function $f(x)=(x-5)^n$ where n is a real number.

    a. What is the root r of f ?
    b. What is the multiplicity of the root?
    c. Assume an algorithm has provided the approximation $x_r = 5.001$ of r. Compute for $n=1, 2, 4$ and 0.5 the forward and backward error and present the results in a table. What do you observe?

17. We want to provide an approximation x_r of a root of the function $f(x)$ with a relative error less than 10^{-5}. Among the following proposed approximations x_r which ones satisfy this condition?

    a. $f(x)=e^x+2^{-x}+2\cos x-6$ and $x_r=1.82938543$
    b. $f(x)=\ln(x-1)+\cos(x-1)$ and $x_r=1.3977486157$
    c. $f(x)=3x^2-e^x$ and $x_r=0.91091849$
    d. $f(x)=(x-2)^2-\ln x$ and $x_r=3.0571034$
    e. $f(x)=x-0.5(\sin x+\cos x)$ and $x_r=0.72287$

18. Find approximations of root(s) of the function f of exercise 2 with an accuracy better than 10^{-6} using the fixed-point method. For this, the root-finding problems must first be reformulated as fixed-point problems with a function $g(x)$ such that the fixed-point algorithm will converge.

    Note: do not use the true root r to estimate the errors. You can use the true root to validate your calculations.

19. Find approximations of root(s) of the function f of Exercise 2 with an accuracy better than 10^{-12} using Newton-Raphson's method.

    Note: do not use the true root r to estimate the errors. You can use the true root to validate your calculations.

20. Find an approximation of the root meeting the specifications:

    a. A root of $2+\cos(e^x-2)=e^x$ in the interval $[0.5, 1.5]$ with an absolute error less than 10^{-14}.
    b. A root of $2x\cos(2x)-(x-2)^2=0$ in the interval $[2, 3]$ with an absolute error less than 10^{-10}.
    c. A root of $e^x+2^{-x}+2\cos x-6=0$ in the interval $[1, 2]$ with an absolute error less than 10^{-10}.
    d. The positive root of $\sqrt{x}=e^{-2x}$ with an absolute error less than 10^{-10}.
    e. A root of $\sin(e^{3x}+4)=0.5$ in the interval $[0, 0.5]$ with an absolute error less than 10^{-12}.

NOTES

1 Joanna M. Papakonstantinou, Richard A. Tapia "Origin and Evolution of the Secant Method in One Dimension", The American Mathematical Monthly, Vol. 120, No. 6 (June–July 2013), pp. 500–518.

2 The reader may argue that one could verify that $x_r = 1$ is indeed the searched root by verifying that $f(x_r) = 0$. This way to proceed, unfortunately, has some major flaws that will be discussed soon in this chapter.

3 Isaac Newton (1642–1727) English mathematician, physicist, astronomer, alchemist, and theologian.

4 Joseph Raphson (1668–1715), English mathematician.

5 If the reader has difficulties visualizing this statement, it is useful to try out an example where $E_i \cong 0.1$ and compute a few errors E_{i+1}, E_{i+2}, ... with $\alpha = 2$ using Equation (3.32). Regardless of λ, the sequence will decrease very similarly. The effect will be more pronounced for larger α.

6 As g and g' are continuous, the various limits and function evaluations can be interchanged.

7 Graphically, such cases can be identified as a function crossing the x-axis with a none zero slope but presenting an inflection point. A typical example is $f(x) = \sin x$.

8 During the calculations, one will have to simplify by $p(x)$ the fraction. Here the fact that $p(r) \neq 0$ is essential to guarantee that $g(r)$ is differentiable (in order to satisfy the hypotheses of Theorem 3.5).

4 Systems of Linear Equations

4.1 INTRODUCTION

Let us start by outlining some notations and terminology that will be useful for the remainder of the text. We will deal with systems of linear equations that are arranged such that all unknowns are on the left-hand side and constant terms are on the right-hand-side of the equations:

$$\begin{cases} a_{11}x_1 + a_{12}x_2 + \ldots + a_{1n}x_n &= b_1 \\ a_{21}x_1 + a_{22}x_2 + \ldots + a_{2n}x_n &= b_2 \\ \vdots & \vdots \\ a_{n1}x_1 + a_{n2}x_2 + \ldots + a_{nn}x_n &= b_n \end{cases} \tag{4.1}$$

The system has n unknowns labeled as x_i with $i = 1,\ldots,n$ and n equations. The case where the number of equations is larger than the number of unknowns will be discussed in detail in Chapter 6. The coefficients in front of the unknowns are labeled as a_{ij} where we use the convention that the first index i is the number of the equation and the second index j is the number of the unknown. Finally, the constant terms are denoted as b_i where the index i is the number of the equation.

As we will soon realize, a more efficient way to present this equation is to use an equivalent matrix equation instead:

$$\begin{bmatrix} a_{11} & a_{12} & \ldots & a_{1n} \\ a_{21} & a_{22} & \ldots & a_{2n} \\ \vdots & \vdots & & \vdots \\ a_{n1} & a_{n2} & \ldots & a_{nn} \end{bmatrix} \begin{bmatrix} x_1 \\ x_2 \\ \vdots \\ x_n \end{bmatrix} = \begin{bmatrix} b_1 \\ b_2 \\ \vdots \\ b_n \end{bmatrix} \tag{4.2}$$

Here, the unknowns are grouped into the column vector $\underline{x} = [x_1 \cdots x_n]^T$, termed the vector of unknowns. The matrix $A = [a_{ij}]$ is called the coefficient matrix, and the column vector $\underline{b} = [b_1 \cdots b_n]^T$ is referred to as the right-hand side vector. Using these notations, Equation (4.2) can be written as

$$A\underline{x} = \underline{b} \tag{4.3}$$

DOI: 10.1201/9781003262121-5

It is straightforward to verify that Equations (4.1) and (4.2) are indeed equivalent. By multiplying the left-hand side of Equation (4.2), one ends up with the equality of two column vectors. Two column vectors are equal if all their components are equal. This condition translates directly into Equation (4.1).

Using a matrix equation rather than a traditional system of equations presents tremendous benefits. It results in a more compact equation and makes it possible to use the vast theory of real-valued vector spaces. For example, the solution of Equation (4.3) can be simply written as

$$\underline{x} = A^{-1}\underline{b} \tag{4.4}$$

as long as the inverse A^{-1} of the coefficient matrix A exists.[1] From a purely mathematical point of view, the problem is essentially solved. However, computing the inverse of a matrix is a lengthy process. The algorithms we will discuss in the remainder of this chapter will avoid this calculation and find the solution of a linear system in an efficient way.

4.2 GAUSS ELIMINATION ALGORITHMS

An impressive arsenal of methods to solve systems of linear equations exists. In the frame of this book, we selected one popular example we will use as a prototype algorithm for solving linear equations. The general discussion about the control and estimations of errors will be independent of the specific algorithm used to produce the solution of the system of equations. We will, however, use the family of algorithms presented in this section for illustrations and examples.

Gauss elimination algorithms are a family of algorithms in which one aims to transform the system (4.1), respectively (4.2), into a system presenting a form facilitating its solution finding. This form is an upper diagonal system of equations. Written as a matrix equation:

$$\begin{bmatrix} u_{11} & u_{12} & u_{13} & \dots & u_{1n} \\ 0 & u_{22} & u_{23} & \dots & u_{2n} \\ 0 & 0 & u_{33} & \dots & u_{3n} \\ \vdots & \vdots & \vdots & & \vdots \\ 0 & 0 & 0 & \dots & u_{nn} \end{bmatrix} \begin{bmatrix} x_1 \\ x_2 \\ x_3 \\ \vdots \\ x_n \end{bmatrix} = \begin{bmatrix} d_1 \\ d_2 \\ d_3 \\ \vdots \\ d_n \end{bmatrix} \tag{4.5}$$

Subsequently, solving a linear system of equations now requires identifying an upper diagonal matrix U and a right-hand side vector $\underline{d}$, where both U and d are derived from the initial system of equation $A\underline{x} = b$ and such that both systems of linear equations (4.5) and (4.2) are equivalent (i.e., have the same solution). This process is called forward elimination. Once the upper diagonal system is found, the new system of equations (4.5) is solved by back substitution. Back substitution starts with the last equation in order to find the last unknown as $x_n = d_n / u_{nn}$. Once x_n is known, the second last equation is solved for x_{n-1}:

$$x_{n-1} = \frac{d_{n-1} - u_{n-1,n} \cdot x_n}{u_{n-1,n-1}} \tag{4.6}$$

Proceeding the same way with the following equations allows us to obtain efficiently the values of all unknowns:

$$x_i = \frac{d_i - \sum_{k=i+1}^{n} u_{ik} \cdot x_k}{u_{ii}} \tag{4.7}$$

The task remaining is to describe an efficient way to perform forward elimination. Two possible algorithms will be presented thereafter. Regardless of the particular implementation of forward elimination, all methods are based on a combination of operations performed on a system of equations as described in Theorem 4.1.

Theorem 4.1. Operations on Systems of Linear Equations

A system of linear equations is transformed into an equivalent system, i.e., a system having the same solution set as the untransformed one, by any of the following operations:

1. Swap two equations
2. Add or subtract a multiple of one equation from another
3. Multiply an equation by a none zero constant

It is convenient to introduce a notation to formalize the operations described in Theorem 4.1. The following notation is used. For example,

$$r_2 \leftarrow R_2 - 2R_1$$

means that we replace the second equation (second row r_2) in the new system with the equation created by subtracting two times the first equation (first row R_1) from the second equation (second row R_2) of the initial system of equations.

4.2.1 Naïve Gauss Elimination

Naïve Gauss elimination is an algorithm to perform forward elimination. The simplest way to explain the algorithm is through an example.

Example 4.1. Naïve Gauss Elimination

Consider the following system of linear equations:

$$\begin{bmatrix} 2 & 1 & -3 \\ -4 & -1 & 10 \\ 2 & 4 & -5 \end{bmatrix} \begin{bmatrix} x_1 \\ x_2 \\ x_3 \end{bmatrix} = \begin{bmatrix} 4 \\ -13 \\ 17 \end{bmatrix} \tag{4.8}$$

Naïve Gauss elimination uses operations from Theorem 4.1 in order to progressively transform the system of equations into an upper diagonal form.

In the first step, the algorithm keeps the first equation as is. This equation is referred to in the algorithm as the pivot equation, and the coefficient a_{11} as the pivot element.[2] The algorithm aims then to transform all equations below the pivot equation into equations where the coefficients of x_1 are null using operations involving the pivot equation and one of the equations below it. There isn't a unique way to achieve this. Naïve Gauss elimination *subtracts* a well-chosen number of times an equation from the pivot equation. For the second equation of the system, this is:

- Subtract (-2) times the second equation from equation one: $r_2 \leftarrow R_2 - (-2) R_1$

  and for the third equation:

- Subtract $(+1)$ times the third equation from equation one: $r_3 \leftarrow R_3 - (+1) R_1$

The reason for this choice, which may seem not the most obvious one, will become clear once we discuss further algorithms based on Gauss elimination. Executing these two operations on the system of equations (4.8) results in

$$\begin{bmatrix} 2 & 1 & -3 \\ 0 & 1 & 4 \\ 0 & 3 & -2 \end{bmatrix} \begin{bmatrix} x_1 \\ x_2 \\ x_3 \end{bmatrix} = \begin{bmatrix} 4 \\ -5 \\ 13 \end{bmatrix} \tag{4.9}$$

The Naïve Gauss elimination algorithm continues with the same strategy, but this time using the second equation as the pivot equation. The aim is now to transform the equations below the second equation into equations having the coefficients of the unknowns x_1 and x_2 equal to zero. Again, the algorithm does this by subtracting these equations from the pivot equation an adequate number of times. In our example, this results in the operation $r_3 \leftarrow R_3 - (+3) R_2$:

$$\begin{bmatrix} 2 & 1 & -3 \\ 0 & 1 & 4 \\ 0 & 0 & -14 \end{bmatrix} \begin{bmatrix} x_1 \\ x_2 \\ x_3 \end{bmatrix} = \begin{bmatrix} 4 \\ -5 \\ 28 \end{bmatrix} \tag{4.10}$$

This completes the forward elimination in the case of our example. If the system contains more equations, the algorithm will proceed further, using the third equation as a pivot equation and so on and so forth.

The final step is back substitution. Starting with the last equation, the system is solved using successively Equation (4.7):

$$\begin{bmatrix} x_1 \\ x_2 \\ x_3 \end{bmatrix} = \begin{bmatrix} -2.5 \\ 3 \\ -2 \end{bmatrix} \tag{4.11}$$

As observed from Example 4.1, Naïve Gauss elimination involves relatively few calculations. We don't discuss the efficiency or complexity of algorithms in the frame of this book. Let us simply mention that Naïve Gauss elimination is one of the more efficient algorithms used in solving systems of linear equations.

Implementation in Octave is relatively simple. The Octave function that follows gives a possible implementation of the forward elimination step.[3] Based on a matrix A and right-hand side vector $\underline{b}$, the algorithm computes the upper diagonal matrix U and the modified right-hand side vector $\underline{d}$.

```
function [U d] = naive_gauss(A, b)
  n = length(b);
  for j = 1:n-1
    for i = j+1:n
      g = A(i,j)/A(j,j);
      for k = j:n
        A(i,k) = A(i,k) - g*A(j,k);
      endfor
      b(i) = b(i) - g*b(j);
    endfor
  endfor
  U = A;
  d = b;
endfunction
```

Similarly, the back substitution step can be implemented using Equation (4.7):

```
function x = back_substitution(U, d)
  n = length(d);
  for i = n:-1:1
```

```
    for j = i+1:n
      d(i) = d(i) - U(i,j)*x(j);
    endfor
    x(i) = d(i)/U(i,i);
  endfor
  x = x';
endfunction
```

The main interest in these two code snippets is in the opportunity to become familiar with for-loop structures and to illustrate how summation equations, such as Equation (4.7), can be coded using such loops. From a mathematical point of view, we will soon see in Section 4.2.2 a better algorithm that will address some issues that Naïve Gauss elimination presents.

Example 4.2. Naïve Gauss Elimination With Octave

Consider again the system of equations (4.8) from Example 4.1. Let us put in action the two Octave functions `naive_gauss` and `back_substitution` outlined previously. Firstly, one has to define the coefficient matrix A and the right-hand-side vector $\underline{b}$:

```
>> A = [2 1 -3; -4 -1 10; 2 4 -5];
>> b = [4 -13 17]';
```

Secondly, we deploy our two functions. We start with the forward elimination step:

```
>> [U d] = naive_gauss(A,b)
U =
    2   1   -3
    0   1    4
    0   0  -14
d =
     4
    -5
    28
```

Followed by back substitution:

```
>> back_substitution(U,d)
ans =
   -2.5000
    3.0000
   -2.0000
```

The Naïve Gauss elimination algorithm can be presented in an operator form. The idea is to replace the row operations used in Naïve Gauss elimination with a well-chosen matrix G that will perform the needed operations onto matrix A, i.e., such that matrix GA is the matrix obtained after applying the desired row operation. This is possible because elimination is based on linear combinations only. The starting point is the simple identity $A = IA$ where I is the identity matrix. To implement an operation such as $r_2 \leftarrow R_2 - gR_1$, the identity matrix needs only a minor modification:

$$\begin{bmatrix} 1 & 0 & 0 \\ -g & 1 & 0 \\ 0 & 0 & 1 \end{bmatrix} \begin{bmatrix} a_{11} & a_{12} & a_{13} \\ a_{21} & a_{22} & a_{23} \\ a_{31} & a_{32} & a_{33} \end{bmatrix} = \begin{bmatrix} a_{11} & a_{12} & a_{13} \\ a_{21} - ga_{11} & a_{22} - ga_{12} & a_{23} - ga_{13} \\ a_{31} & a_{32} & a_{33} \end{bmatrix} \quad (4.12)$$

Similar transformation matrices G can be identified for other Gauss elimination steps. We will use the notation $G_{ij}(g)$ for such matrices where the indexes i and j indicate the position of the element g inside the matrix G_{ij}. Note the negative sign in front of g in the matrix G_{ij}. The two indexes i and j also refer to the equation on which Gauss elimination is performed and the pivot equation, respectively.

Using an operator formulation of the algorithm presents some advantages. From a mathematical point of view, it is much simpler to handle matrices than verbal descriptions of row operations. From an implementation point of view, it can have advantages as well, as many data analysis software packages allow the use of Gauss elimination packages, thereby eliminating the need to code nested for-loops. For example, in the case of Octave, a matrix multiplication will be executed much faster than a series of nested for-loops.[4]

Example 4.3. Naïve Gauss Elimination in Operator Form

We consider once more the system of equations (4.8) from Example 4.1. The first Gauss elimination step is the row operation $r_2 \leftarrow R_2 - (-2)R_1$. This can be represented by the matrix

$$G_{21}(-2) = \begin{bmatrix} 1 & 0 & 0 \\ 2 & 1 & 0 \\ 0 & 0 & 1 \end{bmatrix} \quad (4.13)$$

Note the sign used in the matrix G_{21}. The coefficient to put in position $ij = 21$ is $-g$, which is $+2$ in our case. The second Gauss elimination step is $r_3 \leftarrow R_3 - (+1)R_1$ which corresponds to the matrix

$$G_{31}(1) = \begin{bmatrix} 1 & 0 & 0 \\ 0 & 1 & 0 \\ -1 & 0 & 1 \end{bmatrix} \quad (4.14)$$

We can verify that cascading these two operations eliminates the coefficients of the unknown x_1 in the second and third equation:

$$\begin{bmatrix} 1 & 0 & 0 \\ 0 & 1 & 0 \\ -1 & 0 & 1 \end{bmatrix}\begin{bmatrix} 1 & 0 & 0 \\ 2 & 1 & 0 \\ 0 & 0 & 1 \end{bmatrix}\begin{bmatrix} 2 & 1 & -3 \\ -4 & -1 & 10 \\ 2 & 4 & -5 \end{bmatrix} = \begin{bmatrix} 2 & 1 & -3 \\ 0 & 1 & 4 \\ 0 & 3 & -2 \end{bmatrix} \tag{4.15}$$

Finally, the last Gauss elimination step is the row operation $r_3 \leftarrow R_3 - (+3)R_2$, which is represented by the matrix

$$G_{32}(3) = \begin{bmatrix} 1 & 0 & 0 \\ 0 & 1 & 0 \\ 0 & -3 & 1 \end{bmatrix} \tag{4.16}$$

Cascading all operations results in the desired upper diagonal matrix U:

$$U = \begin{bmatrix} 1 & 0 & 0 \\ 0 & 1 & 0 \\ 0 & -3 & 1 \end{bmatrix}\begin{bmatrix} 1 & 0 & 0 \\ 0 & 1 & 0 \\ -1 & 0 & 1 \end{bmatrix}\begin{bmatrix} 1 & 0 & 0 \\ 2 & 1 & 0 \\ 0 & 0 & 1 \end{bmatrix}\begin{bmatrix} 2 & 1 & -3 \\ -4 & -1 & 10 \\ 2 & 4 & -5 \end{bmatrix} = \begin{bmatrix} 2 & 1 & -3 \\ 0 & 1 & 4 \\ 0 & 0 & -14 \end{bmatrix} \tag{4.17}$$

The beauty of the operator form of the naïve Gauss elimination algorithm is that it allows us to compute the vector $\underline{d}$ of Equation (4.5). Indeed, multiplying the original system of equations $A\underline{x} = \underline{b}$ by the Gauss elimination operators results in:

$$G_{32}(3)G_{31}(1)G_{21}(-2)A\underline{x} = U\underline{x} = G_{32}(3)G_{31}(1)G_{21}(-2)\underline{b} \tag{4.18}$$

Comparing (4.18) with equation $U\underline{x} = \underline{d}$ shows that $\underline{d} = G_{32}(3)G_{31}(1)$ $G_{21}(-2)\underline{b}$. Indeed:

$$\underline{d} = \begin{bmatrix} 1 & 0 & 0 \\ 0 & 1 & 0 \\ 0 & -3 & 1 \end{bmatrix}\begin{bmatrix} 1 & 0 & 0 \\ 0 & 1 & 0 \\ -1 & 0 & 1 \end{bmatrix}\begin{bmatrix} 1 & 0 & 0 \\ 2 & 1 & 0 \\ 0 & 0 & 1 \end{bmatrix}\begin{bmatrix} 4 \\ -13 \\ 17 \end{bmatrix} = \begin{bmatrix} 4 \\ -5 \\ 28 \end{bmatrix} \tag{4.19}$$

The reader may have noted that the multiplication of the three Gauss elimination matrices results in a lower diagonal matrix:

$$\begin{bmatrix} 1 & 0 & 0 \\ 0 & 1 & 0 \\ 0 & -3 & 1 \end{bmatrix}\begin{bmatrix} 1 & 0 & 0 \\ 0 & 1 & 0 \\ -1 & 0 & 1 \end{bmatrix}\begin{bmatrix} 1 & 0 & 0 \\ 2 & 1 & 0 \\ 0 & 0 & 1 \end{bmatrix} = \begin{bmatrix} 1 & 0 & 0 \\ 2 & 1 & 0 \\ -7 & -3 & 1 \end{bmatrix} \tag{4.20}$$

This is not an accident. We will investigate this property in more detail in Section 4.3.1.

4.2.2 Gauss Elimination Algorithm With Partial Pivoting

The naïve Gauss elimination algorithm presents a particular drawback whereby it may fail even if the system of equations one aims to solve has a unique solution. Consider, for example, the following Gauss elimination operation:

$$\begin{bmatrix} a_{11} & a_{12} & \cdots & a_{1n} \\ a_{21} & a_{22} & \cdots & a_{2n} \\ \vdots & \vdots & & \vdots \\ a_{n1} & a_{n2} & \cdots & a_{nn} \end{bmatrix} \begin{bmatrix} x_1 \\ x_2 \\ \vdots \\ x_n \end{bmatrix} = \begin{bmatrix} b_1 \\ b_2 \\ \vdots \\ b_n \end{bmatrix} \xrightarrow{r_2 \leftarrow R_2 - gR_1} \begin{bmatrix} a_{11} & a_{12} & \cdots & a_{1n} \\ 0 & a'_{22} & \cdots & a'_{2n} \\ a_{31} & a_{32} & \cdots & a_{3n} \\ \vdots & \vdots & & \vdots \\ a_{n1} & a_{n2} & \cdots & a_{nn} \end{bmatrix} \begin{bmatrix} x_1 \\ x_2 \\ x_3 \\ \vdots \\ x_n \end{bmatrix} = \begin{bmatrix} b_1 \\ b'_2 \\ b_3 \\ \vdots \\ b_n \end{bmatrix} \tag{4.21}$$

where the elimination coefficient g is calculated as

$$g = \frac{a_{21}}{a_{11}} \tag{4.22}$$

If the pivot element a_{11} is equal to zero, this operation can not be performed. This doesn't mean the system of equation has no solution. In fact, switching the pivot equation with any other equation where the coefficient a_{i1} is none zero solves the issue. In the context of Gauss elimination, the operation of switching two equations is called pivoting.

The naïve Gauss elimination algorithm is also prone to failure when the pivot element is a number very close to zero. In such a case, the Gauss elimination coefficient g, such as the one computed by Equation (4.22), becomes very sensitive to round-off errors. To minimize the risk of dividing by small numbers, we must avoid using a pivot equation that has a small number as the pivot element. Both issues, a pivot element equal to zero or a pivot element being a small number, can be addressed by adequate pivoting strategies.

Several strategies exist. A simple one is partial pivoting on rows. For this, all column elements of the coefficient matrix below the pivot element are scanned. We then look for the largest, in absolute value, among them. If this element is larger than the absolute value of the pivot element, the pivot equation and the row containing this largest element are switched. Switching two equations does not change the solution set of the system of equations. It, however, resolves the issue of a small or vanishing pivot element. In this way, the pivot element is always the largest possible value for our given system of equations.

Similarly, it is possible to develop pivoting strategies involving scanning row elements to the right of the pivot element rather than the column elements below it. Switching columns is allowed as it is the same as renumbering the unknowns. Finally, it is possible to combine both strategies, which are referred to as full pivoting strategies. Example 4.4 illustrates the deployment of the partial pivoting strategy by scanning the column elements below the pivoting element.

Example 4.4. Gauss Elimination With Partial Pivoting

We consider the system of equations (4.8) from Example 4.1. The algorithm starts by determining the first pivot element and scanning the column below it. In our example, the first pivot element is, in absolute value, $|a_{11}| = 2$. Scanning the column below it finds the largest entry, in absolute value, in the second row as $|a_{21}| = 4$. As $|a_{21}|$ is larger than the pivot element, the two rows (row 1 and row 2) have to be switched, resulting in the following system of equations to be used for naïve Gauss elimination:

$$\begin{bmatrix} -4 & -1 & 10 \\ 2 & 1 & -3 \\ 2 & 4 & -5 \end{bmatrix} \begin{bmatrix} x_1 \\ x_2 \\ x_3 \end{bmatrix} = \begin{bmatrix} -13 \\ 4 \\ 17 \end{bmatrix} \tag{4.23}$$

Applying naïve Gauss elimination results in:

$$\begin{bmatrix} -4 & -1 & 10 \\ 0 & 0.5 & 2 \\ 0 & 3.5 & 0 \end{bmatrix} \begin{bmatrix} x_1 \\ x_2 \\ x_3 \end{bmatrix} = \begin{bmatrix} -13 \\ -2.5 \\ 10.5 \end{bmatrix} \tag{4.24}$$

Before proceeding with naïve Gauss elimination on the second column, one has to determine if pivoting is required or not. Now, the pivot element is $|a_{22}| = 0.5$. Scanning the column below results in the maximal element as being $|a_{32}| = 3.5$. As this element is larger than the pivot element, the two rows have to be swapped, resulting in the following system of equations:

$$\begin{bmatrix} -4 & -1 & 10 \\ 0 & 3.5 & 0 \\ 0 & 0.5 & 2 \end{bmatrix} \begin{bmatrix} x_1 \\ x_2 \\ x_3 \end{bmatrix} = \begin{bmatrix} -13 \\ 10.5 \\ -2.5 \end{bmatrix} \tag{4.25}$$

Proceeding with naïve Gauss elimination results in:

$$\begin{bmatrix} -4 & -1 & 10 \\ 0 & 3.5 & 0 \\ 0 & 0 & 2 \end{bmatrix} \begin{bmatrix} x_1 \\ x_2 \\ x_3 \end{bmatrix} = \begin{bmatrix} -13 \\ 10.5 \\ -4 \end{bmatrix} \tag{4.26}$$

Compared to Example 4.1, we obtain a different upper triangular system of linear equations. However, the solution to both systems is the same. We can verify this by applying back substitution to system (4.26).

Similarly to naïve Gauss elimination, Gauss elimination with partial pivoting can be written in an operator form. This requires encoding the pivoting operation in a matrix. We can achieve this using permutation matrices.

Definition 4.1. Permutation Matrix

A permutation matrix is a $n \times n$ matrix consisting of all zeros except for a single 1 in every row and column.

From Definition 4.1 it follows that the unity matrix is a permutation matrix. Further permutation matrices can be built by swapping rows and columns of a unity matrix.

Example 4.5. Permutation Matrices

The following six matrices are all possible 3×3 permutation matrices:

$$\begin{bmatrix} 1 & 0 & 0 \\ 0 & 1 & 0 \\ 0 & 0 & 1 \end{bmatrix}, \begin{bmatrix} 0 & 1 & 0 \\ 1 & 0 & 0 \\ 0 & 0 & 1 \end{bmatrix}, \begin{bmatrix} 0 & 0 & 1 \\ 0 & 1 & 0 \\ 1 & 0 & 0 \end{bmatrix}, \begin{bmatrix} 1 & 0 & 0 \\ 0 & 0 & 1 \\ 0 & 1 & 0 \end{bmatrix}, \begin{bmatrix} 0 & 0 & 1 \\ 1 & 0 & 0 \\ 0 & 1 & 0 \end{bmatrix}, \begin{bmatrix} 0 & 1 & 0 \\ 0 & 0 & 1 \\ 1 & 0 & 0 \end{bmatrix}$$

They were all constructed by swapping rows and columns of the 3×3 identity matrix.

Permutation matrices can be used to encode exchanges of rows within a matrix, as stated in Theorem 4.2.

Theorem 4.2. Permutation Matrices and Row Exchanges

Let P be a permutation matrix formed by a particular set of row exchanges applied to the identity matrix. Then, for any matrix A, PA is the matrix obtained by applying exactly the same set of row exchanges to A.

For example, let us exchange the last two rows of the 3×3 identity matrix. This results in the following permutation matrix:

$$P = \begin{bmatrix} 1 & 0 & 0 \\ 0 & 0 & 1 \\ 0 & 1 & 0 \end{bmatrix} \tag{4.27}$$

Applying this matrix to a general 3×3 matrix results indeed in the exchange of the last two rows:

$$P=\begin{bmatrix}1&0&0\\0&0&1\\0&1&0\end{bmatrix}\begin{bmatrix}a_{11}&a_{12}&a_{13}\\a_{21}&a_{22}&a_{23}\\a_{31}&a_{32}&a_{33}\end{bmatrix}=\begin{bmatrix}a_{11}&a_{12}&a_{13}\\a_{31}&a_{32}&a_{33}\\a_{21}&a_{22}&a_{23}\end{bmatrix} \tag{4.28}$$

Let us note that the same result applies to a general three-dimensional column vector. With the concept of permutation matrices, one can formulate the Gauss elimination algorithm with partial pivoting in an operator form.

Example 4.6. Gauss Elimination With Partial Pivoting in Operator Form

Let us show how the various steps of Example 4.4 can be written in operator form. We start with the pivoting step. Revisiting Example 4.4 shows that we had to exchange rows 1 and 2 of matrix A. This can be achieved with a permutation matrix obtained by exchanging the first two rows of the identity matrix:

$$\begin{bmatrix}0&1&0\\1&0&0\\0&0&1\end{bmatrix}\begin{bmatrix}2&1&-3\\-4&-1&10\\2&4&-5\end{bmatrix}=\begin{bmatrix}-4&-1&10\\2&1&-3\\2&4&-5\end{bmatrix} \tag{4.29}$$

Next, we perform the elimination steps to remove the coefficient of the unknown x_1 in the last two equations. We already know how to encode these operations with a matrix:

$$\begin{bmatrix}1&0&0\\0&1&0\\0.5&0&1\end{bmatrix}\begin{bmatrix}1&0&0\\0.5&1&0\\0&0&1\end{bmatrix}\begin{bmatrix}-4&-1&10\\2&1&-3\\2&4&-5\end{bmatrix}=\begin{bmatrix}-4&-1&10\\0&0.5&2\\0&3.5&0\end{bmatrix} \tag{4.30}$$

Then, we apply the pivoting step for the second row. In our case, as $|3.5|>|0.5|$, we have to exchange the last two rows. This can be achieved with a permutation matrix obtained by exchanging the last two rows of the identity matrix:

$$\begin{bmatrix}1&0&0\\0&0&1\\0&1&0\end{bmatrix}\begin{bmatrix}-4&-1&10\\0&0.5&2\\0&3.5&0\end{bmatrix}=\begin{bmatrix}-4&-1&10\\0&3.5&0\\0&0.5&2\end{bmatrix} \tag{4.31}$$

By applying the last naïve Gauss elimination step on column 2, we obtain our upper diagonal matrix U as in Example 4.4:

$$\begin{bmatrix}1&0&0\\0&1&0\\0&-1/7&1\end{bmatrix}\begin{bmatrix}-4&-1&10\\0&3.5&0\\0&0.5&2\end{bmatrix}=\begin{bmatrix}-4&-1&10\\0&3.5&0\\0&0&2\end{bmatrix} \tag{4.32}$$

At this stage, there seems to be very little, if any, benefit to encoding the row exchanges in permutation matrices. We will see in Section 4.3.3 how this encoding presents some significant benefits.

4.3 MATRIX DECOMPOSITIONS

In Section 4.2, we discussed how Gauss elimination algorithms can be represented in operator form. Once the adequate transformation matrix is available, the application of the algorithm becomes straightforward and is simply a task of executing a matrix product. This can be particularly beneficial when we have to solve several systems of equations having all the same coefficient matrix A but different right-hand side vectors $\underline{b}$. Once the Gauss elimination algorithm is encoded in a matrix G, we can simply apply the operations by multiplying the various systems of equations by G to bring them into an upper diagonal form.

Before showing in detail how this can be done, we have to revisit the Gauss elimination algorithm in operator form in order to find a methodology that can construct this transformation matrix G in a simple and efficient way. It turns out that mathematically, it is simpler to construct the inverse of this transformation matrix. We start by explaining how this matrix can be constructed.

4.3.1 LU DECOMPOSITION OF A MATRIX

Consider Example 4.4. In this example the coefficient matrix A was:

$$A = \begin{bmatrix} -4 & -1 & 10 \\ 2 & 1 & -3 \\ 2 & 4 & -5 \end{bmatrix} \tag{4.33}$$

Recall that the naïve Gauss elimination algorithm transformed the matrix A into the following upper diagonal matrix U:

$$U = \begin{bmatrix} 2 & 1 & -3 \\ 0 & 1 & 4 \\ 0 & 0 & -14 \end{bmatrix} \tag{4.34}$$

In Example 4.4, we found the following transformation matrix G that transforms the original system $A\underline{x} = \underline{b}$ into the upper diagonal system $U\underline{x} = \underline{d}$ (the reader is invited to verify once more that $U = GA$):

$$G = \begin{bmatrix} 1 & 0 & 0 \\ 2 & 1 & 0 \\ -7 & -3 & 1 \end{bmatrix} \tag{4.35}$$

Now, it turns out that the following equation holds:

$$A = \begin{bmatrix} -4 & -1 & 10 \\ 2 & 1 & -3 \\ 2 & 4 & -5 \end{bmatrix} = \begin{bmatrix} 1 & 0 & 0 \\ -2 & 1 & 0 \\ 1 & 3 & 1 \end{bmatrix} \begin{bmatrix} 2 & 1 & -3 \\ 0 & 1 & 4 \\ 0 & 0 & -14 \end{bmatrix} \tag{4.36}$$

The matrix A is factored into the product of two matrices. A first lower diagonal matrix L and a second upper diagonal matrix which is the matrix U constructed by the naïve Gauss elimination algorithm. Interestingly, matrix L is the inverse of matrix G. This is straightforward to prove. As $U = GA$, assuming G^{-1} exists, then $A = G^{-1}U$ and consequently $L = G^{-1}$. The not straightforward fact is that G^{-1} turns out to be a lower diagonal matrix. We will come back on this shortly. For now, let us investigate matrix $L = G^{-1}$.

Matrix L has all its diagonal elements equal to 1. If we inspect the remaining non-zero entries, we find that they are exactly equal to the elimination coefficients g used in the naïve Gauss elimination algorithm (compare with Example 4.1 and Example 4.4). This is not an accident, as stated by Theorem 4.3:

Theorem 4.3. LU Decomposition of a Square Matrix

Any square matrix A for which the naïve Gauss elimination algorithm can be successfully executed can be decomposed into the product

$$A = LU \tag{4.37}$$

where U is the upper diagonal matrix computed by the naïve Gauss elimination algorithm, and L is a lower diagonal matrix, with all its diagonal elements equal to 1 and its lower diagonal elements equal to the Gauss elimination coefficients g used during the execution of the naïve Gauss elimination algorithm.

Before giving a sketch of the proof of Theorem 4.3, let us note that if one does not force matrix U to be an upper diagonal matrix calculated by the naïve Gauss elimination algorithm but allows U to be any upper diagonal matrix, then the LU decomposition is not unique. Indeed, we can simply multiply U by a number $k \neq 0$ and the matrix L by k^{-1} to find two new matrices $U' = kU$ and $L' = k^{-1}L$, which are still upper and lower diagonal matrices, and their product is still equal to A.

Let us now explain why Theorem 4.3 holds. We know already that $L = G^{-1}$. What is not obvious at this stage, though, is why G^{-1} is always a lower diagonal matrix and why its lower diagonal elements are equal to the Gauss elimination coefficients g used during the execution of the naïve Gauss elimination algorithm. Both properties result from a few simple facts. We explain them in the case of a 3×3 matrix A, but the results hold for a general $n \times n$ matrix. Firstly, we write down explicitly the inverse of G. Recall that in the case of a 3×3 coefficient matrix A, the Gauss elimination operator G encoding the eliminations steps writes as the product of three individual eliminations steps:

$$G = G_{32}\left(g_{32}\right)G_{31}\left(g_{31}\right)G_{21}\left(g_{21}\right) \tag{4.38}$$

with g_{ij} the respective Gauss elimination coefficients used. From Equation (4.38) it follows:

$$G^{-1} = G_{21}^{-1}\left(g_{21}\right)G_{31}^{-1}\left(g_{31}\right)G_{32}^{-1}\left(g_{32}\right) \tag{4.39}$$

It turns out that the inversion of a Gauss elimination operator is very simple. For example:

$$G_{21}^{-1}\left(g_{21}\right) = \begin{bmatrix} 1 & 0 & 0 \\ -g_{21} & 1 & 0 \\ 0 & 0 & 1 \end{bmatrix}^{-1} = \begin{bmatrix} 1 & 0 & 0 \\ g_{21} & 1 & 0 \\ 0 & 0 & 1 \end{bmatrix} = G_{21}\left(-g_{21}\right) \tag{4.40}$$

This can be verified by checking that indeed the product $G_{21}^{-1}\left(g_{21}\right)G_{21}\left(g_{21}\right)$ is equal to the identity matrix. But by thinking about how a Gauss elimination matrix is built and how the inverse of the given Gauss elimination step looks like (i.e., which row operation must be conducted to reverse the operation), it is quite obvious that the inverse of a Gauss elimination matrix is simply computed by changing the sign of the Gauss elimination coefficient.

Using this fact, Equation (4.39) can be computed as

$$G^{-1} = \begin{bmatrix} 1 & 0 & 0 \\ g_{21} & 1 & 0 \\ 0 & 0 & 1 \end{bmatrix}\begin{bmatrix} 1 & 0 & 0 \\ 0 & 1 & 0 \\ g_{31} & 0 & 1 \end{bmatrix}\begin{bmatrix} 1 & 0 & 0 \\ 0 & 1 & 0 \\ 0 & g_{32} & 1 \end{bmatrix} = \begin{bmatrix} 1 & 0 & 0 \\ g_{21} & 1 & 0 \\ g_{31} & g_{32} & 1 \end{bmatrix} \tag{4.41}$$

which is indeed the matrix L of Theorem 4.3.

Example 4.7. LU Decomposition

Let us compute the LU decomposition of the following matrix:

$$A = \begin{bmatrix} -3 & 1 & 4 \\ 6 & 0 & -13 \\ -9 & 7 & 1 \end{bmatrix} \tag{4.42}$$

Applying naïve Gauss elimination on A transforms A into the following upper diagonal matrix U:

$$U = \begin{bmatrix} -3 & 1 & 4 \\ 0 & 2 & -5 \\ 0 & 0 & -1 \end{bmatrix} \tag{4.43}$$

This can be verified using our Octave function `naive_gauss` from Section 4.2.1 implementing the naïve Gauss elimination on a system of equations (note: as our function requires a right-hand side vector $\underline{b}$, which is not relevant for the calculation of U, we provide an arbitrarily chosen vector):

```
>> A = [-3 1 4; 6 0 -13; -9 7 1];
>> naive_gauss(A,[1 1 1]')
ans =
   -3    1    4
    0    2   -5
    0    0   -1
```

During the application of the algorithm, the following Gauss elimination coefficients were used (the reader is invited to modify the function `naive_gauss` to provide these coefficients):

$$g_{21} = -2,\ g_{31} = 3,\ g_{32} = 2. \tag{4.44}$$

which results in the following matrix L:

$$L = \begin{bmatrix} 1 & 0 & 0 \\ -2 & 1 & 0 \\ 3 & 2 & 1 \end{bmatrix} \tag{4.45}$$

One can verify that indeed $A = LU$.

```
>> U = [-3 1 4; 0 2 -5; 0 0 -1];
>> L = [1 0 0; -2 1 0; 3 2 1];
>> L*U
ans =
   -3    1    4
    6    0  -13
   -9    7    1
```

The LU decomposition is an efficient algorithm for computing the determinant of a matrix. This comes from the special structure of the matrix L and U. Determinants of matrices are lengthy to compute. Lower and upper diagonal matrices are an exception. In both cases, the determinant is simply the product of their diagonal elements. This fact is used in is used in Example 4.8.

Example 4.8. Determinant of a Matrix Using LU Decomposition

Consider the following matrix:

$$A = \begin{bmatrix} -3 & 2 & -1 \\ 6 & -6 & 7 \\ 3 & -4 & 4 \end{bmatrix} \tag{4.46}$$

We aim to compute its determinant.

The reader is invited to compute the LU decomposition of A and verify that the result it

$$\begin{bmatrix} -3 & 2 & -1 \\ 6 & -6 & 7 \\ 3 & -4 & 4 \end{bmatrix} = \begin{bmatrix} 1 & 0 & 0 \\ -2 & 1 & 0 \\ -1 & 1 & 1 \end{bmatrix} \begin{bmatrix} -3 & 2 & -1 \\ 0 & -2 & 5 \\ 0 & 0 & -2 \end{bmatrix} \tag{4.47}$$

We next compute the determinant of A as

$$\det(A) = \det(LU) = \det(L) \cdot \det(U) = 1 \cdot 1 \cdot 1 \cdot (-3)(-2)(-2) = -12 \tag{4.48}$$

using the fact that the determinants of lower and upper diagonal matrices are the product of their diagonal elements.

4.3.2 Solving a Linear System of Equations Using LU Decomposition

Once a square matrix A has been decomposed into a product $A = LU$, it becomes possible to efficiently solve any system of equations $A\underline{x} = \underline{b}$, regardless of the value of the right-hand side vector $\underline{b}$. To understand the principle, start with:

$$A\underline{x} = LU\underline{x} = \underline{b} \tag{4.49}$$

Multiplying from the left by L^{-1} results in

$$U\underline{x} = L^{-1}\underline{b} \tag{4.50}$$

Defining $\underline{d} = L^{-1}\underline{b}$ leads directly to the steps to follow to solve the system of equations.

Algorithm 4.1. Solving a System of Linear Equations by LU Decomposition

Given a linear system of equations $A\underline{x} = \underline{b}$, the following steps lead to find its solution:

1. Decompose the coefficient matrix A into $A = LU$.
2. Solve for $\underline{d}$ the system $L\underline{d} = \underline{b}$ by forward substitution.
3. Solve for $\underline{x}$ the system $U\underline{x} = \underline{d}$ by back substitution.

One may wonder why the procedure is of interest as, compared to the original system $A\underline{x} = \underline{b}$, we now have two systems, $L\underline{d} = \underline{b}$ and $U\underline{x} = \underline{d}$, to solve. To understand the advantage of the LU decomposition, we start by noting that, when conducting naïve Gauss elimination, all the elements required to build the matrices L and U are computed. As such, the LU decomposition comes at no cost. Next, we note that this decomposition can be conducted by considering only the coefficient matrix A. This operation is totally independent of the particular right-hand side vector $\underline{b}$ of the system. Consequently, Algorithm 4.1 is of interest if several systems with the same coefficient matrix A have to be solved. In other words, the bulk of the work is encoded in the LU decomposition of the matrix A and afterward, for a given right-hand side vector $\underline{b}$, remains to solve the two systems $L\underline{d} = \underline{b}$ and $U\underline{x} = \underline{d}$, which are both straightforward to solve, as we will see in the next example.

Example 4.9. Solving a System of Linear Equations by LU Decomposition

Consider the following system of linear equations:

$$\begin{bmatrix} -3 & 1 & 4 \\ 6 & 0 & -13 \\ -9 & 7 & 1 \end{bmatrix} \begin{bmatrix} x_1 \\ x_2 \\ x_3 \end{bmatrix} = \begin{bmatrix} 13 \\ 25 \\ 40 \end{bmatrix} \tag{4.51}$$

From Example 4.7, we already know the LU decomposition of A:

$$A = \begin{bmatrix} -3 & 1 & 4 \\ 6 & 0 & -13 \\ -9 & 7 & 1 \end{bmatrix} = \begin{bmatrix} 1 & 0 & 0 \\ -2 & 1 & 0 \\ 3 & 2 & 1 \end{bmatrix} \begin{bmatrix} -3 & 1 & 4 \\ 0 & 2 & -5 \\ 0 & 0 & -1 \end{bmatrix} \tag{4.52}$$

We find the vector $\underline{d}$ by solving

$$\begin{bmatrix} 1 & 0 & 0 \\ -2 & 1 & 0 \\ 3 & 2 & 1 \end{bmatrix} \begin{bmatrix} d_1 \\ d_2 \\ d_3 \end{bmatrix} = \begin{bmatrix} 13 \\ 25 \\ 40 \end{bmatrix} \tag{4.53}$$

This can be done in a similar way to back substitution but starting with the first equation and moving downwards. This process is referred to as forward substitution:

$$\begin{bmatrix} d_1 \\ d_2 \\ d_3 \end{bmatrix} = \begin{bmatrix} 13 \\ 51 \\ -101 \end{bmatrix} \tag{4.54}$$

The last step is to compute the vector $\underline{x}$ by solving:

$$\begin{bmatrix} -3 & 1 & 4 \\ 0 & 2 & -5 \\ 0 & 0 & -1 \end{bmatrix} \begin{bmatrix} x_1 \\ x_2 \\ x_3 \end{bmatrix} = \begin{bmatrix} 13 \\ 51 \\ -101 \end{bmatrix} \tag{4.55}$$

Using back substitution results in:

$$\begin{bmatrix} x_1 \\ x_2 \\ x_3 \end{bmatrix} = \begin{bmatrix} 223 \\ 278 \\ 101 \end{bmatrix} \tag{4.56}$$

4.3.3 PA=LU Decomposition of a Matrix

The LU decomposition of a matrix A is essentially the operator version of the naïve Gauss elimination algorithm. The matrix L keeps track of the various row operations that were done to transform A into the upper diagonal matrix U. However, as discussed in Section 4.2.2, naïve Gauss elimination can fail or become quite sensitive to round-off errors if the various pivot elements are close to zero. Pivoting strategies avoid these issues. The various switching operations that must be executed to avoid small pivot elements can be encoded in permutation matrices we introduced in Section 4.2.2. By grouping them all together in a single matrix P, one obtains the so-called $PA = LU$ decomposition of a matrix A.

To understand how to construct the $PA = LU$ decomposition, let us revisit Example 4.6 and Example 4.4. During the execution of the Gauss elimination algorithm with partial pivoting, two switching operations were identified. The first operation consisted of switching row 1 and row 2. The second consisted of switching row 2 and row 3.

Using permutation matrices, these two operations can be encoded and afterward applied to the coefficient matrix A. If P_1 is the permutation matrix switching row 1 and row 2, and P_2 is the permutation matrix switching row 2 and row 3, the matrix $P = P_2 P_1$ will execute these operations in the correct order on the matrix A. Note the order of executing the multiplication between P_1 and P_2. As the operations encoded in P_1 must be applied first to A, P_1 appears at the utmost right in the multiplication:

$$P = P_2 P_1 = \begin{bmatrix} 1 & 0 & 0 \\ 0 & 0 & 1 \\ 0 & 1 & 0 \end{bmatrix} \begin{bmatrix} 0 & 1 & 0 \\ 1 & 0 & 0 \\ 0 & 0 & 1 \end{bmatrix} = \begin{bmatrix} 0 & 1 & 0 \\ 0 & 0 & 1 \\ 1 & 0 & 0 \end{bmatrix} \tag{4.57}$$

Knowing which rows must be switched in which order allows the calculation of the matrix that has to be decomposed into LU instead of the original one:[5]

$$PA = \begin{bmatrix} 0 & 1 & 0 \\ 0 & 0 & 1 \\ 1 & 0 & 0 \end{bmatrix} \begin{bmatrix} 2 & 1 & -3 \\ -4 & -1 & 10 \\ 2 & 4 & -5 \end{bmatrix} = \begin{bmatrix} -4 & -1 & 10 \\ 2 & 4 & -5 \\ 2 & 1 & -3 \end{bmatrix} \tag{4.58}$$

Proceeding with LU decomposition on the matrix PA, gives the desired decomposition:

$$PA = \begin{bmatrix} 0 & 1 & 0 \\ 0 & 0 & 1 \\ 1 & 0 & 0 \end{bmatrix} \begin{bmatrix} -4 & -1 & 10 \\ 2 & 1 & -3 \\ 2 & 4 & -5 \end{bmatrix} = \begin{bmatrix} 1 & 0 & 0 \\ -0.5 & 1 & 0 \\ -0.5 & 1/7 & 1 \end{bmatrix} \begin{bmatrix} -4 & -1 & 10 \\ 0 & 3.5 & 0 \\ 0 & 0 & 2 \end{bmatrix} \tag{4.59}$$

Octave implements the function `lu` to perform the $PA = LU$ decomposition. The reader is invited to read more about this function using the command `help lu` in Octave. The Octave function `lu` performs several matrix decompositions, and among them is the LU decomposition. However, the implemented decomposition is not the same as presented in Section 4.3.1. To perform the $\boldsymbol{PA} = \boldsymbol{LU}$ decomposition, the following steps are required:

```
>> A = [2 1 -3; -4 -1 10; 2 4 -5];
>> [L U P] = lu(A)
L =
   1.00000   0.00000   0.00000
  -0.50000   1.00000   0.00000
  -0.50000   0.14286   1.00000
U =
  -4.00000  -1.00000  10.00000
   0.00000   3.50000   0.00000
   0.00000   0.00000   2.00000
P =
Permutation Matrix
   0   1   0
   0   0   1
   1   0   0
```

Let us recall the reason for using pivoting strategies in Gauss elimination algorithms. Pivoting strategies, though adding some overhead to the algorithm compared to naïve Gauss elimination, are required to ensure the algorithm does not fail in case a pivot element is equal to zero. It also helps reduce the effects of round-off errors in the case of small pivot elements.

4.3.4 Solving Systems of Linear Equations Using PA=LU Decomposition

Once the $PA = LU$ decomposition on a matrix A is performed, any system of linear equations $A\underline{x} = \underline{b}$ can be solved for an arbitrary right-hand side vector $\underline{b}$. The method is essentially the same as in the case of LU decomposition. However, we must not forget to apply the various row-switching operations required by the pivoting strategy onto the right-hand side vector, too. This is only a matter of multiplying the vector $\underline{b}$ by the permutation matrix P.

Starting from the equation $A\underline{x} = \underline{b}$ multiplied from the left side by P, we obtain:

$$PA\underline{x} = LU\underline{x} = P\underline{b} \tag{4.60}$$

Multiplying from the left by L^{-1} results in

$$U\underline{x} = L^{-1}P\underline{b} \tag{4.61}$$

Defining $\underline{d} = L^{-1}P\underline{b}$ leads directly to the steps to follow to solve the system of equations.

Algorithm 4.2. Solving a System of Linear Equations by PA=LU Decomposition

Given a linear system of equations $A\underline{x} = \underline{b}$, the following steps lead to find its solution:

1. Decompose the coefficient matrix A into $PA = LU$
2. Solve for $\underline{d}$ the system $L\underline{d} = P\underline{b}$ by forward substitution
3. Solve for $\underline{x}$ the system $U\underline{x} = \underline{d}$ by back substitution

---

Octave has a dedicated operator to solve a system of linear equations using the $PA = LU$ decomposition: the left division operator. The following example illustrates its usage:

```
>> A = [2 1 -3; -4 -1 10; 2 4 -5];
>> b = [4 -13 17]';
>> A\b
ans =
   -2.5000
    3.0000
   -2.0000
```

Conceptually, the left division `A\b` is the same as computing `inv(A)*b`. However, Octave avoids the calculation of the inverse and uses, in the case of a $n \times n$ system of equations, the $PA = LU$ decomposition to solve the system $A\underline{x} = \underline{b}$ instead.

Example 4.10. Solving a System of Linear Equations With the PA=LU Decomposition

We revisit the system of Equation (4.51) form Example 4.9.

The first step is to find the PA=LU decomposition of the coefficient matrix A of the system. Following the method described in Section 4.3.3, one finds:

$$\begin{bmatrix} 0 & 0 & 1 \\ 0 & 1 & 0 \\ 1 & 0 & 0 \end{bmatrix} \begin{bmatrix} -3 & 1 & 4 \\ 6 & 0 & -13 \\ -9 & 7 & 1 \end{bmatrix} = \begin{bmatrix} 1 & 0 & 0 \\ -2/3 & 1 & 0 \\ 1/3 & -2/7 & 1 \end{bmatrix} \begin{bmatrix} -9 & 7 & 1 \\ 0 & 14/3 & -37/3 \\ 0 & 0 & 1/7 \end{bmatrix} \tag{4.62}$$

The same result can be obtained using Octave:

```
>> A = [-3 1 4; 6 0 -13; -9 7 1];
>> [L U P] = lu(A);
```

Next, we solve $L\underline{d} = P\underline{b}$ for $\underline{d}$. In Octave:

```
>> b = [13; 25; 40];
>> d = L\(P*b)
d =
    40.000
    51.667
    14.429
```

And as a last step, we solve $U\underline{x} = \underline{d}$ for $\underline{x}$. In Octave:

```
>> x = U\d
x =
    223
    278
    101
```

In Octave, the answer could be found directly with

```
>> x = A\b
```

which will use the PA=LU decomposition to solve the system of Equation (4.51).

4.4 CONDITIONING

It might seem unnecessary to discuss errors in the context of systems of linear equations given that the algorithms presented for solving these systems are free of truncation errors. However, the total numerical error has two components: truncation and round-off errors. Even if an algorithm is free of truncation errors, round-off errors cannot be avoided. This is why it is still required to conduct a thorough error analysis. We will see next that systems of linear equations, despite their mathematical simplicity, can become very sensitive to round-off errors. In fact, there are systems of linear equations that are very sensitive to small changes in their input. For example, changing only slightly the right-hand side vector $\underline{b}$ can lead to a very large change in the solution $\underline{r}$. Such systems will be very challenging to solve numerically, as was the case with equations in one variable presenting multiple roots.

4.4.1 Measuring Errors for Systems of Equations

In Chapter 1, we discussed how to measure errors. We defined the absolute error as being the difference, in absolute value, between the true and approximated solution of a mathematical problem. This definition applies in the case our numerical solution is a number. In the context of systems of equations, where the solution is no longer a single number, this definition needs to be generalized. In fact, there are different ways of measuring errors, and during the course of this text, we will encounter several of them. Measuring errors is linked to the concept of similarity, which aims to quantify how similar two mathematical objects are. At this stage, we have only to handle vectors as mathematical objects. To measure the error, or in other words, the difference between a true and an approximated solution, we use the concept of vector norm (norms will be discussed more in-depth in Chapter 5).

Definition 4.2. True Errors of System Of Equations

The true absolute error of a numerical approximation $\underline{x}_r$ of a true solution $\underline{r}$ of a system of equations is the quantity

$$E(\underline{x}_r) = \|\underline{x}_r - \underline{r}s\| \tag{4.63}$$

The true relative error of a numerical approximation $\underline{x}_r$ of a true solution $\underline{r}$ of a system of equations is the quantity

$$e(\underline{x}_r) = \frac{\|\underline{x}_r - \underline{r}\|}{\|\underline{r}\|} \tag{4.64}$$

Note that sometimes we will use $\underline{x}_r$ in the denominator of Equation (4.64) instead of $\underline{r}$.

Any norm can be used to measure errors. The context in which a numerical solution has to be found will usually favor a norm over others. For the remainder of this chapter, we will use the infinity norm as it is particularly simple to compute from a numerical point of view. Furthermore, as will be illustrated in the following examples, the infinity norm presents some other interesting properties that often make it a well-suited choice.

Example 4.11. Vector Norms

Octave implements the function `norm` that allows the calculation of norms belonging to the family of the p-norm. The p-norm is defined as:

$$\left\| \underline{x}_p \right\| = \sqrt[p]{|x_1|^p + |x_2|^p + \ldots + |x_n|^p} \tag{4.65}$$

with $p \geq 1$ a real number.

Let us illustrate its usage with the vector:

$$\underline{x} = \begin{bmatrix} -2 \\ 4 \\ -8 \\ 3 \\ 5 \end{bmatrix} \tag{4.66}$$

Let us define the vector in Octave:

```
>> x = [-2; 4; -8; 3; 5];
```

The following code computes the Euclidian norm:

```
>> norm(x)
ans = 10.863
```

Alternatively, we can use the syntax norm(x,2), where the second parameter tells Octave to use $p = 2$ for the calculation of the p-norm, which is indeed the Euclidian norm.

The following two lines of Octave code, compute successively the 1-norm, or Manahan norm of the vector, and the infinity norm:

```
>> norm(x,1)
ans = 22
>> norm(x,inf)
ans = 8
```

As expected, the value achieved will vary depending on the norm used. The usage of norms to measure errors allows the definition of backward and forward errors for systems of equations.

The usage of norms to measure errors allows the definition of backward and forward errors for systems of equations.

Definition 4.3. Forward and Backward Errors of Systems of Equations

The absolute forward error of an approximation $\underline{x}_r$ of a true solution $\underline{r}$ of a system of equations is the true error $\|\underline{x}_r - \underline{r}\|$ of $\underline{x}_r$ and the relative forward error is the true relative error $\|\underline{x}_r - \underline{r}\| / \|\underline{r}\|$ of $\underline{x}_r$.

The absolute backward error of an approximation $\underline{x}_r$ of a true solution $\underline{r}$ of a system of linear equations $A\underline{x} = \underline{b}$ is the quantity $\|A\underline{x}_r - \underline{b}\|$ and the relative backward error is the quantity $\|A\underline{x}_r - \underline{b}\| / \|\underline{b}\|$.

The following Example 4.12 illustrates Definition 4.3.

Example 4.12. Forward and Backward Errors

Consider the system of linear equations:

$$\begin{bmatrix} 3 & 2 \\ -1 & 2 \end{bmatrix} \begin{bmatrix} x_1 \\ x_2 \end{bmatrix} = \begin{bmatrix} 18 \\ 2 \end{bmatrix} \tag{4.67}$$

The true solution of this system is $\underline{x} = \begin{bmatrix} 4 & 3 \end{bmatrix}^T$. Let us compute the forward and backward errors of the approximation $\underline{x}_r = \begin{bmatrix} 3 & 5 \end{bmatrix}^T$. Using Octave, we start by defining the required variables:

```
>> r = [4 3]';
>> xr = [3 5]';
>> A = [3 2; -1 2];
>> b = [18 2]';
```

Using the infinity norm to measure errors, the absolute and relative forward errors are given by:

```
>> norm(xr-r, inf)
ans = 2
>> norm(xr-r, inf)/norm(r, inf)
ans = 0.50000
```

The absolute and relative backward errors evaluate to:

```
>> norm(A*xr-b, inf)
ans = 5
>> norm(A*xr-b, inf) / norm(b, inf)
ans = 0.27778
```

The usage of the infinity norm, besides being particularly simple to evaluate numerically, presents another advantage: it provides the largest error present in the components of the error vector, for example, in the vector $\underline{e} = \underline{x}_r - \underline{r}$. Some applications may indeed require keeping the error of each component of a vector below a given accuracy. In situations where each component represents a different physical quantity, this may be different. Imagine the first component represents a distance that has to be estimated to an accuracy below 1 mm, and the second component represents a velocity that has to be controlled to an accuracy below 5 km/h. In such a case, the infinity norm may not be the most appropriate norm to use. An alternative way is to use relative errors, which are independent of the physical unit.

In order to conduct error analysis on a linear system of equations, one more tool is needed. It will often be required to estimate quantities like $\|A\underline{x}\|$, where A is a square matrix and $\underline{x}$ is a column vector. Note that $A\underline{x}$ is a column vector, and, consequently, the quantity $\|A\underline{x}\|$ is well-defined. If instead of a matrix A, the expression is written with a scalar a, then the task is straightforward. In this case, one can write $\|a\underline{x}\| = |a|\|\underline{x}\|$. Our aim is to define a quantity $\|A\|$, which will be called the norm of a matrix, such that some similar calculations can be done.[6] Unfortunately, it is not possible to define a matrix norm such that $\|A\underline{x}\| = \|A\|\|\underline{x}\|$ holds for any vector $\underline{x}$. It is only possible to define a matrix norm such that the relation $\|A\underline{x}\| \leq \|A\|\|\underline{x}\|$ holds.

It is instructive to develop an intuitive understanding of why it is not possible to define a matrix norm that behaves in a simple way, as is the case when a vector is multiplied by a number. For this, let us imagine that matrix A has two eigenvalues, $\lambda_1$and λ_2, associated with the two eigenvectors $\underline{v}_1$ and $\underline{v}_2$. We can write:

$$\|A\underline{v}_1\| = \|\lambda_1\underline{v}_1\| = |\lambda_1|\|\underline{v}_1\| \quad (4.68)$$

Similarly, using the second eigenvector:

$$\|A\underline{v}_2\| = \|\lambda_2\underline{v}_2\| = |\lambda_2|\|\underline{v}_2\| \quad (4.69)$$

If the two eigenvalues are, then we could define $\|A\|$ as being equal to $|\lambda_1|$. But this is, of course, not the case. Generally, a matrix' eigenvalues are not all equal, and comparing Equations (4.68) and (4.69) leads directly to the conclusion that it is not possible to define a quantity $\|A\|$ for which all vectors $\underline{x}$ satisfy the relation $\|A\underline{x}\| = \|A\|\|\underline{x}\|$. However, it is possible to define $\|A\|$ such that $\|A\underline{x}\| \leq \|A\|\|\underline{x}\|$ holds. For example, one could choose $\|A\|$ as being the largest eigenvalue in absolute value.

There is indeed a way such a matrix norm can be defined. Let us now define the norm of a square matrix. We will afterward show that this definition indeed leads to the desired property that $\|A\underline{x}\| \leq \|A\|\|\underline{x}\|$ for any vector $\underline{x}$.

Definition 4.4. Matrix Norm

The matrix norm $\|A\|$ of a $n \times n$ matrix is the quantity:

$$\|A\| = \sup\left\{\|A\underline{u}\| : \underline{u} \in \mathbb{R}^n \text{ with} \|\underline{u}\| = 1\right\} \tag{4.70}$$

Definition 4.4 says that the matrix norm is the largest norm of the vector $\|A\underline{u}\|$ where one computes this norm over all unity vectors $\underline{u}$. It is worth noting that the definition is independent of the specific vector norm used. More precisely, the matrix norm, as defined in Definition 4.4, is called the matrix norm induced by a vector norm. Induced means that the definition is based on a given vector norm. It is possible to define matrix norms in a more general way, not induced on vector norms. In the frame of the present text, only matrix norms induced by vector norms are used. Theorem 4.4 follows directly from Definition 4.4.

Theorem 4.4. Submultiplicative Property of Matrix Norms

Consider a $n \times n$ matrix A. For any vector $\underline{x} \in \mathbb{R}^n$ we have

$$\|A\underline{x}\| \leq \|A\|\|\underline{x}\| \tag{4.71}$$

The proof is straightforward. Consider $\underline{x} \in \mathbb{R}^n$ We can write $\underline{x} = \|\underline{x}\|\underline{u}$, with $\underline{u}$ a unity vector, it follows, together with Definition 4.4, that:

$$\|A\underline{x}\| = \|\underline{x}\| \cdot \|A\underline{u}\| \leq \|A\|\|\underline{x}\| \tag{4.72}$$

Definition 4.4 is not very helpful in computing the matrix norm. There is no general expression that allows computing a matrix norm. But, for a given vector norm, such an expression can be derived. We present it here for the case of the infinity norm.

Consider a general unity vector $\underline{u}$ and a $n \times n$ matrix A. The component i of $A\underline{u}$ is given by:

$$\left(Au\right)_i = a_{i1}u_1 + a_{i2}u_1 + \ldots a_{in}u_n \tag{4.73}$$

The infinity norm of the vector $A\underline{u}$ is the largest, in absolute value, component of the vector. Keeping in mind that $\|\underline{u}\| = 1$, it follows that $|u_i| \leq 1$ for $i = 1,\ldots,n$. Consequently, the largest component of $A\underline{u}$, in absolute value, is:

$$\max_{i=1,\ldots,n} \left|a_{i1}u_1 + a_{i2}u_1 + \ldots a_{in}u_n\right| = \max_{i=1,\ldots,n}\left[\left|a_{i1}\right| + \left|a_{i2}\right| + \ldots + \left|a_{in}\right|\right] \tag{4.74}$$

From this observation, we can deduce that computing the matrix infinity norm can be done as follows:

Theorem 4.5. Matrix Infinity Norm

The infinity norm of a $n \times n$ matrix A is given by

$$\|A\|_\infty = \max_{i=1,\ldots,n}\left[\left|a_{i1}\right| + \left|a_{i2}\right| + \ldots + \left|a_{in}\right|\right] \tag{4.75}$$

The theorem means that the infinity norm of a $n \times n$ matrix A is the maximal absolute row sum of all rows of the matrix. Let us illustrate the usage of Theorem 4.5 with an example.

Example 4.13. Matrix Norm

Consider the following matrix:

$$A = \begin{bmatrix} 3 & 2 & 2 \\ -2 & -7 & 4 \\ 4 & 1 & -3 \end{bmatrix} \tag{4.76}$$

The absolute row sums are equal to $|3|+|2|+|2| = 7$ for row 1, $|-2|+|-7|+|4| = 13$ for row 2, and $|4|+|1|+|-3| = 8$ for row 3. Consequently,

$$\|A\|_\infty = 13 \tag{4.77}$$

Octave implements the `norm` function to compute matrix norms. Its usage is identical for matrices and vectors.

```
>> A = [3 2 2; -2 -7 4; 4 1 -3];
>> norm(A,inf)
ans = 13
```

Octave allows computing other norms as well. For example, the following lines of code compute the 1-norm and 2-norm of the matrix A:

```
>> norm(A,1)
ans = 10
>> norm(A,2)
ans = 9.2204
```

4.4.2 A System of Linear Equations With High Error Magnification Factor

In Chapter 3, we discussed how, in the case of equations in one variable, it is possible to have situations in which the backward error is very small while the forward error is very large. Such equations are challenging to solve numerically. The difficulty comes from the fact that once an approximation of the solution of the equation is available, only the backward error can be computed. A small backward error does not guarantee a small forward error, which is the error we are interested in controlling. The concept of error magnification factor formalizes this problematic. In Chapter 3, an estimation for the error magnification factor was provided and allowed for estimating the forward error based on the backward error. As discussed in Chapter 3, equations with multiple roots present high error magnification factors. In the case of linear equations, we never have multiple roots. This indicates that solving a linear equation in one variable is not challenging from a numerical point of view. However, the situation changes as soon as we have more than one equation.

Consider the following system of linear equations:

$$\begin{bmatrix} 1 & 1 \\ 1.0001 & 1 \end{bmatrix} \begin{bmatrix} x_1 \\ x_2 \end{bmatrix} = \begin{bmatrix} 2 \\ 2.0001 \end{bmatrix} \tag{4.78}$$

The true solution of this system is $\underline{r} = \begin{bmatrix} 1 & 1 \end{bmatrix}^T$. Let us consider the approximation $\underline{x}_r = \begin{bmatrix} 10 & -8 \end{bmatrix}^T$. This is a rather rough approximation, as it shows the high value of the relative forward error:

$$\frac{\left\| \underline{r} - \underline{x}_r \right\|_\infty}{\left\| \underline{r} \right\|_\infty} = \frac{9}{1} = 9 \tag{4.79}$$

On the other hand, the relative backward error is surprisingly low:

$$\frac{\left\| A\underline{x}_r - \underline{b} \right\|_\infty}{\left\| \underline{b} \right\|_\infty} = \frac{0.0009}{2.0001} \cong 0.0005 \tag{4.80}$$

Equations (4.78) are almost satisfied when using the approximation $\underline{x}_r$. The backward error is approximatively equal to zero. This can be quantified using the relative error magnification factor, which is the ratio between the relative forward and backward error:

$$m = \frac{9}{0.0005} = 18'000 \tag{4.81}$$

This example demonstrates that for systems of linear equations, it is possible to have a very low backward error while having a very large forward error. Equivalently, we can say that the system has a high error magnification factor. In the case of equations (4.78), the high error magnification factor is due to the fact that the two equations are almost identical. In fact, if we attempt to plot them and determine the solution graphically, we observe that it is impossible to distinguish the two equations (Figure 4.1). A high error magnification factor occurs whenever two or more equations are almost linearly dependent on each other. But, it is not the only case leading to high error magnification factors.

4.4.3 Error Magnification Factor of a Linear System of Equations

In the case of linear systems of equations, it is possible to give an estimation of the error magnification factor of the system. Consider a system of linear equations $A\underline{x} = \underline{b}$ with its true solution $\underline{r}$. If $\underline{x}_r$ approximates the true solution, then

$$\|\Delta \underline{b}\| = \|A\underline{x}_r - \underline{b}\| \tag{4.82}$$

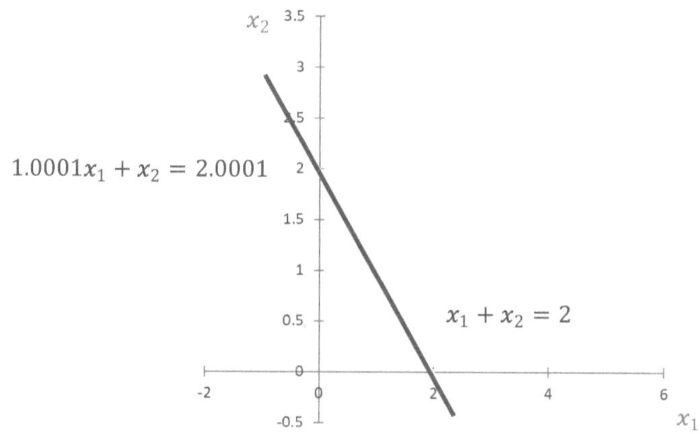

FIGURE 4.1 Graphical solution of the system of equations (4.78). The two equations are almost identical. It becomes very difficult to locate graphically the intersection of the two equations and find the solution of (4.78). This is an example of a system of equations with a very high error magnification factor.

is the backward error, and

$$\|\Delta \underline{r}\| = \|\underline{x}_r - \underline{r}\| \tag{4.83}$$

is the forward error. Using the notions of (4.83) and (4.82), we can write:

$$A(\underline{r} + \Delta \underline{r}) = A\underline{x}_r = \underline{b} + \Delta \underline{b} \tag{4.84}$$

From which follows:

$$A\Delta \underline{r} = \Delta \underline{b} \tag{4.85}$$

The relation is rather intuitive. If instead of a system of linear equations we were to consider a single equation, relation (4.85) is simple to understand. In fact, it can be seen as the reformulation of Theorem 3.8 for systems of linear equations.

Equation (4.85) allows the estimation of the forward error in function of the backward error. To obtain a lower limit, we calculate with the help of Theorem 4.4:

$$\|\Delta \underline{b}\| = \|A\Delta \underline{r}\| \leq \|A\| \cdot \|\Delta \underline{r}\| \tag{4.86}$$

Consequently:

$$\|\Delta \underline{r}\| \geq \frac{\|\Delta \underline{b}\|}{\|A\|} \tag{4.87}$$

To obtain an upper limit, which is the one required to make conservative error estimations, we rearrange Equation (4.85) and use again Theorem 4.4:

$$\|\Delta \underline{r}\| = \|A^{-1}\Delta \underline{b}\| \leq \|A^{-1}\| \cdot \|\Delta \underline{b}\| \tag{4.88}$$

This proves Theorem 4.6:

Theorem 4.6. Absolute Forward Error of a System of Linear Equations

Consider a system of linear equations $A\underline{x} = \underline{b}$ with true solution $\underline{r}$ and an approximation $\underline{x}_r$ of $\underline{r}$. The absolute forward error of $\underline{x}_r$ is bounded by

$$\frac{\|A\underline{x}_r - \underline{b}\|}{\|A\|} \leq \|\underline{r} - \underline{x}_r\| \leq \|A^{-1}\| \cdot \|A\underline{x}_r - \underline{b}\| \tag{4.89}$$

Example 4.14 demonstrates how Theorem 4.6 can be applied.

Example 4.14. Absolute Forward Error of a System of Linear Equations

Recall the system of linear equations (4.78). Using Theorem 4.6, we can compute the lower limit of the forward error of our approximation:

```
>> A = [1 1; 1.0001 1];
>> b = [2; 2.0001];
>> xr = [10; -8];
>> norm(A*xr-b,inf)/norm(A,inf)
ans = 0.00044998
```

which tells us that the forward error is at least equal to 0.00044998.

On the other hand, the upper limit of the forward error is given by:

```
>> norm(inv(A),inf)*norm(A*xr-b)
ans = 18.001
```

which allows us to state that 18.001 is a conservative estimation of the true forward error. This is indeed the case, as the true forward error is equal to 9:

```
>> norm([1;1]-xr,inf)
ans = 9
```

In practice, the relative forward error is often more relevant than the absolute forward error. Using the same calculations that led to Theorem 4.6, we derive a lower and upper limit of the relative forward error in function of the relative backward error.

From Equation (4.87), we can write, using the relation $\underline{r} = A^{-1}\underline{b}$ and Theorem 4.4 and assuming $\underline{r} \neq \underline{0}$:

$$\frac{\|\Delta \underline{b}\|}{\|A\|} \leq \|A^{-1}\underline{b}\| \frac{\|\Delta \underline{r}\|}{\|\underline{r}\|} \leq \|A^{-1}\| \cdot \|\underline{b}\| \frac{\|\Delta \underline{r}\|}{\|\underline{r}\|} \tag{4.90}$$

After rearranging, a lower limit for the forward relative error is found while assuming $\underline{b} \neq \underline{0}$:

$$\frac{1}{\|A^{-1}\| \cdot \|A\|} \cdot \frac{\|\Delta \underline{b}\|}{\|\underline{b}\|} \leq \frac{\|\Delta \underline{r}\|}{\|\underline{r}\|} \tag{4.91}$$

To obtain the more relevant upper limit, we rearrange Equation (4.88) using the relation $\underline{b} = A\underline{r}$ and Theorem 4.4, assuming $\underline{b} \neq \underline{0}$:

$$\|\Delta \underline{r}\| \le \|A^{-1}\| \cdot \|\underline{b}\| \frac{\|\Delta \underline{b}\|}{\|\underline{b}\|} = \|A^{-1}\| \cdot \|A\underline{r}\| \frac{\|\Delta \underline{b}\|}{\|\underline{b}\|} \le \|A^{-1}\| \cdot \|A\| \cdot \|\underline{r}\| \frac{\|\Delta \underline{b}\|}{\|\underline{b}\|} \quad (4.92)$$

After rearranging, an upper limit for the forward relative error is found as long as $\underline{r} \neq \underline{0}$:

$$\frac{\|\Delta \underline{r}\|}{\|\underline{r}\|} \le \|A^{-1}\| \cdot \|A\| \frac{\|\Delta \underline{b}\|}{\|\underline{b}\|} \quad (4.93)$$

Considering Equations (4.91) and (4.93), it appears that the product $\|A^{-1}\| \cdot \|A\|$ plays an important role. This justifies the introduction of the following definition:

Definition 4.5. Matrix Condition Number

The condition number $\text{cond}(A)$ of a $n \times n$ matrix A is given by:

$$\text{cond}(A) = \|A^{-1}\| \cdot \|A\| \quad (4.94)$$

Octave implements the function `cond` to compute the condition number of a matrix. In the next example, its usage will be demonstrated.

Using the definition of the matrix condition number and Equations (4.91) and (4.93), the following theorem can be formulated:

Theorem 4.7. Relative Forward Error of a System of Linear Equations

The relative forward error of a system of equations $A\underline{x} = \underline{b}$ with true solution $\underline{r}$ and an approximation $\underline{x}_r$ of $\underline{r}$ is bounded by:

$$\frac{1}{\text{cond}(A)} \cdot \frac{A\underline{x}_r - \underline{b}}{\|\underline{b}\|} \le \frac{\|\underline{r} - \underline{x}_r\|}{\|\underline{r}\|} \le \text{cond}(A) \frac{\|A\underline{x}_r - \underline{b}\|}{\|\underline{b}\|} \quad (4.95)$$

under the conditions that $\underline{r} \neq \underline{0}$ and $\underline{b} \neq \underline{0}$.

In the light of Theorem 4.7, the condition number of a matrix can be defined as the maximal relative error magnification factor we can obtain when solving any system of equations $A\underline{x} = \underline{b}$.

Theorem 4.8. Maximal Relative Error Magnification Factor

The maximal relative error magnification factor m one obtains when solving a system of linear equations $A\underline{x} = \underline{b}$, is given by the condition number of the matrix A.

Indeed, using Theorem 4.7, the relative error magnification factor is bounded by

$$m = \frac{\left\|\underline{r} - \underline{x}_r\right\|}{\left\|\underline{r}\right\|} / \frac{\left\|A\underline{x}_r - \underline{b}\right\|}{\left\|\underline{b}\right\|} \leq \text{cond}(A) \tag{4.96}$$

The proof of Theorem 4.8 comes from the fact that Theorem 4.7 holds for any system of linear equations $A\underline{x} = \underline{b}$, regardless of the right-hand-side vector $\underline{b}$ and the approximation $\underline{x}_r$ considered. Furthermore, the condition number of A is independent of $\underline{b}$ and $\underline{x}_r$.

Example 4.15. Relative Forward Error of a System of Linear Equations

Let us consider again the system of linear equations (4.78). Using Theorem 4.7, we can compute the lower limit of the relative forward error of the numerical approximation:

```
>> A = [1 1; 1.0001 1];
>> b = [2; 2.0001];
>> xr = [10; -8];
>> 1/cond(A,inf)*norm(A*xr-b,inf)/norm(b,inf)
ans = 0.000000011248
```

which tells us that the forward error is at least equal to 0.000000011248. The Octave function `cond` was used to compute the condition number of the coefficient matrix A. Its usage is similar to the `norm` function. In the second argument, we must provide the type of norm we want to use for the calculation of the condition number. In our case, we used the infinity norm.

On the other hand, the upper limit of the relative forward error is given by:

```
>> cond(A,inf)*norm(A*xr-b)/norm(b,inf)
ans = 18.001
```

which allows us to state that 18.001 is a conservative estimation of the true relative forward error. This is indeed the case, as the true relative forward error is equal to 9:

```
>> norm([1;1]-xr,inf)/norm([1;1],inf)
ans = 9
```

In the present example, the maximal error magnification factor we can expect, as given by the condition number of the coefficient matrix A is:

```
>> cond(A,inf)
ans = 40004.00010
```

This is rather a very large relative error magnification factor, which explains why even though the relative backward error is very small (i.e., the system of linear equations is almost satisfied for the given approximation $\underline{x}_r$), the relative forward error is very large.

4.4.4 Conditioning of Linear Systems of Equations

The numerical error of an approximation $\underline{x}_r$ of a true solution $\underline{r}$ has two contributions: truncation and round-off errors. In the case of linear systems of equations, algorithms that are free of truncation errors, such as the ones described in this chapter, can be deployed. However, round-off errors remain. With Theorem 4.6 and Theorem 4.7, we have two tools to estimate the forward error of a computed approximation $\underline{x}_r$. As demonstrated by the following example, even though solving a system of linear equations using the $PA = LU$ decomposition is free of truncation errors, its result can still be quite wrong. There can even be extreme cases in which the backward error evaluates to zero; despite this, the forward error is still large. In such cases, the application of Theorem 4.6 and Theorem 4.7 no longer help us to obtain a correct error estimation, and we will require a different approach.

Example 4.16. Estimating Forward Errors in Linear Systems of Equations

Consider the following coefficient matrix A:

$$A = \begin{bmatrix} 1.0 & 58.1 \\ 5.5 & 321.8 \end{bmatrix} \tag{4.97}$$

Let us construct a system of linear equations with a known true solution based on this coefficient matrix. We compute a right-hand side vector $\underline{b}$ such that the vector $\underline{r} = \begin{bmatrix} 1 & 15 \end{bmatrix}^T$ is a true solution:

$$\underline{b} = \begin{bmatrix} 1.0 & 58.1 \\ 5.5 & 321.8 \end{bmatrix} \begin{bmatrix} 1 \\ 15 \end{bmatrix} = \begin{bmatrix} 872.5 \\ 4832.5 \end{bmatrix} \tag{4.98}$$

Let us now solve this system using the Octave left division (i.e., Octave solves the system by using the $PA = LU$ decomposition):

```
>> A=[1.0 58.1; 5.5 321.8];
>> r=[1;15];
>> b=A*r;
>> xr=A\b
ans =
    1.0000
   15.0000
```

The solution $\underline{x}_r$, up to the displayed digits, is correct. Plugging back our solution into the original system seems to confirm that the approximation should be correct since the backward error is equal to zero:

```
>> norm(A*xr-b,inf)
ans =
    0
```

But, Octave was not able to find the true solution, as seen in the following calculation:

```
>> xr-r
ans =
    1.1079e-11
   -1.9007e-13
```

In fact, the true relative forward error is equal to:

```
>> norm(xr-r,inf)/norm(r,inf)
ans = 7.3862e-13
```

The problem in this example is that if we don't know the true solution, we will not be able to get an estimation of the forward error. Indeed, the backward error numerically evaluates exactly to zero, which makes the predictions of Theorem 4.6 and Theorem 4.7 unusable.

The problematic that Example 4.16 points to, is that the computation of the backward error is subject to round-off errors, too, as is the computation of the numerical approximation $\underline{x}_r$ of the solution of the system of linear equations. If the accumulated round-off error in the calculation of the backward error becomes too large, Theorem 4.6 and Theorem 4.7 become unusable in practice. Luckily, the situation is not quite as dramatic as it looks. These theorems can still be used but need some revisiting to address the problematic of large round-off errors in the backward error.

Theorem 4.6 and Theorem 4.7 are formulated in terms of error magnification factor. They are useful if one is confident that the computed backward error has round-off errors that are small enough compared to the computed backward error. If we want to develop a method of estimating the forward error independently of the backward error, we must consider the problem in the context of conditioning. The aim is to find how a small error in the input of the linear system of equations, for example, round-off errors in the right-hand-side vector $\underline{b}$, translates into the forward error of the solution of the system of equations.

For this, let us consider a system of linear equations $A\underline{x} = \underline{b}$ with true solution $\underline{r}$. Assume we have some error $\Delta\underline{b}$, for example, round-off errors, in the right-hand side vector $\underline{b}$. In this case, the system of equations an algorithm attempts to solve is no longer $A\underline{x} = \underline{b}$ but

$$A\underline{x} = \underline{b} + \Delta\underline{b} \tag{4.99}$$

The true solution of this new system is no longer $\underline{r}$ but $\underline{r}' = \underline{r} + \Delta\underline{r}$. The error $\Delta\underline{b}$ in $\underline{b}$ translates in a change of $\Delta\underline{r}$ in the true solution $\underline{r}$ of the original system. We can write:

$$A(\underline{r} + \Delta\underline{r}) = \underline{b} + \Delta\underline{b} \tag{4.100}$$

Since $A\underline{r} = \underline{b}$, Equation (4.100) is identical to Equation (4.85), which was the starting point leading to Theorem 4.6 and Theorem 4.7. The only difference is the interpretation of $\Delta\underline{b}$. In Equation (4.85), $\Delta\underline{b}$ was the backward error, while in Equation (4.100), $\Delta\underline{b}$ is the error in the right-hand-side vector $\underline{b}$, such as round-off errors. In the light of these explanations, the following theorem results:

Theorem 4.9. Round-Off Errors in Solving a S ystem of Linear Equations

Consider a system of linear equations $A\underline{x} = \underline{b}$ with true solution $\underline{r}$. If the right-hand side vector $\underline{b}$ has a round-off error of $\Delta\underline{b}$, then a truncation error-free algorithm will find an approximation $\underline{x}_r$ of the solution with an error $\|\underline{r} - \underline{x}_r\|$ such that

$$\frac{\|\underline{r} - \underline{x}_r\|}{\|\underline{r}\|} \leq \text{cond}(A)\frac{\|\Delta\underline{b}\|}{\|\underline{b}\|} \tag{4.101}$$

under the conditions that $\underline{r} \neq \underline{0}$ and $\underline{b} \neq \underline{0}$.

Since the condition number of a coefficient matrix gives the largest error magnification factor that can occur over all possible right-hand-side vectors $\underline{b}$, Theorem 4.9 gives an estimation for the worst-case situation.

Revisiting Example 4.15 with Theorem 4.9 allows us to estimate correctly the forward error of the linear system of equations. The condition number of the coefficient matrix A is:

```
>> cond([1.0 58.1; 5.5 321.8], inf)
ans = 55262.78666666607
```

The relative round-off error of the right-hand-side vector $\underline{b}$ is given by the machine epsilon. Consequently, using Theorem 4.9, the largest forward error to be expected when solving the system $A\underline{x} = \underline{b}$ is:

```
>> cond([1.0 58.1; 5.5 321.8], inf)*eps
ans = 1.227080363245616e-11
```

This is indeed a conservative estimation of the true relative forward error. Consequently, Theorem 4.9 warrants that a given approximation has a relative forward error below a certain value. Explicitly computing the backward error is not necessary.

Theorem 4.9 can further be used to design the solution process of a system of linear equations, as illustrated in Example 4.17.

Example 4.17. Number of Digits Required to Find a Solution to a Desired Accuracy

Consider the system of Equations (4.77). We want to know how many correct significant digits we can expect by solving this system numerically with floating-point numbers when using a truncation error-free algorithm, such as the PA=LU decomposition, for example.

The condition number of the coefficient matrix is

```
>> A = [1 1; 1.0001 1];
>> cond(A, inf)
ans = 4.0004e+04
```

When solving numerically with floating-point numbers, the system of Equations (4.77), approximately 16 significant digits are used. The relative round-off error is about the machine epsilon. From Theorem 4.9, it follows that the expected relative forward error, when solving the system, is about

```
>> cond(A, inf)*eps
ans = 8.8827e-12
```

which is about 10^{-11}. In other words, only about 11 digits will be correct in the produced answer. We can verify this numerically. The numerical approximation of the solution of the system (4.77) is:

```
>> b = [2; 2.0001];
>> xr = A\b;
```

As the true solution is $\underline{r} = \begin{bmatrix} 1 & 1 \end{bmatrix}^T$, the true relative forward error is:

```
>> norm(xr-[1 1]', inf)/norm([1 1]', inf)
ans = 1.9442e-12
```

Our estimation of 10^{-11} was, as stated by Theorem 4.9, a conservative estimation of the relative forward error.

With this result, one can design the solution process. For a desired accuracy, the required machine epsilon can be computed (i.e., how many digits must be used at least in the calculations to reach the desired accuracy). For example, if a relative forward error of maximum 10^{-6} is needed, the required relative backward error (i.e., the required machine epsilon) is:

$$10^{-6} / \text{cond}(A, \infty) \cong 2 \cdot 10^{-11} \tag{4.102}$$

which is well below the machine epsilon when using floating-point numbers. In other words, the desired accuracy would be reached.

On the other hand, if a relative forward error of maximum 10^{-14} is required, one would need a machine epsilon of

$$10^{-14} / \text{cond}(\text{A}, \infty) \cong 2 \cdot 10^{-19} \tag{4.103}$$

This value is below the machine epsilon corresponding to floating-point numbers. A higher number of digits would be required (in fact, more than 19 significant digits).

In real problems, the round-off error is present in the right-hand-side vector $\underline{b}$ and in the coefficient matrix A. Theorem 4.9 can be generalized to the following theorem:

Theorem 4.10. Relative Forward Error Estimation

Consider a system of linear equations $A\underline{x} = \underline{b}$ with true solution $\underline{r}$. If the coefficient matrix A has a round-off error of ΔA and the right-hand side vector $\underline{b}$ has a round-off error of $\Delta \underline{b}$, then a truncation error-free algorithm will find an approximation $\underline{x}_r$ of the solution with an error $\|\underline{r} - \underline{x}_r\|$ such that

$$\frac{\|\underline{r} - \underline{x}_r\|}{\|\underline{r}\|} \leq \frac{\text{cond}(A)}{1 - \text{cond}(A)\dfrac{\|\Delta A\|}{\|A\|}} \left(\frac{\|\Delta A\|}{\|A\|} + \frac{\|\Delta \underline{b}\|}{\|\underline{b}\|} \right) \tag{4.104}$$

under the conditions that $\underline{r} \neq \underline{0}$, $\underline{b} \neq \underline{0}$ and $\|\Delta A\| < 1 / \|\Delta A^{-1}\|$.

The proof can, for example, be found in *A first course in numerical analysis* by A. Ralston and Ph. Rabinowitz. As the relative round-off error is typically in the order of magnitude of the machine epsilon ε_M, Equation (4.104) can be rewritten as:

$$\frac{\left\|\underline{r}-\underline{x}_r\right\|}{\left\|\underline{r}\right\|} \leq \frac{2\text{cond}(A)\varepsilon_M}{1-\text{cond}(A)\varepsilon_M} \tag{4.105}$$

4.5 LAB: SOLVING SYSTEMS OF LINEAR EQUATIONS

Octave is particularly well suited for solving systems of linear equations. In this lab, we show two examples, including how to estimate the numerical errors of the computed approximated solutions.

4.5.1 Solving a System of Linear Equations

Let us solve numerically the following system of linear equations and estimate the error of the computed approximation:

$$\begin{bmatrix} 4 & 7 & 2 & 9 \\ 3 & 8 & 1 & 3 \\ 9 & 3 & 7 & 2 \\ 7 & 4 & 7 & 2 \end{bmatrix} \begin{bmatrix} x_1 \\ x_2 \\ x_3 \\ x_4 \end{bmatrix} = \begin{bmatrix} 2 \\ 3 \\ 6 \\ 7 \end{bmatrix} \tag{4.106}$$

We start by defining the coefficient matrix A and right-hand side vector $\underline{b}$:

```
>> A = [4 7 2 9; 3 8 1 3; 9 3 7 2; 7 4 7 2];
>> b = [2; 3; 6; 7];
```

The numerical solution of system (4.106) is computed with the left division `A\b`. In order to display all the digits used by Octave, we further use `format long`:

```
>> format long
>> A\b
ans =
   -0.284741144414169
    0.430517711171662
    1.104904632152589
   -0.231607629427793
```

The error due to round-off errors in the calculations can be estimated using Theorem 4.9. For this, we need to compute the conditioning number of the coefficient matrix:

```
>> cond(A, inf)
ans = 23.16893732970026
```

Following the discussion of Example 4.17, this means that about two digits out of the initial 16 will be lost during the solution process.

As the relative round-off error on the right-hand-side vector is about the magnitude of the machine epsilon, we can estimate, according to Theorem 4.9, the relative forward error on the provided solution of system (4.106) as:

```
>> cond(A, inf) * eps
ans = 5.144537535906105e-15
```

and the absolute forward error $\|x_r - r\|$ will be about

```
>> norm(A\b, inf) * cond(A, inf) * eps
ans = 5.684223353705520e-15
```

Considering the definition of the infinity norm (and this is one of the main reasons why we used it here), it follows that the maximal absolute error of each x_i is lower than about $6 \cdot 10^{-15}$.

4.5.2 Vandermond Matrix

In this second example, we solve a system of linear equations involving a matrix with a high conditioning number. As an example, we use a system in which the coefficient matrix belongs to the family of the Vandermond[7] matrices. Solving linear systems of equations involving Vandermond matrices is a problem we will encounter in Chapter 6 when we will discuss polynomial regression problems. The system we want to solve is

$$\begin{bmatrix} 1 & 3 & 9 & 27 \\ 1 & 4 & 16 & 64 \\ 1 & 2 & 4 & 8 \\ 1 & 9 & 81 & 729 \end{bmatrix} \begin{bmatrix} x_1 \\ x_2 \\ x_3 \\ x_4 \end{bmatrix} = \begin{bmatrix} 5 \\ 7 \\ 3 \\ 2 \end{bmatrix} \tag{4.107}$$

Note the particular form of the coefficient matrix. The first column is made out of all 1. The second column contains some numbers. The third column is the second column to the power of two, and the fourth column is the second column to the power of three.

Octave gives the following numerical approximation of the solution of system (4.107):

```
>> A = [1 3 9 27; 1 4 16 64; 1 2 4 8; 1 9 81 729];
>> b = [5; 7; 3; 2;];
>> A\b
ans =
   7.142857142857073e-01
   1.428571428571483e-01
```

```
   6.428571428571416e-01
  -7.142857142857134e-02
```

However, the conditioning number of the coefficient matrix is very high:

```
>> cond(A, inf)
ans = 20687.42857142853
```

which means we will lose about five significant digits during the solution process.

According to Theorem 4.9, the relative forward error on the provided solution of system (4.107) is:

```
>> cond(A, inf)*eps
ans = 4.593531904057653e-12
```

and the absolute forward error $\|x_r - r\|$ will be about

```
>> norm(A\b, inf)*cond(A, inf)*eps
ans = 3.281094217184005e-12
```

This means that the absolute error of each x_i is lower than about $3 \cdot 10^{-12}$. Despite this relatively high forward error (considering we use 16 significant digits in our calculations), the backward error is very small (i.e., all the equations seem to be satisfied correctly):

```
>> xr=A\b;
>> A*xr
ans =
   5
   7
   3
   2
```

This is once more an example of a mathematical problem that is ill-conditioned. The provided approximation seems to satisfy the problem, even though the error is actually quite large. This comes from the round-off errors while evaluating the quantity $A\underline{x}_r$. If we use an $\underline{x}_r$ in which we alter the last digits (highlighted in bold), we still get a very small backward error:

```
>> xr=[7.142857142999999e-01
   1.428571428999999e-01
   6.428571428999999e-01
  -7.142857142999999e-02];
>> A*xr
```

```
ans =
    5
    7
    3
    2
```

The system (4.107) still seems satisfied with this different $\underline{x}_r$. The reader is invited to try other values by altering the last digits of $\underline{x}_r$.

4.6 EXERCISES

4.6.1 Introduction

1. Rewrite the following systems of equations as a matrix equation

a. $$\begin{cases} 3x_1 + 6x_2 + 2x_3 & = & 7 \\ -5x_1 + 9x_2 - 4x_3 & = & -2 \\ x_1 + 5x_2 & = & 4 \end{cases}$$

b. $$\begin{cases} -3x_2 + 4x_3 & = & 1 \\ 3x_1 - 4x_2 + 5x_3 & = & -7 \\ -x_1 + 3x_2 + 9x_3 & = & 2 \end{cases}$$

c. $$\begin{cases} 8x_1 + 2x_2 - 5x_3 & = & 2 \\ 6x_1 - 7x_2 - 2x_3 & = & 5 \\ 7x_1 - 5x_2 + 2x_3 & = & 3 \end{cases}$$

d. $$\begin{cases} 2x_1 + 4x_2 + 7x_3 & = & 6 \\ 9x_1 - 8x_2 - 4x_3 & = & 1 \\ -3x_1 + x_2 + 3x_3 & = & 0 \end{cases}$$

4.6.2 Gauss Elimination Algorithms

2. Solve the systems of Exercise 1 using naïve Gauss elimination. If the algorithm fails, explain why.

3. Rewrite the various row operations used in Exercise 2 in matrix form following example 4.3

4. Solve the systems of Exercise 1 using Gauss elimination with partial pivoting.

5. Rewrite the various row operations used in Exercise 4 in matrix form following Example 4.6.

4.6.3 Matrix Decompositions

6. For each coefficient matrix of Exercise 1, find the LU decomposition. If no decomposition exists, explain why.

7. Solve the systems of Exercise 1 using LU decomposition if the LU decomposition exists.

8. For each coefficient matrix of Exercise 1, find the PA=LU decomposition.

9. Solve the systems of Exercise 1 using PA=LU decomposition

4.6.4 Conditioning

10. This exercise illustrates that, depending on the choice of the vector norm, two vectors can be ordered differently according to their norm.

    Provide some examples of two vectors $\underline{x}_1$ and $\underline{x}_2$ for which the following conditions hold:

    a. $\|\underline{x}_1\|_2 < \|\underline{x}_2\|_2$ and $\|\underline{x}_1\|_\infty < \|\underline{x}_2\|_\infty$.

    b. $\|\underline{x}_1\|_2 < \|\underline{x}_2\|_2$ and $\|\underline{x}_1\|_\infty > \|\underline{x}_2\|_\infty$

    By using vectors with two components, explain intuitively this behavior on a figure.

11. Consider the following systems of equations and corresponding approximated solutions $\underline{x}_r$:

    a. $$\begin{bmatrix} 5 & -3 & 2 \\ 9 & 4 & -1 \\ -4 & -7 & 0 \end{bmatrix}\begin{bmatrix} x_1 \\ x_2 \\ x_3 \end{bmatrix} = \begin{bmatrix} 4 \\ -2 \\ 5 \end{bmatrix} \qquad \underline{x}_r = \begin{bmatrix} 7.0 \\ -4.7 \\ -2.2 \end{bmatrix}$$

    b. $$\begin{bmatrix} 3 & 2 & 7 \\ -2 & 4 & -5 \\ 1 & 0 & 4 \end{bmatrix}\begin{bmatrix} x_1 \\ x_2 \\ x_3 \end{bmatrix} = \begin{bmatrix} -7 \\ -44 \\ 7 \end{bmatrix} \qquad \underline{x}_r = \begin{bmatrix} 0.7 \\ -13 \\ 2.5 \end{bmatrix}$$

    For each of them, compute the absolute and relative forward and backward errors using the infinity norm. After computing the infinity norm of the coefficient matrix, discuss your results.

12. For all coefficient matrixes A from Exercise 1, compute their conditioning number using the infinity norm.

13. For the following system of linear equations, find the number of digits required to use in the calculations to achieve a precision of 12 significant figures in the numerical solution (use the infinity norm in your analysis):

a. $$\begin{bmatrix} 5 & 2 & -1 \\ 1 & 4 & 5 \\ -3 & -2 & 7 \end{bmatrix}\begin{bmatrix} x_1 \\ x_2 \\ x_3 \end{bmatrix} = \begin{bmatrix} 2 \\ 1 \\ 7 \end{bmatrix}$$

b. $$\begin{bmatrix} 1 & -6 & 7 & -9 \\ 1 & -5 & 0 & 0 \\ 0 & 1 & -5 & 0 \\ 0 & 0 & 1 & -5 \end{bmatrix}\begin{bmatrix} x_1 \\ x_2 \\ x_3 \\ x_4 \end{bmatrix} = \begin{bmatrix} -44.1 \\ -6.8 \\ 12.9 \\ -13.3 \end{bmatrix}$$

c. $$\begin{bmatrix} 4 & 7 & 39 & 145 \\ 7 & 39 & 145 & 723 \\ 39 & 145 & 723 & 3337 \\ 145 & 723 & 3337 & 16419 \end{bmatrix}\begin{bmatrix} x_1 \\ x_2 \\ x_3 \\ x_4 \end{bmatrix} = \begin{bmatrix} -44.1 \\ -6.8 \\ 12.9 \\ -13.3 \end{bmatrix}$$

d. $$\begin{bmatrix} 2 & 2 & -2 \\ -1 & 3 & 5 \\ -2 & -4 & 4 \end{bmatrix}\begin{bmatrix} x_1 \\ x_2 \\ x_3 \end{bmatrix} = \begin{bmatrix} 3 \\ 1 \\ -7 \end{bmatrix}$$

NOTES

1 As the reader certainly knowns, linear systems of equations can have either a unique solution, a sub-space as solution or no solution, depending on the coefficient matrix A.

2 The reason for the wording “pivot” will become clear in Section 0.

3 This implementation does not attempt to be efficient. There are ways to make the implementation more efficient by reducing mathematical operations and variable copying.

4 This comes from the fact that Octave is an interpreted language. Executing for loops is very time consuming. Matrix multiplication are already available in compiled versions and, consequently, will execute significantly faster.

5 The various row switching operations cannot be identified by simple inspection of the matrix A. One has to conduct the Gauss elimination algorithm with partial pivoting to identify them.

6 Note that we use the same symbol $\| \|$ for the norm of a vector and the norm of a matrix. In calculations, one has to be careful to understand well if the vector or the matrix norm has to be applied by inspecting the object on which the norm is applied.

7 Alexandre-Théophile Vandermonde (1735–1796) French mathematician, musician, and chemist.

5 Orthogonality

The concept of orthogonality plays an important role in several domains of mathematics and engineering, especially in analytical geometry, where it becomes possible to replace geometrical constructions and geometrical theorems by algebraic calculations. The key, as first proposed by René Descartes,[1] is to introduce a coordinate system, in particular an orthogonal coordinate system, which simplifies calculations significantly. A generalization of this idea is the concept of abstract vector spaces. Real-valued vector spaces can be used to represent a large variety of mathematical objects as abstract vectors and handle them in similar ways one handles geometrical vectors.

We will learn that several numerical methods algorithms take advantage of the existence of orthogonal vectors and orthogonal vector bases. For example, in the context of linear equations, algorithms that are less sensitive to round-off error can be developed. It becomes possible, as will be discussed later, to give a precise meaning to the solution of a system of linear equations containing more equations than unknowns.

An example of this concept is the development of function approximation algorithms. Even in domains that seem very far from vector spaces, orthogonality can play a key role. For instance, we'll explore a suite of algorithms used to compute numerical approximations of definite integrals, leveraging the concept of orthogonal functions.

This chapter starts by introducing the concept of orthogonality in the frame of real-valued vector spaces. Next, we introduce algorithms capable of constructing sets of orthogonal vectors. These algorithms will serve as invaluable tools throughout our exploration of numerical methods.

5.1 INNER SPACES

In the frame of the present text, inner spaces are real-valued vector spaces on which a new operation is defined: the inner product of two vectors.

5.1.1 NOTATIONS

Consider a real-valued vector space V. The elements of V will be called vectors and designated by the symbol $|x\rangle$. This symbol, first introduced by Pau Dirac,[2] is used to designate vectors from a general real-valued vector space. An important particular case of a real-valued vector space is $\mathbb{R}^n$ The vectors of $\mathbb{R}^n$ are called column vectors. For the particular case of $\mathbb{R}^n$ we will use the notation $\underline{x}$ to designate a vector, as we did in Chapter 4.

DOI: 10.1201/9781003262121-6

Vector bases introduced in V are designated by $\mathcal{B} = \{|e_1\rangle, \ldots, |e_n\rangle\}$ A vector $|x\rangle$ of V can then be decomposed in this base according

$$|x\rangle = \sum_{i=1}^{n} x_i |e_i\rangle \tag{5.1}$$

An important base of $\mathbb{R}^n$is the following base $\mathcal{B} = \{e_1, \ldots, e_n\}$ which will be referred to as the natural base of $\mathbb{R}^n$

Definition 5.1. Natural Vector Base of $\mathbb{R}^n$

The natural vector base $\mathcal{B} = \{\underline{e}_1, \ldots, \underline{e}_n\}$ of $\mathbb{R}^n$is:

$$\underline{e}_1 = \begin{bmatrix} 1 \\ 0 \\ \vdots \\ 0 \end{bmatrix}, \underline{e}_2 = \begin{bmatrix} 0 \\ 1 \\ 0 \\ \vdots \\ 0 \end{bmatrix}, \ldots, \underline{e}_n = \begin{bmatrix} 0 \\ \vdots \\ 0 \\ 1 \end{bmatrix} \tag{5.2}$$

The vector space $\mathbb{R}^n$is of significant practical value as it allows us to perform calculations on abstract vectors using the more familiar, and simpler to handle, vectors of $\mathbb{R}^n$. To do so, a vector $|x\rangle$ from a real-valued vector space V of dimension n is represented by an element $\underline{x}$ of $\mathbb{R}^n$. The representation is obtained by arranging in a column vector the coordinates of the vector $|x\rangle = \sum_{i=1}^{n} x_i |e_i\rangle$ in a chosen base $\mathcal{B} = \{|e_1\rangle, \ldots, |e_n\rangle\}$:

$$|x\rangle = \sum_{i=1}^{n} x_i |e_i\rangle := \begin{bmatrix} x_1 \\ \vdots \\ x_i \\ \vdots \\ x_n \end{bmatrix} \tag{5.3}$$

Note the symbol ":=", which stands for "is represented by". The column vector $\underline{x}$ of $\mathbb{R}^n$ representing the vector $|x\rangle$ of V is often called the coordinate vector or the coordinate representation of $|x\rangle$.

Mathematically, the vector $|x\rangle$ and the element $\underline{x}$ of $\mathbb{R}^n$ are not equal. It is just convenient to use $\underline{x}$ for executing operations such as vector additions, for example. This allows us, in particular, to use Octave to perform calculations in abstract vector spaces. Note that the representation is dependent on the chosen base. If one chooses a different base, $\mathcal{B}'$ instead of $\mathcal{B}$, then the representation becomes a different column vector of $\mathbb{R}^n$ Note that the order in which the vectors of the base in V are chosen is important. Ordering them differently will result in a different representation.

Example 5.1. Representation of Abstract Vectors by Vectors of $\mathbb{R}^n$

Consider the real-valued vector space $P_2[x]$ of the real polynomials of degree 2 or less. The following two polynomials are vectors of $P_2[x]$

$$p(x) = 3x^2 - 2x + 5 \tag{5.4}$$

$$Q(x) = 5x - 4 \tag{5.5}$$

The vector addition in $P_2[x]$ is defined by the polynomial addition. For example, $|P(x)\rangle + |Q(x)\rangle$ will be the vector $|T(x)\rangle$:

$$T(x) = 3x^2 + 3x + 1 \tag{5.6}$$

A convenient way to perform these calculations is to use the representation in $\mathbb{R}^3$ We use $\mathbb{R}^3$ as the real-valued vector space $P_2[x]$ of dimension 3. We choose the natural base $B = \{1, x, x^2\}$ from $P_2[x]$ to represent our vectors:

$$|P(x)\rangle := \underline{P} = \begin{bmatrix} 5 \\ -2 \\ 3 \end{bmatrix} \tag{5.7}$$

and

$$|Q(x)\rangle := \underline{Q} = \begin{bmatrix} -4 \\ 5 \\ 0 \end{bmatrix} \tag{5.8}$$

The addition $|P(x)\rangle + |Q(x)\rangle$ can then be performed in Octave using the representations in $\mathbb{R}^3$of the vectors $|P(x)\rangle$ and $|Q(x)\rangle$:

```
>> p = [5; -2; 3];
>> q = [-4; 5; 0];
>> p+q
```

```
ans =
   1
   3
   3
```

This is indeed the vector $|T(x)\rangle$ represented in $\mathbb{R}^3$

5.1.2 Norms

The norm of a vector plays an important role in several domains. We have already used it extensively in Chapter 4 to measure errors. Let us give a formal definition of a general vector norm of a real-valued vector space.

Definition 5.2. Vector Norm

A norm of a vector $|x\rangle$ from a real-valued vector space V is any function that associates a positive number $\||x\rangle\|$ to the vector $|x\rangle$ such that the following conditions are satisfied

1. *Positive definite*: $\||x\rangle\| \geq 0$ and $\||x\rangle\| = 0$ only and only if $|x\rangle = |0\rangle$
2. *Absolute homogeneous*: $\|a|x\rangle\| = |a||x\rangle$
3. *Triangle inequality*: $\||x\rangle + |y\rangle\| \leq \||x\rangle\| + \||y\rangle\|$

A real-valued vector space equipped with a vector norm is called a normed vector space.

Throughout the remainder of the document, we will adopt a simplified notation, using $\|x\|$ instead of $\||x\rangle\|$.

Example 1 Vector Norm

For a given real-valued vector space V, different norms can be defined. Here a few examples

- For $\mathbb{R}^n$ we can define the p-norm

$$\|\underline{x}\|_p = \sqrt[p]{|x_1|^p + |x_2|^p + \ldots + |x_n|^p}$$

with $p \geq 1$ a real number.

- For the vector space $P_n[x]$ of the real polynomials of degree n or less defined over a finite interval $[a, b]$ the following norm can be defined:

$$\|P_n(x)\| = \left(\int_a^b [P_n(x)]^2\, dx\right)^{\frac{1}{2}}$$

  This norm is often referred to as the L^2-norm.
- The L^2-norm is a particular case of the L^p-norm of the vector space $P_n[x]$ (defined for any $p > 1$):

$$\|P_n(x)\|_p = \left(\int_a^b [P_n(x)]^p\, dx\right)^{\frac{1}{p}}$$

  For $p \to \infty$ one obtains the infinity or maximum-norm:

$$\|P_n(x)\|_\infty = \max_{x\in[a,b]} |P_n(x)|$$

The reader is encouraged to verify that the three conditions of the norm from Definition 5.2 hold true for these examples.

5.1.3 Inner Products

Vector spaces can be equipped with an inner product. Inner products are related, to a certain extent, to norms and are used to define orthogonality between two vectors. Vector spaces equipped with an inner product are termed inner spaces.

Definition 5.3. Inner Product

An inner product between two vectors $|x\rangle$ and $|y\rangle$ of a real-valued vector space V, is a function that associates to these two vectors a positive real number $\langle x \mid y \rangle \in \mathrm{R}$ and satisfies the following three conditions

1. The function is a *bilinear mapping* from

$$V \times V \to \mathrm{R}$$

   For any numbers a and b and vectors $|x\rangle$, $|y\rangle$ and $|z\rangle$ we have:

   a. $\langle ax + by|z\rangle = a\langle x|z\rangle + b\langle y|z\rangle$

   b. $\langle x|ay + bz\rangle = a\langle x|y\rangle + b\langle x|z\rangle$

2. The function is *symmetric*: $\langle x|y\rangle = \langle y|x\rangle$
3. The function is *positive definite*: $\langle x|x\rangle > 0$ for $|x\rangle \in V \setminus \{|0\rangle\}$ and $\langle 0|0\rangle = 0$

Example 5.2 illustrates a variety of inner products.

Example 5.2. Inner Products

As for norms, there exists a wide variety of inner products. Here are a few examples:

- In the vector space $\mathbb{R}^n$ an inner product between two vectors $\underline{x}$ and $\underline{y}$ can be defined as

$$\underline{x}^T \underline{y} = \sum_{i=1}^{n} x_i y_i$$

  This inner product is referred to as the standard inner product of $\mathbb{R}^n$ It's numerical calculation in Octave is particularly simple. If we define two column vectors `x` and `y` in Octave, their inner product is computed as `x'*y`.
- For the vector space $P_n[x]$ of the real polynomials of degree n or less defined over a finite interval $[a, b]$, the following inner product can be defined:

$$\langle P_n(x) | Q_n(x)\rangle = \int_a^b P_n(x) Q_n(x)\, dx$$

The reader may wish to verify that the three conditions of the inner product hold true for these two examples.

There exists a relation between inner products and norms. In fact, any inner product can be used to define a norm, the so-called induced norm. Consequently, any inner space is also a normed vector space equipped with the induced norm.

Definition 5.4. Induced Norm by an Inner Product

The induced norm of an inner product is defined as $\|x\| = \sqrt{\langle x|x\rangle}$.

Using the properties outlined in Definition 5.3, it is possible to show that the quantity $\sqrt{\langle x|x\rangle}$ satisfies indeed all three conditions of a norm.

If any inner product induces a norm, the contrary is not true. There are norms that are not induced by any inner product. Two notable examples are the 1-norm and the infinity-norm.

Induced norms are often used. For example, in $\mathbb{R}^n$ the inner product $\langle x|y\rangle = \underline{x}^T \underline{y}$ induces the 2-norm or Euclidian norm. In the vector space of the real polynomials of degree n, the inner product, as defined in Example 5.2, induces the L^2-norm.

Example 5.3. Induced Norms

Consider the vector $\underline{x}$ from $\mathbb{R}^4$ defined in Octave as follows:

```
>> x = [1 2 3 4]';
```

Its induced norm, from the inner product $\langle x|y\rangle = \underline{x}^T \underline{y}$, is

```
>> sqrt(x'*x)
ans = 5.4772
```

The same answer can be found by computing the Euclidian norm of $\underline{x}$:

```
>> norm(x, 2)
ans = 5.4772
```

Induced norms satisfy an important relation, often considered as one of the most important inequalities in mathematics: the Cauchy-Schwarz inequality.[3,4]

Theorem 5.1. Cauchy-Schwarz Inequality

For an induced norm $\|x\|$ from an inner product $\langle x \mid y\rangle$, the following inequality holds:

$$|\langle x|y\rangle| \leq \|x\| \cdot \|y\| \tag{5.9}$$

This inequality is called the Cauchy-Schwarz inequality.

Let us emphasize that in Inequality (5.9), it is essential to utilize the norm induced by the inner product. The inequality does, in general, not hold if a different norm is employed instead of the induced norm. Throughout this text, we adhere to the convention that whenever inner products and norms appear in an equation, the norm is consistently the induced norm derived from the inner product referenced in that equation.

Example 5.4. Cauchy-Schwarz Inequality

This example illustrates Theorem 5.1. We work in the real-valued vector space $\mathbb{R}^3$. We start by defining two vectors of $\mathbb{R}^3$

```
>> x = [1; 2; 3];
>> y = [3; 4; 5];
```

We use the inner product defined by $\langle x|y\rangle = \underline{x}^T \underline{y}$. In our example, this results in:

```
>> x'*y
ans = 32
```

The induced norm of the inner product $\underline{x}^T \underline{y}$ is the 2-norm. Using this norm, Theorem 5.1 is indeed satisfied:

```
>> norm(x,2)*norm(y,2)
ans = 32.833
```

However, if the infinity norm is used, then Theorem 5.1 is no longer satisfied:

```
>> norm(x,inf)*norm(y,inf)
ans = 18
```

The reader is invited to try out other norms. Some will satisfy Theorem 5.1 some not. Only the induced norm is guaranteed to always satisfy Theorem 5.1.

5.1.4 Generalized Angle and Orthogonality

An inner product allows to introduce geometrical concepts into a general real-valued vector space V. In particular, it allows the definition of the generalized angle between two vectors $|x\rangle$ and $|y\rangle$. Let us emphasize that the concepts presented in this text are only valid for induced norms, and all equations involve the inner product together with its induced norm.

We begin by providing some intuitive context. An angle serves to quantify the degree of alignment between two segments' orientations. If we interpret vectors from V as segments, then we can say that the vectors $|x\rangle$ and $a|x\rangle$ have an angle of zero for positive numbers a (the two vectors are of the same orientation) and an angle of π for negative numbers a (the two vectors are of opposite orientation). Calculating the inner product between these vectors yields $a\langle x|x\rangle = a\|x\|^2$. The result is of the same sign as the number a. Therefore, a positive inner product indicates that vectors are of similar orientations, whereas negative numbers indicate that they are of opposite orientations.

We can take this concept a step further by extending it to two arbitrary vectors. We use an important consequence of the Cauchy-Schwarz inequality:

$$-1 \leq \frac{\langle x|y\rangle}{\|x\|\cdot\|y\|} \leq +1 \tag{5.10}$$

As the ratio $\langle x|y\rangle$ over $\|x\|\cdot\|y\|$ is between -1 and +1, we can define a quantity θ such that

$$\cos\theta = \frac{\langle x|y\rangle}{\|x\|\cdot\|y\|} \tag{5.11}$$

Applying Equation (5.11) to our initial example of the inner product between the two vectors $|x\rangle$ and $a|x\rangle$, results in $\theta = 0$ for positive numbers a and $\theta = \pi$ for negative numbers a. In that sense, θ can be seen as an angle, or a measure, of how similar the orientations of two vectors are.

Definition 5.5. Generalized Angle

The generalized angle θ between the vectors $|x\rangle$ and $|y\rangle$ is defined by $\cos\theta = \frac{\langle x|y\rangle}{\|x\|\cdot\|y\|}$

According to Definition 5.5, if two vectors are such that their inner product is equal to zero, then their generalized angle is equal to $\frac{\pi}{2}$. In a geometric picture, they are orthogonal. Vectors such that their inner product is equal to zero play a key role in a large variety of applications, which justifies the introduction of a dedicated definition.

Definition 5.6. Orthogonal Vectors

Two vectors $|x\rangle$ and $|y\rangle$ are orthogonal if $\langle x|y\rangle = 0$

Vector bases composed of orthogonal vectors hold particular significance. Moreover, if these vectors additionally have a unit norm, they are referred to as orthonormal vectors.

Definition 5.7. Orthonormal Base

A vector base $\mathcal{B} = \{|e_1\rangle, \ldots, |e_n\rangle\}$ is called an orthonormal vector base if $\langle e_i | e_j \rangle = \delta_{ij}$, where δ_{ij} is the Kronecker[5] symbol ($\delta_{ij} = 1$ if $i = j$ and 0 otherwise).

The terminology "orthonormal" comes from the following two facts. First, we note that the norm of all vectors $|e_i\rangle$ is equal to one as $\|e_i\| = \sqrt{\langle e_i | e_i \rangle} = 1$. We say the base vectors are normalized. Second, the generalized angle between $|e_i\rangle$ and $|e_j\rangle$, for $i \neq j$ is equal to a right angle. We say the vectors are orthogonal.

We will see later that for every real-valued vector space, one can construct an orthonormal vector base using the Gram-Schmidt algorithm.[6,7]

So far, the definition of orthogonality may seem somewhat arbitrary. The next theorem allows us to give a justification for this concept.

Theorem 5.2. Generalized Pythagoras' Theorem

Consider two vectors $|x\rangle$ and $|y\rangle$ of a real-valued inner space V. If these vectors are orthogonal, i.e. $\langle x | y \rangle = 0$, then

$$\|x + y\|^2 = \|x\|^2 + \|y\|^2 \tag{5.12}$$

where $\|x\|$ is the induced norm in the inner space V.

The proof is straightforward using the definition of orthogonality:

$$\|x + y\|^2 = \langle x + y | x + y \rangle = \langle x | x \rangle + \langle x | y \rangle + \langle y | x \rangle + \langle y | y \rangle = \|x\|^2 + \|y\|^2 \tag{5.13}$$

Theorem 5.2 shows that if two vectors are orthogonal, then the norm of their sum can be calculated as we are accustomed to elementary geometry. In that sense, it can be seen as a generalization of Pythagoras'[8] theorem.

5.2 ORTHOGONAL PROJECTIONS

From elementary geometry, it is known that the shortest distance between a vector and a sub-space (for example, a line) is given by the norm of orthogonal projection of the vector on that sub-space. The concept of orthogonality introduced by Definition 5.6 allows the formulation of this idea in a general abstract inner space.

5.2.1 Orthogonal Projects in a General Inner Space

Theorem 5.3 shows that the shortest distance between a vector and a sub-space is given by the norm of the vector orthogonal to this sub-space constructed in the same way as we know from elementary geometry. This property plays key roles in several algorithms across this book.

Theorem 5.3. Orthogonal Projection

Consider a sub-space U of a real-valued inner space V. Consider $|x\rangle$ a vector of V and let $\mathcal{B}=\{|e_1\rangle,\ldots,|e_m\rangle\}$ be an orthonormal base of U. Define $P_U|x\rangle$ the vector of U that is the closest to the vector $|x\rangle$, i.e., such that $\left\| |x\rangle - P_U|x\rangle \right\|$ is minimal, where $\|x\|$ is the induced norm in the inner vector space V. Then

1. $P_U|x\rangle = \sum_{i=1}^{m} \langle e_i|x\rangle |e_i\rangle$
2. $|x\rangle - P_U|x\rangle$ is orthogonal to any vector of U

The vector $P_U|x\rangle$ is called the orthogonal projection of $|x\rangle$ onto U.

The proof is relatively straightforward and serves as an excellent exercise to familiarize oneself with the various concepts and calculation techniques associated with inner products and induced norms.

To find the vector $P_U|x\rangle$, the orthogonal projection of $|x\rangle$ onto U, we search for which vector of the sub-space U the norm $\left\| |x\rangle - P_U|x\rangle \right\|$ is minimal. First, we note that any vector $|u\rangle$ of U can be decomposed in the base B Let us calculate, keeping in mind that $\mathcal{B}=\{|e_1\rangle,\ldots,|e_m\rangle\}$ is an orthonormal base:

$$
\begin{aligned}
\left\| |x\rangle - \sum_{i=1}^{m} u_i |e_i\rangle \right\|^2 &= \left\langle x - \sum_{i=1}^{m} u_i e_i \middle| x - \sum_{i=1}^{m} u_i e_i \right\rangle \\
&= \langle x|x\rangle - \sum_{i=1}^{m} u_i \langle e_i|x\rangle - \sum_{i=1}^{m} u_i \langle x|e_i\rangle + \sum_{i=1}^{m} u_i^2 \\
&= \|x\|^2 - 2\sum_{i=1}^{m} u_i \langle e_i|x\rangle + \sum_{i=1}^{m} u_i^2 \\
&= \|x\|^2 - \sum_{i=1}^{m} \langle e_i|x\rangle^2 + \sum_{i=1}^{m} \left(u_i - \langle e_i|x\rangle\right)^2
\end{aligned}
$$

Consequently, the norm $\left\| |x\rangle - |u\rangle \right\|$, for a general vector $|u\rangle$ of U, is minimal if $u_i = \langle e_i | x \rangle$ for all $i = 1,\ldots,m$, which proves the first part of Theorem 5.3.

To prove the second part of Theorem 5.3, we calculate

$$\begin{aligned}\langle x - P_U x | e_j \rangle &= \langle x | e_j \rangle - \langle P_U x | e_j \rangle = \langle x | e_j \rangle - \sum_{i=1}^{m} \langle e_i | x \rangle \langle e_i | e_j \rangle \\ &= \langle x | e_j \rangle - \langle e_j | x \rangle = 0\end{aligned} \tag{5.14}$$

As $|x - P_U x\rangle$ is orthogonal to any base vector of U, the vector $|x - P_U x\rangle$ is orthogonal to any vector of U. This proves Theorem 5.3.

Let us add a remark. If the base $B = \{|e_1\rangle,\ldots,|e_m\rangle\}$ of the sub-space U is not normalized but only an orthogonal base, then the orthogonal projection of $|x\rangle$ onto the sub-space U writes:

$$P_U |x\rangle = \sum_{i=1}^{m} \frac{\langle e_i | x \rangle}{\langle e_i | e_i \rangle} |e_i\rangle \tag{5.15}$$

Equation (5.15) shows that it is important to always carefully check if an orthogonal or an orthonormal base is used in the calculations.

5.2.2 Orthogonal Projection Matrix

As already mentioned, orthogonal projections play an important role in several fields of engineering and science. With Equation (5.15), the projection of a vector $|x\rangle$ onto a sub-space U can be computed. We want to give here, for the particular case of $\mathbb{R}^n$ a slightly different reformulation of Equation (5.15) particularly well-suited for calculations in Octave. Rewriting Equation (5.15) in the particular case of $\mathbb{R}^n$ results in:

$$P_U \underline{x} = \sum_{i=1}^{m} \frac{\underline{e}_i^T \underline{x}}{\underline{e}_i^T \underline{e}_i} \underline{e}_i \tag{5.16}$$

where $\{\underline{e}_1,\ldots,\underline{e}_m\}$ is an orthogonal base of the sub-space U. Let us have a closer look at the products $\underline{e}_i^T \underline{x}\, \underline{e}_i$. The first product $\underline{e}_i^T \underline{x}$ is a number. Consequently, we can write:

$$\left(\underline{e}_i^T \underline{x}\right) \underline{e}_i = \underline{e}_i \left(\underline{e}_i^T \underline{x}\right) \tag{5.17}$$

All columns, respectively, line vectors appearing in this expression can be seen as matrices. As the matrix product is associative, we can further write

$$\left(\underline{e}_i^T \underline{x}\right) \underline{e}_i = \underline{e}_i \left(\underline{e}_i^T \underline{x}\right) = \left(\underline{e}_i \underline{e}_i^T\right) \underline{x} \tag{5.18}$$

Using this result, Equation (5.16) can be rewritten as:

$$P_U \underline{x} = \sum_{i=1}^{m} \frac{\underline{e}_i \underline{e}_i^T}{\underline{e}_i^T \underline{e}_i} \underline{x} \tag{5.19}$$

Note that the quantity $\left(\underline{e}_i \underline{e}_i^T\right)$ is a $n \times n$ matrix. This allows the definition of a matrix P_U which, when multiplied by $\underline{x}$, projects the vector $\underline{x}$ on the sub-space U. This matrix is called the orthogonal projection matrix.

Definition 5.8. Orthogonal Projection Matrix

The matrix P_U defined by

$$P_U = \sum_{i=1}^{m} \frac{\underline{e}_i \underline{e}_i^T}{\underline{e}_i^T \underline{e}_i} \tag{5.20}$$

is called the orthogonal projection matrix onto the sub-space U spanned by the orthogonal vectors $\{\underline{e}_1, \ldots, \underline{e}_m\}$.

The following example illustrates how using the projection matrix is efficient in Octave.

Example 5.5. Orthogonal Projections in $\mathbb{R}^3$

In this example, we work in $\mathbb{R}^3$ to obtain an intuitive understanding of the various operations. In all calculations, we use the natural base of $\mathbb{R}^3$ to express the coordinates of the vectors. We start by considering the sub-space U spanned by the following two vectors:

```
>> x1 = [1; 0; 0];
>> x2 = [0; 1; 0];
```

The two vectors span the x-y plane in a geometrical representation of $\mathbb{R}^3$ Consequently, the projection of any vector of $\mathbb{R}^3$onto that plane will be the vector formed by the two x and y components of the vector. Let's verify if Equation (5.20) will reach the same conclusion.

The orthogonal projection matrix onto this sub-space is:[9]

```
>> Pu = (x1*x1')/(x1'*x1) + (x2*x2')/(x2'*x2)
   1   0   0
   0   1   0
   0   0   0
```

The projection of any vector $\underline{x}$ onto the sub-space U can then be calculated as:

```
>> x = [1; 2; 3];
>> Pu*x
ans =
     1
     2
     0
```

This is what we expected. Observe how the orthogonal projection matrix selects the first two components of the vector $\underline{x}$.

As a second example, let us consider the projection onto the sub-space U formed by the following two vectors:

```
>> x3 = [0; 0; 1];
>> x4 = [1; 1; 0];
```

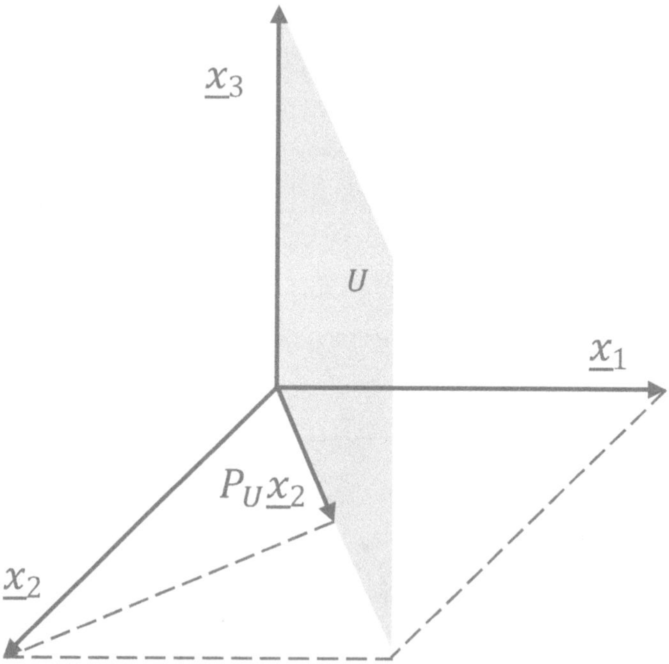

FIGURE 5.1 Illustration of the vector space spanned by the two vectors x3 = [0; 0; 1] and x4 = [1; 1; 0] and the projection of the vector x2 = [0; 1; 0] onto this space.

The orthogonal projection matrix onto this sub-space U is:

```
>> Pu = (x1*x1')/(x1'*x1) + (x2*x2')/(x2'*x2)
Pu =
   0.50000 0.50000 0.00000
   0.50000 0.50000 0.00000
   0.00000 0.00000 1.00000
```

The projection of the vector $\underline{x}_2$ onto that sub-space is:

```
>> Pu*x2
ans =
   0.50000
   0.50000
   0.00000
```

This aligns with the expected geometric interpretation (refer to Figure 5.1). It's important to note that the coordinates of the projected vector are represented in the same vector base as the vectors spanning the subspace U. In this example, it corresponds to the natural base of $\mathbb{R}^3$

The proof of Theorem 5.3 gives a first feeling as to why orthonormal vector bases are convenient choices to execute calculations in vector spaces. In fact, when using orthonormal bases, several results familiar from geometry can directly be translated into counterparts in real-valued vector spaces. For example, consider $|x\rangle = \sum_{i=1}^{n} x_i |e_i\rangle$ and $|y\rangle = \sum_{i=1}^{n} y_i |e_i\rangle$ two vectors of a real-valued vector space. Then, their inner product writes

$$\langle x|y\rangle = \sum_{i,j=1}^{n} x_i y_j \langle e_i|e_j\rangle \tag{5.21}$$

If we use an orthonormal base $\mathcal{B} = \{|e_1\rangle, \ldots, |e_m\rangle\}$ Equation (5.21) implies that the inner product $\langle x|y\rangle$ can be calculated by the simpler relation $\underline{x}^T \underline{y}$ using the $\mathbb{R}^n$ representation of the involved vectors, as now $\langle e_i|e_j\rangle = \delta_{ij}$. This is indeed the expression we expected from elementary geometry.

This observation can be summarized by the following theorem:

Theorem 5.4. Inner Products and Norms Using $\mathbb{R}^n$ Representations

Consider two vectors $|x\rangle$ and $|y\rangle$ and an orthonormal base $\mathcal{B} = \{|e_1\rangle, \ldots, |e_n\rangle\}$ of a real-valued vector space of dimension n. If all involved vectors are represented by elements of $\mathbb{R}^n$ then we have

$$\langle x|y\rangle = \|x\|\cdot\|y\|\cdot\cos\theta = \underline{x}^T\cdot\underline{y} \tag{5.22}$$

$$\|x\| = \sqrt{\langle x|x\rangle} = \sqrt{\underline{x}^T\cdot\underline{x}} = \sqrt{x_1^2+\cdots+x_n^2} \tag{5.23}$$

with θ representing the generalized angle between the vectors $|x\rangle$ and $|y\rangle$. This angle is equal to the geometrical angle between the vectors $\underline{x}$ and $\underline{y}$ of $\mathbb{R}^n$

Theorem 5.4 shows the benefit of using orthonormal bases to represent vectors. In such bases, the calculations become very similar to the well-known rules in $\mathbb{R}^n$. Theorem 5.4 gives a further justification of the definition of the generalized angle. The generalized angle between two vectors $|x\rangle$ and $|y\rangle$ measures, if orthonormal bases are used to represent the involved vectors, the same angle as the angle between their representations $\underline{x}$ and $\underline{y}$ in $\mathbb{R}^n$.

Example 5.6. Illustration of Theorem 5.4

Let us first start with a counter-example to illustrate that when we are not using an orthonormal base, Theorem 5.4 does not apply.

Consider the vector space $P_2[x]$ of the real polynomials of degree 2 or less. We choose the natural base $\mathcal{B} = \{1, x, x_2\}$ and the following inner product

$$\langle P(x)|Q(x)\rangle = \int_{-1}^{+1} P(x)Q(x)\,dx \tag{5.24}$$

The two polynomials $P(x) = 2+3x-x^2$ and $Q(x) = 3-x+3x^2$ are represented in $\mathbb{R}^3$ by

$$|P(x)\rangle := \underline{P} = \begin{bmatrix} 2 \\ 3 \\ -1 \end{bmatrix} \tag{5.25}$$

and

$$|Q(x)\rangle := \underline{Q} = \begin{bmatrix} 3 \\ -1 \\ 3 \end{bmatrix} \tag{5.26}$$

As an exercise, the reader is invited to verify that $\langle P(x)|Q(x)\rangle = \frac{54}{5} \neq \underline{P}^T\underline{Q} = 0$.

Let us switch to using an orthonormal base for $p_2[x]$ instead of the natural base. Such orthonormal bases exist, and one example is inspired by Legendre polynomials.[10] We won't delve into how to find it just yet, but here's an orthonormal basis for the vector space $p_2[x]$ It is

$$\mathcal{B}=\left\{E_1(x)=\frac{1}{\sqrt{2}},E_2(x)=\sqrt{\frac{3}{2}}x,E_3(x)=\sqrt{\frac{45}{8}}\left(x^2-\frac{1}{3}\right)\right\} \tag{5.27}$$

As an exercise, the reader is invited to verify that this base is orthonormal for the inner product (5.24) by checking that, indeed, $\langle E_i(x)|E_j(x)\rangle=\delta_{ij}$.

In this base, one has

$$|P(x)\rangle=\frac{5\sqrt{2}}{3}|E_1(x)\rangle+3\sqrt{\frac{2}{3}}|E_2(x)\rangle-\sqrt{\frac{8}{45}}|E_3(x)\rangle \tag{5.28}$$

and

$$|Q(x)\rangle=4\sqrt{2}|E_1(x)\rangle-\sqrt{\frac{2}{3}}|E_2(x)\rangle+3\sqrt{\frac{8}{45}}|E_3(x)\rangle \tag{5.29}$$

Consequently, in this base, the representations in R^3 of $P(x)=2+3x-x^2$ and $Q(x)=3-x+3x^2$ become

$$|P(x)\rangle:=\underline{P}=\begin{bmatrix}\frac{5\sqrt{2}}{3}\\ 3\sqrt{\frac{2}{3}}\\ -\sqrt{\frac{8}{45}}\end{bmatrix}\text{ and }|Q(x)\rangle:=\underline{Q}=\begin{bmatrix}4\sqrt{2}\\ -\sqrt{\frac{2}{3}}\\ 3\sqrt{\frac{8}{45}}\end{bmatrix} \tag{5.30}$$

As an exercise the reader is invited to verify that this time $\langle P(x)|Q(x)\rangle=\underline{P}^T\underline{Q}=\frac{54}{5}$.

Orthonormal bases have other advantages. For example, in orthonormal vector bases, the calculation of the components of a vector is particularly simple. Consider a vector $|x\rangle$ from a real-valued vector space V, and let us calculate the coordinates x_i of $|x\rangle$ in an orthonormal base $\mathcal{B}=\{|e_1\rangle,\ldots,|e_n\rangle\}$

In the base B the vector $|x\rangle$ writes $|x\rangle=\sum_{i=1}^{n}x_i|e_i\rangle$. As we have

$$\langle e_j|x\rangle=\sum_{i=1}^{n}x_i\langle e_j|e_i\rangle=x_j \tag{5.31}$$

the coordinate x_j can be calculated by taking the inner product between the base vector $|e_j\rangle$ and the vector $|x\rangle$. This is a familiar result of geometry.

5.3 GRAM-SCHMIDT ORTHOGONALIZATION

There are several fields in mathematics where it is desirable to create a set of orthogonal vectors. We will discuss some of these applications later. For now, we want to present an algorithm known as Gram-Schmidt orthogonalization,[11] which is able to construct such sets. To explain the algorithm, we use our knowledge of orthogonal projections from Theorem 5.3.

Consider $k+1$ given vectors $|x_i\rangle$ of a real-valued vector space which we number from 0 to k. We chose these vectors linearly independent, which means they form a vector base of the sub-space they span. Consequently, any vector $|y\rangle$ of this sub-space can be written as

$$\left|y\right\rangle = \sum_{i=0}^{k} a_i \left|x_i\right. \tag{5.32}$$

with unique real numbers a_i.

Our goal is to find an orthogonal vector base $\mathcal{B} = \{|z_o\rangle, |z_1\rangle, \ldots, |z_k\rangle\}$ of this sub-space. In this case, we can write

$$|y\rangle = \sum_{i=0}^{k} a_i |x_i\rangle = \sum_{i=0}^{k} \tilde{a}_i |z_i\rangle \tag{5.33}$$

with some unique real numbers $\tilde{a}_i$. Keep in mind that, in order to define orthogonality, an inner product $\langle x|y\rangle$ must have been defined on this real-valued vector space.

The gram-Schmidt orthogonalization algorithm proceeds iteratively to construct this orthogonal base $\mathcal{B}$. Imagine we found a set $\mathcal{B}_i = \{|z_o\rangle, |z_1\rangle, \ldots, |z_i\rangle\}$ of $i+1$ vectors forming an orthogonal base of the sub-space Π_i spanned by the $i+1$ vectors $\{|x_o\rangle, |x_1\rangle \ldots, |x_i\rangle\}$. We want to add a vector $|z_{i+1}\rangle$ to $\mathcal{B}_i$ to form an orthogonal base $\mathcal{B}_{i+1}$ of the sub-space Π_{i+1} spanned by the $i+2$ vectors $\{|x_o\rangle, \ldots, |x_{i+1}\rangle\}$. In other words, we want to find a vector $|z_{i+1}\rangle$ orthogonal to the sub-space Π_i. We can do this by calculating

$$|z_{i+1}\rangle = |x_{i+1}\rangle - P_{\Pi_i} |x_{i+1}\rangle \tag{5.34}$$

where $P_{\Pi_i} |x_{i+1}\rangle$ is the orthogonal projection of $|x_{i+1}\rangle$ on the sub-space Π_i (Figure 5.2). The orthogonal projection can be calculated using Equation (5.15):

$$|z_{i+1}\rangle = |x_{i+1}\rangle - \sum_{k=0}^{i} \frac{\langle z_k | x_{i+1}\rangle}{\langle z_k | z_k\rangle} |z_k\rangle \tag{5.35}$$

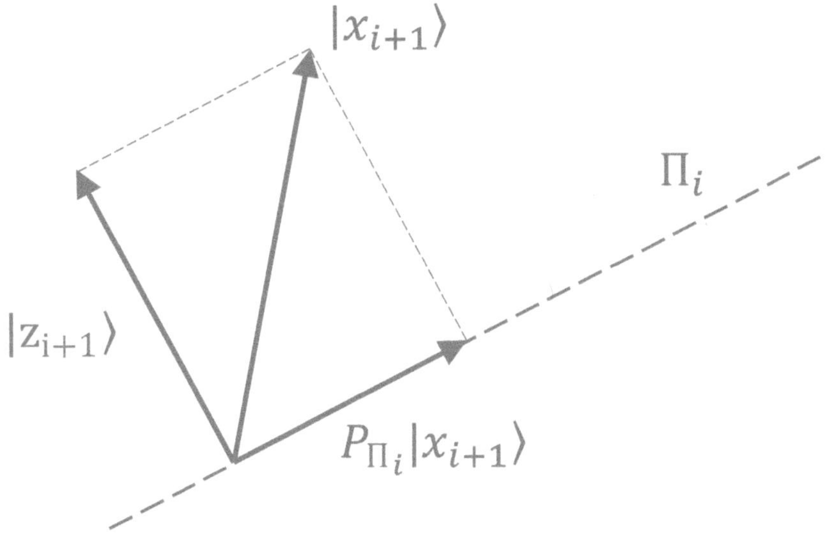

FIGURE 5.2 Constructing a vector $|z_{i+1}\rangle$ orthogonal to the sub-space π_i spanned by the $i+1$ vectors $\{|x_o\rangle, \langle x_1|, \ldots, |x_i\rangle\}$. This vector is $z \in \mathbb{R}^{n\times(k+1)}$, where $P_{\Pi_i}|x_{i+1}\rangle$ is the orthogonal projection of $|x_{i+1}\rangle$ on the sub-space π_i.

To use the algorithm, we need to start it. For this we choose $|z_o\rangle = |x_o\rangle$. Using Equation (5.35), we can then iteratively construct the base $\mathcal{B}_i = \{|z_o\rangle, |z_1\rangle, \ldots, |z_k\rangle\}$.

This strategy can be summarized in the following algorithm:

Algorithm 5.1. Gram-Schmidt Orthogonalization

Consider a set of linearly independent vectors $\{|x_o\rangle, |x_1\rangle, ..., |x_k\rangle\}$ of a real-valued inner space. Then, we can construct an orthogonal vector base $\mathcal{B}_i = \{|z_o\rangle, |z_1\rangle, \ldots, |z_k\rangle\}$ of the sub-space spanned by $\{|x_o\rangle, |x_1\rangle, ..., |x_k\rangle\}$ using the following procedure

1. Set $|z_o\rangle = |x_o\rangle$
2. For $i = 1, 2, \ldots, k$ compute successively

$$|z_i\rangle = |x_i\rangle - \sum_{j=0}^{i-1} \lambda_{j,i} |z_j\rangle$$

with the i numbers $\lambda_{j,i} = \dfrac{\langle z_j | x_i \rangle}{\langle z_j | z_j \rangle}$.

Let us illustrate the application of Algorithm 5.1 with some examples.

Example 5.7. Gram-Schmidt Orthogonalization

Consider the three vectors

$$\underline{x}_0 = \begin{bmatrix} 4 \\ 0 \\ 3 \end{bmatrix}, \underline{x}_1 = \begin{bmatrix} 8 \\ 2 \\ 6 \end{bmatrix} \text{ and } \underline{x} = \begin{bmatrix} 1 \\ -2 \\ 7 \end{bmatrix} \tag{5.36}$$

from R^3. Let us verify that they form a vector base of $\mathbb{R}^3$and construct, based on these three vectors, an orthogonal base using Gram-Schmidt's algorithm according to the standard inner product $\underline{x}^T \underline{y}$.

To confirm the linear independence of the three vectors, we validate that

$$\begin{vmatrix} 4 & 8 & 1 \\ 0 & 2 & -2 \\ 3 & 6 & 7 \end{vmatrix} = 50 \neq 0 \tag{5.37}$$

Let us build an orthogonal vector base $\mathcal{B} = \{z_o, z_1, z_2\}$ following Gram-Schmidt's method. We start by setting

$$\underline{z}_0 = \underline{x}_0 = \begin{bmatrix} 4 \\ 0 \\ 3 \end{bmatrix} \tag{5.38}$$

The vector $\underline{z}_1$ is calculated as

$$\underline{z}_1 = \underline{x}_1 - \frac{\underline{z}_0^T \underline{x}_1}{\underline{z}_0^T \underline{z}_0} \underline{z}_0 = \begin{bmatrix} 8 \\ 2 \\ 6 \end{bmatrix} - \frac{50}{25} \begin{bmatrix} 4 \\ 0 \\ 3 \end{bmatrix} = \begin{bmatrix} 0 \\ 2 \\ 0 \end{bmatrix} \tag{5.39}$$

The vector $\underline{z}_2$ becomes

$$\underline{z}_2 = \underline{x}_2 - \frac{\underline{z}_0^T \underline{x}_2}{\underline{z}_0^T \underline{z}_0} \underline{z}_0 - \frac{\underline{z}_1^T \underline{x}_2}{\underline{z}_1^T \underline{z}_1} \underline{z}_1 = \begin{bmatrix} 1 \\ -2 \\ 7 \end{bmatrix} - \frac{25}{25} \begin{bmatrix} 4 \\ 0 \\ 3 \end{bmatrix} - \frac{(-4)}{4} \begin{bmatrix} 0 \\ 2 \\ 0 \end{bmatrix} = \begin{bmatrix} -3 \\ 0 \\ 4 \end{bmatrix} \tag{5.40}$$

The reader is invited to verify that, indeed, $\mathcal{B} = \{\underline{z}_o, \underline{z}_1, \underline{z}_2\}$forms an orthogonal vector base.

The Gram-Schmidt algorithm does not have to be applied on the full vector space (as we did in Example 5.7) but can be applied to any sub-spaces. The next example illustrates this.

Example 5.8. Gram-Schmidt Orthogonalization on a Sub-Space

Consider the two vectors

$$\underline{x}_0 = \begin{bmatrix} 4 \\ 0 \\ 3 \end{bmatrix} \text{ and } \underline{x}_1 = \begin{bmatrix} 8 \\ 2 \\ 6 \end{bmatrix} \tag{5.41}$$

from $\mathbb{R}^3$ Let us verify that they are linearly independent and construct, based on these vectors, an orthogonal base of the sub-space spanned by these two vectors using Gram-Schmidt's algorithm according to the standard inner product $x^T y$.

As $\underline{x}_0$ and $\underline{x}_1$ are not proportional to each other, they are indeed linearly independent and will span a sub-space of dimension two of $\mathbb{R}^3$ Let us build an orthogonal vector base $\mathcal{B} = \{\underline{z}_o, \underline{z}_1\}$ of this sub-space following Gram-Schmidt's method. We start by setting

$$\underline{z}_0 = \underline{x}_0 = \begin{bmatrix} 4 \\ 0 \\ 3 \end{bmatrix} \tag{5.42}$$

The vector $\underline{z}_1$ is calculated as:

$$\underline{z}_1 = \underline{x}_1 - \frac{\underline{z}_0^T \underline{x}_1}{\underline{z}_0^T \underline{z}_0} \underline{z}_0 = \begin{bmatrix} 8 \\ 2 \\ 6 \end{bmatrix} - \frac{50}{25} \begin{bmatrix} 4 \\ 0 \\ 3 \end{bmatrix} = \begin{bmatrix} 0 \\ 2 \\ 0 \end{bmatrix} \tag{5.43}$$

The reader is invited to verify that, indeed, $\mathcal{B} = \{\underline{z}_o, \underline{z}_1\}$ forms an orthogonal vector base.

We can efficiently conduct these computations on Octave. We first define $\underline{x}_0$ and $\underline{x}_1$:

```
>> xo = [4 0 3]';
>> x1 = [8 2 6]';
```

The vectors $\underline{z}_o$ and $\underline{z}_1$ are computed using the Gram-Schmidt orthogonalization procedure:

```
>> zo = xo;
>> z1 = x1-zo'*x1/(zo'*zo)*zo
z1 =
   0
   2
   0
```

The two vectors $\underline{z}_0$ and $\underline{z}_1$ form indeed an orthogonal base:

```
>> zo'*z1
ans = 0
```

Gram-Schmidt's algorithm can be applied to any real-valued vector space. The next example illustrates how sets of orthogonal polynomials can be constructed.

Example 5.9. Legendre Polynomials

Construct an orthogonal vector base of the real-valued vector space $|z_1\rangle$ of the polynomials of degree 3 or less. To define orthogonality, use the following inner product

$$\langle P(x)|Q(x)\rangle = \int_{-1}^{+1} P(x)Q(x)\,dx \tag{5.44}$$

We start by choosing a vector base of the vector space $P_3[x]$. We choose the canonical base:

$$|P_0\rangle = |1\rangle, |P_1\rangle = |x\rangle, |P\rangle = |x^2\rangle \text{ and } |P_3\rangle = |x^3\rangle$$

We want to construct an orthogonal vector base $B = \{|z_o\rangle, |z_1\rangle, |z_2\rangle, |z_3\rangle\}$ following Gram-Schmidt's method. We start by setting

$$|z_0\rangle = |1\rangle \tag{5.45}$$

The vector $|z_1\rangle$ is then

$$|z_1\rangle = |P_1\rangle - \frac{\langle z_0|P_1\rangle}{\langle z_0|z_0\rangle}|z_0\rangle \tag{5.46}$$

As the inner product

$$\langle z_0|P_1\rangle = \int_{-1}^{+1} 1\cdot x\,dx = 0 \tag{5.47}$$

it follows that $|z_1\rangle = |P_1\rangle = |x\rangle$.

The vector $|z_2\rangle$ is calculated as

$$\left|z_2\right\rangle = \left|P_2\right\rangle = \frac{\left\langle z_0 \middle| P_2\right\rangle}{\left\langle z_0 \middle| z_0\right\rangle}\left|z_0\right\rangle - \frac{\left\langle z_1 \middle| P_2\right\rangle}{\left\langle z_1 \middle| z_1\right\rangle}\left|z_1\right\rangle \tag{5.48}$$

The needed inner products are

$$\left\langle z_0 \middle| P_2\right\rangle = \int_{-1}^{+1} 1 \cdot x^2\, dx = \frac{2}{3} \tag{5.49}$$

$$\left\langle z_0 \middle| z_0\right\rangle = \int_{-1}^{+1} 1 \cdot 1\, dx = 2 \tag{5.50}$$

$$\left\langle z_1 \middle| P_2\right\rangle = \int_{-1}^{+1} x \cdot x^2\, dx = 0 \tag{5.51}$$

It follows that

$$\left|z_2\right\rangle = \left|x^2\right\rangle - \frac{2/3}{2}\left|1\right\rangle = \left|x^2 - \frac{1}{3}\right\rangle \tag{5.52}$$

Finally, the vector $\left|z_2\right\rangle$ is calculated as

$$\left|z_3\right\rangle = \left|P_3\right\rangle - \frac{\left\langle z_0 \middle| P\right\rangle_3}{\left\langle z_0 \middle| z_0\right\rangle}\left|z_0\right\rangle - \frac{\left\langle z_1 \middle| P_3\right\rangle}{\left\langle z_1 \middle| z_1\right\rangle}\left|z_1\right\rangle - \frac{\left\langle z_2 \middle| P_3\right\rangle}{\left\langle z_2 \middle| z_2\right\rangle}\left|z_2\right\rangle \tag{5.53}$$

After calculating the various inner products, one finds:

$$\left|z_3\right\rangle = \left|x^3\right\rangle - \frac{0}{\left\langle z_0 \middle| z_0\right\rangle}\left|z_0\right\rangle - \frac{2/5}{2/3}\left|z_1\right\rangle - \frac{0}{\left\langle z_2 \middle| z_2\right\rangle} z_2 = \left|x^3 - \frac{3}{5}x\right\rangle \tag{5.54}$$

In summary, the base formed by the following polynomials is an orthogonal base of the vector space of polynomials of degree 3 or less:

$$\left\{z_0(x) = 1,\, z_1(x) = x,\, z_2(x) = x^2 - \frac{1}{3},\, z_3(x) = x^3 - \frac{3}{5}x\right\} \tag{5.55}$$

The reader is invited to verify explicitly that all four polynomials are orthogonal to each other. The polynomials $z_i(x)$ we found belong to the family of the Legendre polynomials. As we will learn later, they have important applications in developing accurate formulas to approximate definite integrals.

A version of this algorithm called the Gram-Schmidt orthonormalization procedure exists. In this version of the algorithm, the produced base is an orthonormal base (all vectors are orthogonal to each other and of norm one). To achieve this result, the algorithm normalizes the vectors $|z_i\rangle$ computed by the Gram-Schmidt algorithm in each step.

Gram-Schmidt orthonormalization starts with $|q_o\rangle$, instead of $|x_o\rangle$, defined as

$$|q_o\rangle = \frac{|x_o\rangle}{\sqrt{\langle x_o|x_o\rangle}} \tag{5.56}$$

Further, Equation (5.35) is replaced by

$$|z_{i+1}\rangle = |x_{i+1}\rangle - \sum_{j=0}^{i-1} r_{ji}|q_j\rangle \text{ with } r_{ji} = \langle q_j|x_i\rangle \tag{5.57}$$

followed by

$$|q_{i+1}\rangle = \frac{|z_{i+1}\rangle}{r_{i+1,i+1}} \text{ with } r_{i+1,i+1} = \sqrt{\langle z_{i+1}|z_{i+1}\rangle} \tag{5.58}$$

in order to keep the vectors normalized.

It is possible to conduct these calculations efficiently using the following pseudo-code, which is often referred to as the classical Gram-Schmidt orthonormalization algorithm:

Algorithm 5.2. Classical Gram-Schmidt Orthonormalisation Algorithm

Input: $k+1$ linearly independent vectors $|x_i\rangle$ of a real-valued inner space of dimension n

Output: $k+1$ orthonormal vectors $|q_i\rangle$ forming a base of the sub-space spanned by the vectors $|x_i\rangle$

Algorithm:

for $j = 0,1,\ldots,k$

$|y\rangle = |x_j\rangle$

for $i = 0, 1,\ldots, j-1$

$r_{ij} = \langle q_i|x_j\rangle$

$|y\rangle = |y\rangle - r_{ij}|q_j\rangle$

end

$r_{jj} = \sqrt{\langle y|y\rangle}$

$|q_j\rangle = \frac{1}{r_{jj}}|y\rangle$

end

The following example illustrates its usage.

Example 5.10. Gram-Schmidt Orthonormalization

Consider the two vectors $\underline{x}_0$ and $\underline{x}_1$ from Example 5.8. Let us construct an orthonormal base of the sub-space spanned by these two vectors.

We first define $\underline{x}_0$ and $\underline{x}_1$:

```
>> xo = [4 0 3]';
>> x1 = [8 2 6]';
```

Gram-Schmidt's orthonormalization algorithm starts by computing the vector $|q_o\rangle$:

```
>> qo = 1/sqrt(xo'*xo)*xo
qo =
   0.8000
        0
   0.6000
```

Vector $|q_1\rangle$ is computed as:

```
>> z1 = x1-qo'*x1*qo;
>> r11 = sqrt(z1'*z1);
>> q1 = z1/r11
q1 =
        0
   1.0000
  -0.0000
```

The two vectors $|q_0\rangle$ and $|q_1\rangle$ form an orthonormal base as

```
>> qo'*qo
ans = 1.0000
>> q1'*q1
ans = 1.0000
```

And further

```
>> qo'*q1
ans = -2.6645e-16
```

Note that, due to round-off errors, the two vectors are not exactly orthogonal to each other.

In fact, the Gram-Schmidt orthonormalization algorithm will produce a sequence of equations like these:

$$\begin{aligned}
&|x_0\rangle = r_{00}|q_0\rangle \\
&\|x_1\rangle = r_{01}|q_0\rangle r_{11}|q_1\rangle \\
&|x_2\rangle = r_{02}|q_0\rangle + r_{12}|q_1\rangle + r_{22}|q_2\rangle \\
&\cdots
\end{aligned} \tag{5.59}$$

This suggests that we should be able to translate this algorithm into a matrix equation taking the form $A = QR$ where the columns of A are the coordinate vectors of the vectors $|x_i\rangle$, the columns of Q are the coordinate vectors of the vectors $|q_i\rangle$ and the matrix R is an upper diagonal matrix constructed from the coefficients r_{ij}. This can indeed be done and is the topic of Section 5.4.

5.4 QR DECOMPOSITION

As it is possible to transform the Gauss elimination algorithms into matrix equations using the LU and PA=LU decomposition, it is also possible to transform the Gram-Schmidt algorithm into a matrix equation. This leads to the so-called reduced QR decomposition of a matrix.

We work in a real-valued inner space V of dimension n. Consider a subspace of V spanned by the $k+1$ linearly independent vectors $|x_i\rangle$. We aim to find an orthogonal base of the sub-space using Gram-Schmidt's Algorithm 5.1, but this time using Matrix equations. The key idea is to represent step 2 from Gram-Schmidt's Algorithm 5.1 with a matrix equation. This is possible if we use representations of the vectors $|x_i\rangle$ with elements from $\mathbb{R}^n$. We define the matrix $A \in \mathbb{R}^{n\times(k+1)}$

$$A = [x_0, x_1, \cdots, x_k] \tag{5.60}$$

in which the $k+1$ columns are the vector coordinate representations $\underline{x}_i$ of the $k+1$ vectors $|x_i$ in a base of the vector space V. It is important to note that this vector base must be an orthonormal base as, otherwise, we will not have a match between the inner product of V and $\mathbb{R}^n$ (i.e., that $\langle x \,|\, y\rangle = x^T y$). The reason for this important constraint will become apparent in what follows.

Similarly, we define the matrix $z \in \mathbb{R}^{n\times(k+1)}$

$$z = [\underline{z}_0, \underline{z}_1, \ldots, \underline{z}_k] \tag{5.61}$$

where $\underline{z}_i$ are the representations in $\mathbb{R}^n$ of the $k+1$ orthogonal vectors $|z_i$ constructed by Gram-Schmidt's Algorithm 5.1. Additionally, we define the upper triangular matrix $A \in \mathbb{R}^{n\times(k+1)}$ filled with the coefficients λ_{ij} from Gram-Schmidt's Algorithm 5.1 as follows:

$$\Lambda = \begin{bmatrix} 1 & \lambda_{0,1} & \lambda_{0,2} & \cdots & \lambda_{0,k} \\ 0 & 1 & \lambda_{1,2} & \cdots & \lambda_{1,k} \\ 0 & 0 & 1 & \cdots & \vdots \\ \vdots & \vdots & & \ddots & \lambda_{k-1,k} \\ 0 & 0 & 0 & & 1 \end{bmatrix} \tag{5.62}$$

With these definitions, one can verify that the equation

$$A = Z\Lambda \tag{5.63}$$

is equivalent to

$$|x_i\rangle = \sum_{j=0}^{i-1} \lambda_{j,i} z_j + |z_i\rangle \tag{5.64}$$

Equation (5.64) is identical to Equation (5.35) of step 2 of Gram-Schmidt's Algorithm 5.1. Note that the two Equations (5.63) and (5.64) are only equivalent if the inner products of V and $\mathbb{R}^n$ match. This is the case as we use an orthonormal base for the representations in $\mathbb{R}^n$ of the involved vectors. Equation (5.63) is the matrix representation of Gram'Schmidt's orthogonalization algorithm. Let us add some remarks:

1. Contrary to the matrix Λ, the matrices A and Z are, in general, not square matrices. Their number of lines is equal to the dimension n of the inner space V in which we work, and their number of columns is $k+1$, the dimension of the sub-space spanned by the vectors $|x_i\rangle$, respectively $|z_i\rangle$ (which is equal to the number $k+1$ of these vectors).
2. The structure of the matrix Z is such that its columns are all orthogonal vectors. Consequently, we have that $Z^T Z$ is a $(k+1)\times(k+1)$ diagonal matrix.

Let us illustrate with an example how to find the $Z\Lambda$ decomposition of a matrix A using the Gram-Schmidt algorithm.

Example 5.11. $Z\Lambda$ **Decomposition of a Matrix**

Find the $Z\Lambda$ decomposition of the following matrix

$$A = \begin{bmatrix} 4 & 8 \\ 0 & 2 \\ 3 & 6 \end{bmatrix} \tag{5.65}$$

In Example 5.8, we implemented the Gram-Schmidt's algorithm. Using the results found, we can directly write down the $Z\Lambda$ decomposition:

$$A = \begin{bmatrix} 4 & 8 \\ 0 & 2 \\ 3 & 6 \end{bmatrix} = \begin{bmatrix} 4 & 0 \\ 0 & 2 \\ 3 & 0 \end{bmatrix} \begin{bmatrix} 1 & 2 \\ 0 & 1 \end{bmatrix} \tag{5.66}$$

We can verify that, indeed, $Z^T Z$ is a diagonal matrix:

$$\begin{bmatrix} 4 & 0 & 3 \\ 0 & 2 & 0 \end{bmatrix} \begin{bmatrix} 4 & 0 \\ 0 & 2 \\ 3 & 0 \end{bmatrix} = \begin{bmatrix} 25 & 0 \\ 0 & 4 \end{bmatrix} \tag{5.67}$$

which confirms that the columns of the matrix Z are orthogonal vectors.

The classical Gram-Schmidt orthonormalization Algorithm 5.2 can also be rewritten as a matrix equation. To this end, we introduce the diagonal matrix D, in which the diagonal elements are the norms of the vectors $|z_i\rangle$ found by the algorithm described in Algorithm 5.1, i.e. $d_{ii} = \|z_{i+1}\|$, to write:

$$A = ZD^{-1}D\Lambda = QR \tag{5.68}$$

with $Q = ZD^{-1}$ and $R = D\Lambda$. The matrix R is an upper diagonal matrix. The matrix Q is a matrix in which columns are orthonormal vectors. Consequently,

$$Q^T Q = I_{k+1} \tag{5.69}$$

where I_{k+1} is the $(k+1)\times(k+1)$ identity matrix. The matrix Q can be seen as a generalized orthogonal matrix. We use the term "generalized" to distinguish from the specific definition of an orthogonal matrix, which typically denotes square matrices whose columns consist of orthonormal vectors. In our case, Q is not necessarily a square matrix.

Decomposition (5.68) is usually referred to as the reduced QR factorization. There exist other versions of this factorization that result in a matrix Q, which is an orthogonal matrix, not a generalized orthogonal matrix as in our case. This version of the algorithm is usually termed QR factorization. Note that Octave implements the QR decomposition with its function `qr` and not the reduced QR decomposition.

Definition 5.9. Reduced QR Factorisation of a Matrix

The reduced QR decomposition of a matrix A is the decomposition of the matrix A into a product of a generalized orthogonal matrix Q and an upper diagonal matrix R.

The reduced QR factorization of a matrix can be useful even when the matrix does not refer to the coordinate vectors of a base of a sub-space where an orthonormal base is sought. In that case, the reduced QR decomposition is used to rewrite a matrix A as the product of a generalized orthogonal matrix and an upper diagonal matrix.

Example 5.12. Reduced QR Decomposition

Let us find the reduced QR decomposition of Matrix (5.65) from Example 5.11. To this end, we use the results from the Gram-Schmidt orthonormalization conducted in Example 5.10. The norms of the two vectors $|z_o\rangle$ and $|z_1\rangle$ allow building the matrix D from Equation (5.68):

```
>> D = [norm(zo) 0; 0 norm(z1)]
D =
     5     0
     0     2
```

Together with the two matrices Z and Λ calculated in Example 5.11, we obtain matrices Q and R:

```
>> Z = [4 0; 0 2; 3 0];
>> Lambda = [1 2; 0 1];
>> Q = Z*inv(D)
Q =
    0.8000         0
         0    1.0000
    0.6000         0
>> R = D*Lambda
R =
     5    10
     0     2
```

Multiplying Q and R

```
>> Q*R
ans =
     4     8
     0     2
     3     6
```

yields matrix A. Furthermore, matrix Q is a generalized orthogonal matrix as stated by Equation (5.69):

```
>> Q'*Q
ans =
   1   0
   0   1
```

For reference, let us compute the QR decomposition of matrix A using the Octave function `qr` (as in the case of the `lu` function, the `qr` function computes different versions of the algorithm. The interested reader is encouraged to read the help provided by Octave):

```
>> [Q R] = qr([4 8; 0 2; 3 6])
Q =
  -0.8000         0   -0.6000
        0    1.0000         0
  -0.6000         0    0.8000
R =
   -5   -10
    0     2
    0     0
```

As mentioned before, the function `qr` computes the QR decomposition and not the reduced QR decomposition. In the QR decomposition, matrix Q is an orthogonal matrix, and matrix R is a generalized upper diagonal matrix. But equation $A = QR$ still holds.

5.5 EXERCISES

5.5.1 INNER SPACES

1. Verify that the norms defined in Example 5.1 satisfy all three conditions of the definition of a norm.

2. Verify that the two inner products defined in Example 5.2 satisfy all three conditions of the definition of an inner product.

3. Verify that the base

$$\mathcal{B} = \left\{ E_1(x) = \frac{1}{\sqrt{2}},\ E_2(x) = \sqrt{\frac{3}{2}}x,\ E_3(x) = \sqrt{\frac{45}{8}}\left(x^2 - \frac{1}{3}\right) \right\}$$

of the real-value vector space $P_2[x]$ of the real polynomials of degree 2 or less is an orthonormal vector base for the inner product

$$P(x)|Q(x) = \int_{-1}^{+1} P(x)Q(x)\,dx$$

4. Compute the generalized angle between the following vectors (use the indicated inner products and the corresponding induced norm).

   a. $\underline{x}_1 = \begin{bmatrix} 1 \\ 9 \\ 3 \end{bmatrix}$ and $\underline{x}_2 = \begin{bmatrix} 5 \\ -3 \\ 4 \end{bmatrix}$ with the inner product $\langle x|y\rangle = \underline{x}^T \underline{y}$.

   b. $P(x) = 4x^3 - 2x^2 - 4x + 7$ and $Q(x) = x^2 + 4x - 1$, two polynomials of the vector space $P_3[x]$ of the real polynomials of degree 3 or less, with the inner product

$$\langle P(x)|Q(x)\rangle = \int_{-2}^{4} P(x)Q(x)\,dx.$$

5.5.2 Orthogonal Projections

5. Construct the three orthogonal projection matrices which in the vector space $\mathbb{R}^3$ will compute the first, second, and third components of a general vector $\underline{x}$.

6. Consider the vector space $P_2[x]$ of the real polynomials of degree 2 or less with the orthonormal vector base $\mathcal{B}$ defined in exercise 3. Compute the decomposition of the following vectors $|P(x)\rangle$ in the bases $\mathcal{B}$.

   a. $P(x) = -4 + 2x$

   b. $P(x) = 6 + x - 4x^2$

   c. $P(x) = -2x + 5x^2$

   d. $P(x) = -6 + 7x^2$

7. Consider the vector space $P_2[x]$ of the real polynomials of degree 2 or less with the natural base $B = \{1, x, x^2\}$ and the following inner product

$$\langle P(x)|Q(x)\rangle = \int_{-1}^{+1} P(x)Q(x)\,dx$$

   a. Compute the inner product of the two polynomials $P(x) = 2 + 3x - x^2$ and $Q(x) = 3 - x + 3x^2$.

b. Compute the decomposition of the vectors $|P(x)\rangle$ and $|Q(x)\rangle$ in the bases $\mathcal{B}$ and $\mathcal{B}'$.

5.5.3 Gram-Schmidt Orthogonalization

8. For each of the set of vectors provided, construct an orthogonal base of the sub-space spanned by these vectors using Gram-Schmidt's algorithm. Use the standard inner product $x^T y$ of $\mathbb{R}^n$:

a. $\underline{x}_0 = \begin{bmatrix} -2 \\ 3 \\ 5 \end{bmatrix}$ and $\underline{x}_1 = \begin{bmatrix} 1 \\ -2 \\ 7 \end{bmatrix}$

b. $\underline{x}_0 = \begin{bmatrix} 1 \\ -7 \\ 2 \end{bmatrix}$, $\underline{x}_1 = \begin{bmatrix} 5 \\ 3 \\ 4 \end{bmatrix}$ and $\underline{x}_2 = \begin{bmatrix} 5 \\ -3 \\ 0 \end{bmatrix}$

c. $\underline{x}_0 = \begin{bmatrix} -3 \\ 2 \\ 8 \\ 3 \end{bmatrix}$ and $\underline{x}_1 = \begin{bmatrix} 7 \\ -3 \\ 0 \\ 1 \end{bmatrix}$

9. Repeat exercise 8 using Gram-Schmidt's orthonormalization algorithm.

5.5.4 QR Decomposition

10. For each of the following matrices, find the reduced QR decomposition. Verify your results by checking the product QR of your two matrices and verifying that the matrix Q is a generalized orthogonal matrix by computing $Q^T Q$.

a. $\begin{bmatrix} -2 & 1 \\ 3 & -2 \\ 5 & 7 \end{bmatrix}$

b. $\begin{bmatrix} 1 & 5 & 5 \\ -7 & 3 & -3 \\ 2 & 4 & 0 \end{bmatrix}$

c. $\begin{bmatrix} -3 & 7 \\ 2 & -3 \\ 8 & 0 \\ 3 & 1 \end{bmatrix}$

11. Repeat exercise 10 by computing the QR decomposition using the Octave `qr` function.

NOTES

1 René Descartes (1596–1650), French philosopher, mathematician and scientist.
2 Paul Adrien Maurice Dirac (1902–1984), English theoretical physicist.
3 Augustin-Louis Cauchy (1789–1857), French mathematician, engineer, and physicist.
4 Karl Hermann Amandus Schwarz (1843–1921), German mathematician.
5 Leopold Kronecker (1823–1891), German mathematician.
6 Jørgen Pedersen Gram (1850–1916), Danish actuary and mathematician.
7 Erhard Schmidt (1876–1959), Baltic German mathematician.
8 Pythagoras of Samos (c. 570 B.C.–c.495 B.C.), Ionian Greek philosopher.
9 As in this particular case, the two vectors $\underline{x}_1$ and $\underline{x}_2$ are normalized, one could simplify the calculations and avoid the denominators, as they all evaluate to one.
10 Adrien-Marie Legendre (1752–1833), French mathematician.
11 The method is named after Jørgen Pedersen Gram (1850–1916), Danish mathematician, and Erhard Schmidt (1876–1959), German mathematician. However, Pierre-Simon Laplace (1749–1827), French mathematician, was already familiar with it.

6 Linear Least Square Regression

Chapter 4 discussed various approaches to solving linear systems of equations. Problems involving systems of n equations and n variables were considered. In this chapter, we encounter scenarios where systems of equations possess more equations than variables. Linear regression, our current focus, serves as an illustrative example of this scenario. Regression problems require adjusting some function to fit a set of data points. Generally, the function sought has fewer adjustable parameters than the number of data points available. Such a problem has no solutions (in the sense that it is, in general, impossible to adjust the function parameters to ensure it passes exactly through all points).

This leads to the formulation of systems of linear equations where there are more equations than unknowns. Except for very particular cases, such systems typically have no solution.

We will develop an approach that allows solving such systems "as well as possible". In a first step, we give a precise mathematical definition to the statement "as well as possible". Our approach implies that a vector $\underline{\hat{x}}$ solves the system of linear equations $A\underline{x} = \underline{b}$ "as well as possible" if the backward error $\|A\underline{\hat{x}} - \underline{b}\|$ is the smallest backward error one can obtain over all possible $\|A\underline{x} - \underline{b}\|$.

In this chapter, we introduce an elegant approach to solving this optimization problem, which we apply to solve regression problems. By formulating our method in an abstract vector space, we will be able to extend this approach to function approximation problems, such as, for example, the Fourier[1] approximation.

6.1 REGRESSION PROBLEMS

The aim of a regression problem is to fit a model "as well as possible" to a set of data points. Consider a set of m data points. We will number these points from (x_1, y_1) to (x_m, y_m). Further, consider a function f containing a_i adjustable parameters as a model. We say that f has n model parameters. It is common to start numbering the model parameters with the index 0. Consequently, our model parameters are $a_0, a_1, \ldots, a_{n-1}$. The reader is invited to pay attention to the numbering of the different quantities defined so far. One source of confusion is not paying enough attention to the numbering conventions used in the algorithms to come. This is particularly important when using libraries or comparing results with other reference works.

DOI: 10.1201/9781003262121-7

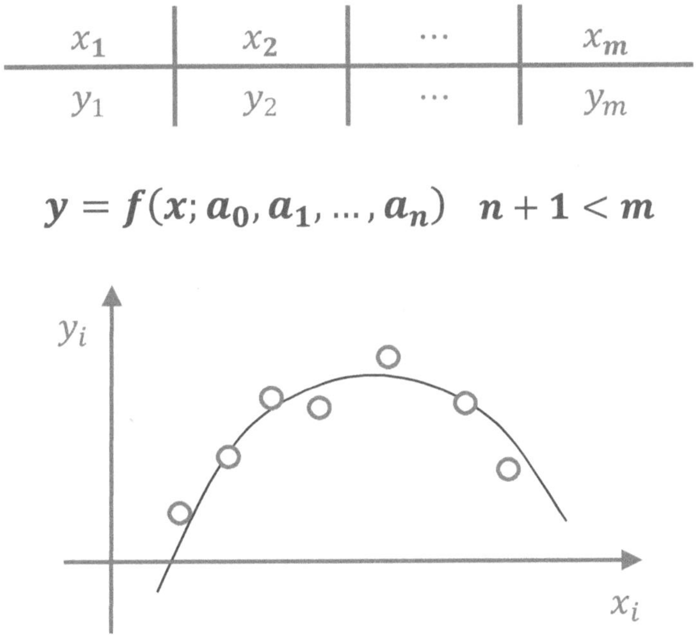

FIGURE 6.1 In a regression problem, the parameters $a_0, a_1, \ldots, a_{n-1}$ of a model $y = f(x; a_0, a_1, \ldots, a_n)$ (continuous line) are tuned in order to represent as well as possible a set of m data points (x_i, y_i) (individual points).

Regressions aim to find model parameters $a_0, a_1, \ldots, a_{n-1}$, such that f represents "as well as possible" the data points (Figure 6.1). In the case of least square regression, "as well as possible" means that the square error is minimal. The square error SE is defined as follows:

Definition 6.1. Square Error

Consider a set of m data points $\{(x_1, y_1), \ldots, (x_m, y_m)\}$ and a model $y = f(x)$. The square error SE is defined as

$$\text{SE} = \sum_{i=1}^{m} \left[f(x_i) - y_i \right]^2 \tag{6.1}$$

and the quantities

$$r_i = f(x_i) - y_i \tag{6.2}$$

are called residual errors, or, in short, residuals.

Since f has parameters a_i that can be freely adjusted, the square error is a function of these model parameters. This function is referred to as the cost function or loss function.

Example 6.1. Square Error

Consider the following set of data points:

x_i	1	−1	1
y_i	2	1	3

and the model

$$f(x) = a_0 + a_1 x \tag{6.3}$$

This model has two parameters that can be chosen. Let us evaluate the square error for two different choices of parameter sets.

In the first case, we choose $a_0 = 1$ and $a_1 = 2$. For this choice, we can compute the model prediction for each data point and its corresponding residual error, as shown in the following table:

x_i	1	−1	1
y_i	2	1	3
$f(x_i)$	3	−1	3
r_i	−1	2	0

Consequently, for this choice, the square error evaluates to

$$SE(1,2) = (-1)^2 + 2^2 + 0^2 = 5 \tag{6.4}$$

In a second scenario, we choose $a_0 = 0$ and $a_1 = 1$. Again, we compute the model predictions and their corresponding residual errors, as shown in the following table:

x_i	1	−1	1
y_i	2	1	3
$f(x_i)$	1	−1	1
r_i	1	2	2

For this choice, the square error evaluates to

$$\mathrm{SE}(0,1) = 1^2 + 2^2 + 2^2 = 9 \tag{6.5}$$

Comparing (6.4) and (6.5), one can conclude that the model for the first choice of parameters fits the data points better than the second choice of parameters. This is visually clear in a figure plotting the data points and the models corresponding to the two choices of the model parameters (Figure 6.2).

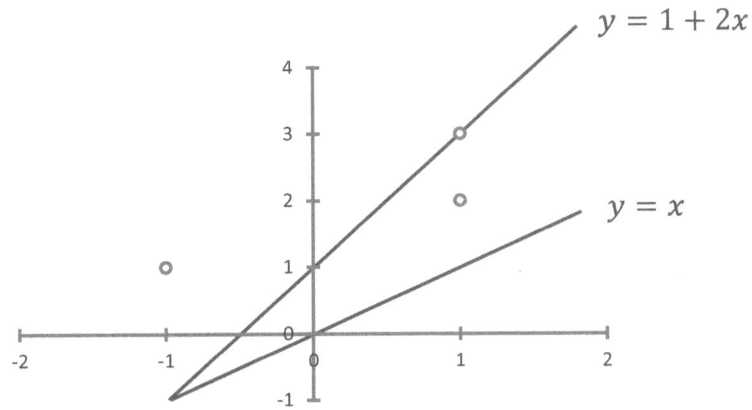

FIGURE 6.2 The data points together with the models corresponding to the two discussed choices for the model parameters.

Considering Figure 6.2, it is not hard to imagine that there must be better choices for the model parameters than the two we proposed.

Example 6.1 provides an initial insight into how model parameters can be objectively adjusted. We must find a set of parameters that minimizes the square error.

Definition 6.2. Least Square Regression

Consider a model $y = f(x)$, having n parameters $a_0, a_1, \ldots, a_{n-1}$, that can be freely adjusted and m points $\{(x_1, y_1), \ldots, (x_m, y_m)\}$, with $m > n$. The parameters $\hat{a}_0, \hat{a}_1, \ldots, \hat{a}_{n-1}$ that minimize the square error are called the optimal parameters in the sense of least square regression.

Different approaches can be developed to find the optimal parameters in the sense of least square regression. One of them, which will be explored in the next section, is to formulate the problem in the form of a set of inconsistent equations (which will be defined in Definition 6.4).

Besides the square error SE, other error measurements are commonly used. Among them are the mean square error MSE and the root mean square error RMSE.

Definition 6.3. Mean Square Error and Root Mean Square Error

Consider a set of m data points $\{(x_1, y_1),\ldots,(x_m, y_m)\}$ and a model $y = f(x)$. The mean square error MSE is defined as

$$\text{MSE} = \frac{1}{m}\sum_{i=1}^{m}\left[f(x_i) - y_i\right]^2 = \frac{\text{SE}}{m} \tag{6.6}$$

and the root mean square error RMSE is defined as

$$\text{RMSE} = \frac{1}{m}\sqrt{\sum_{i=1}^{m}\left[f(x_i) - y_i\right]^2} = \sqrt{\text{MSE}} = \sqrt{\frac{\text{SE}}{m}} \tag{6.7}$$

The mean squared error represents the average squared error across the entire dataset. The root mean squared error is a measure of the typical deviation in each data point. The coming examples will give several practical illustrations.

6.2 GENERAL SOLUTION OF LINEAR LEAST SQUARE REGRESSION

In this section, a general solution for the regression problem for a set of linear equations is derived. The problem is framed as finding the solution to a set of inconsistent equations $A\underline{x} = \underline{b}$. The solution is defined as the vector $\underline{\hat{x}}$ that minimizes the backward error $\|A\underline{x} - \underline{b}\|$.

6.2.1 Definitions

Let us start by defining the concept of inconsistent linear equations.

Definition 6.4. Inconsistent Linear Equations

A system of linear equations having more equations than unknowns is called a system of inconsistent equations.

We number the equations from 1 to m and the unknows from 1 to n. Inconsistent equations are such that $m > n$. Written as a matrix equation:

$$\begin{bmatrix} a_{11} & a_{12} & \cdots & a_{1n} \\ a_{21} & a_{22} & \cdots & a_{2n} \\ \vdots & \vdots & & \vdots \\ a_{m1} & a_{m2} & \cdots & a_{mn} \end{bmatrix} \begin{bmatrix} x_1 \\ x_2 \\ \vdots \\ x_n \end{bmatrix} = \begin{bmatrix} b_1 \\ b_2 \\ \vdots \\ b_m \end{bmatrix} \tag{6.8}$$

Here, we used the same convention as in Chapter 4, where the first index i in the coefficients a_{ij} of the matrix is the number of the equation, and the second index j is the number of the unknown. The reader is invited to take some time to make sure to understand how the various indexes are attributed and how this results indeed in a system of m equations and n unknowns. In particular, note that contrary to Chapter 4, the coefficient matrix is no longer a square matrix. The number of columns corresponds to the number of unknowns x_i, and the number of rows to the number of equations.

Example 6.2. An Inconsistent Equation

To get familiar with the concept of inconsistent equations, we consider an example:

$$\begin{bmatrix} 1 & -1 \\ 1 & 1 \\ 1 & 3 \end{bmatrix} \begin{bmatrix} x_1 \\ x_2 \end{bmatrix} = \begin{bmatrix} 1 \\ 2 \\ 6 \end{bmatrix} \tag{6.9}$$

This is a system with $m = 3$ equations and $n = 2$ unknowns. The reader is invited to verify that this system has no solution (see exercises)

6.2.2 Normal Equations

A system of inconsistent equations, such as Equation (6.8), has no solution in general. However, it is possible to find a vector $\underline{\hat{x}}$ such that the backward error $\left\| A\underline{\hat{x}} - \underline{b} \right\|$ is as small as possible. Throughout the rest of this chapter, unless otherwise specified, we will use the Euclidean norm to evaluate the backward error. When using the Euclidian norm, the vector $\underline{\hat{x}}$ is called a solution in the sense of the least squares.

Definition 6.5. Least Square Solution of Inconsistent Equations

A solution of an inconsistent set of linear equations $A\underline{x} = \underline{b}$, in the sense of least squares, is the vector $\underline{\hat{x}}$, which minimizes the backward error $\left\| A\underline{x} - \underline{b} \right\|$. This solution is called the least square solution of the system $A\underline{x} = \underline{b}$.

Note that in the case of consistent equations (i.e., where the number of equations and unknowns are identical), the solution $\underline{r}$ of $A\underline{x} = \underline{b}$ is also the least squares solution. In this scenario, the backward error is equal to zero. In this sense, Definition 6.5 can be seen as a generalization of the solution of a system of linear equations.

Different approaches exist to find the least square solution. A first approach, which has the advantage of applying it to non-linear equations, too, consists in writing down the backward error as a function of the unknowns $\hat{x}_i$. This function is, in general, called the cost function or loss function $\mathcal{L}$. It is common to use $\mathcal{L} = \|A\underline{x} - \underline{b}\|^2$ instead of $\|A\underline{x} - \underline{b}\|$, as it simplifies the calculations (note that if $\|A\hat{\underline{x}} - \underline{b}\|^2$ is minimal, so will be $\|A\hat{\underline{x}} - \underline{b}\|$):

$$\mathcal{L}(x_1, \ldots, x_n) = \|A\underline{x} - \underline{b}^2\| = \sum_{i=1}^{m} (a_{i1}x_1 + \ldots + a_{im}x_m - b_m)^2 \tag{6.10}$$

The problem in finding the vector $\hat{\underline{x}}$ that minimizes the backward error $\|A\underline{x} - \underline{b}\|$, now becomes a problem of finding the minimum of the loss function $\mathcal{L}$.

A second approach, which we will follow here, avoids the minimization problem of the loss function $\mathcal{L}$ by formulating the problem as a geometrical problem. Both approaches give, in the case of linear equations, the same answer. The calculations in this second approach are, however, simpler and allow us to get familiar with the manipulation of concepts from the theory of vector spaces. This second approach can further be generalized to function approximation, which we will discuss at the end of this chapter.

To explain the approach, let us consider Equation (6.9) from Example 6.2. We start by noting that we can write for any vector $\underline{x} = [x_1\ x_2]^T$:

$$\begin{bmatrix} 1 & -1 \\ 1 & 1 \\ 1 & 3 \end{bmatrix} \begin{bmatrix} x_1 \\ x_2 \end{bmatrix} = \begin{bmatrix} 1 \\ 1 \\ 1 \end{bmatrix} x_1 + \begin{bmatrix} -1 \\ 1 \\ 3 \end{bmatrix} x_2 = x_1\underline{v}_1 + x_2\underline{v}_2 \tag{6.11}$$

where $\underline{v}_1$ and $\underline{v}_2$ are the column vectors formed by the first and second column of the coefficient matrix A. In other words, any vector $A\underline{x}$ will be within the plane spanned by the two vectors $\underline{v}_1$ and $\underline{v}_2$ (Figure 6.3a). This is a general result of the theory of vector spaces. A consequence of Equation (6.11) is that any vector $A\underline{x}$ will belong to the sub-space spanned by the column vectors formed by each column of the matrix A.

Our aim is to find the vector $\hat{\underline{x}}$ which minimizes the quantity $\|A\underline{x} - \underline{b}\|$. Geometrically speaking, this is the vector $\hat{\underline{x}}$, such that $A\hat{\underline{x}}$ is the closest possible to the vector $\underline{b}$. As, in general,[2] the vector $\underline{b}$ is not within the plane spanned by $\underline{v}_1$ and $\underline{v}_2$, the vector $A\hat{\underline{x}}$ which is closest to $\underline{b}$ will be the orthogonal projection of $\underline{b}$ onto the sub-space spanned by $\underline{v}_1$ and $\underline{v}_2$ (Figure 6.3b).

From Theorem 5.3 in Chapter 5, we know that $A\hat{\underline{x}} - \underline{b}$ will be orthogonal to any vector within the plan spanned by $\underline{v}_1$ and $\underline{v}_2$, and, consequently, orthogonal to any vector $A\underline{x}$. This can be formulated as:

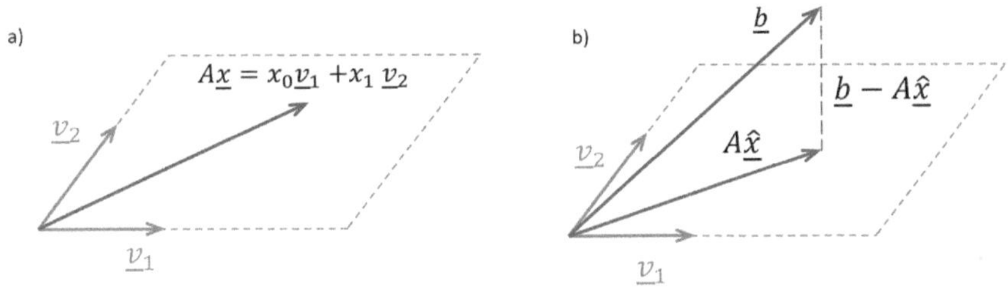

FIGURE 6.3 (a) Any vector $A\underline{x}$ is within the plane spanned by the two vectors $\underline{v}_1$ and $\underline{v}_2$ formed by the first, respectively second, column of the matrix A. (b) The vector $\hat{\underline{x}}$ minimizing $\|A\underline{\hat{x}} - \underline{b}\|$ is such that $A\underline{\hat{x}}$ is the orthogonal projection of the vector $\underline{b}$ onto the plane spanned by the vectors $\underline{v}_1$ and $\underline{v}_2$.

$$\left(A\underline{x}\right)^T \cdot \left(A\underline{\hat{x}} - \underline{b}\right) = 0 \tag{6.12}$$

Rearranging this equation leads to

$$\underline{x}^T A^T \left(A\underline{\hat{x}} - \underline{b}\right) = 0 \tag{6.13}$$

In order for Equation (6.13) to be valid for any vector $\underline{x}$, the following condition must hold:

$$A^T \left(A\underline{\hat{x}} - \underline{b}\right) = 0 \tag{6.14}$$

which can be rearranged as:

$$A^T A\underline{\hat{x}} = A^T\underline{b} \tag{6.15}$$

Equations (6.15) are known as the normal equations. The name stems from the fact that some vectors must be orthogonal (normal) to each other. The reader is invited to verify that Equations (6.15) are, in fact, n linear equations for n unknowns. Their solution gives the vector $\underline{\hat{x}}$ which minimizes the backward error of the system $A\underline{x} = \underline{b}$.

Theorem 6.1. Normal Equations

The vector $\underline{\hat{x}}$ that minimizes the backward error $\|A\underline{x} - \underline{b}\|$ of a linear system of equations $A\underline{x} = \underline{b}$, is given by the solution of the equations:

$$A^T A\underline{\hat{x}} = A^T\underline{b} \tag{6.16}$$

Equations (6.16) are known as the normal equations.

In case the system has the same number of equations and unknowns, the normal equations lead to the solution $\underline{r}$ of the system $A\underline{x} = \underline{b}$ (see exercises).

Note that the matrix A^TA, which is a $n \times n$ matrix, is a symmetric matrix. Indeed, we have:

$$\left(A^TA\right)^T = A^T(A^T)^T = A^TA \tag{6.17}$$

which proves that matrix A^TA is symmetric. This is useful in practical applications, as it allows checking the calculations.

Example 6.3. Normal Equations

Find the vector $\hat{\underline{x}}$ that minimizes the backward error $\|A\underline{x} - \underline{b}\|$ of a linear system of Equations (6.9) from Example 6.2.

To find $\hat{\underline{x}}$, we have to solve the normal equations. Using Octave, we first define the coefficient matrix A and right-hand-side vector $\underline{b}$ of our system:

```
>> A = [1 -1; 1 1; 1 3];
>> b = [1; 2; 6];
```

Note that, indeed, A^TA is a symmetric matrix:

```
>> A'*A
ans =
    3   3
    3  11
```

To find $\hat{\underline{x}}$, we solve the normal equation:

```
>> x=A'*A\A'*b
ans =
    1.7500
    0.2500
```

The backward error we obtain is the smallest possible backward error one can get for any vector $\underline{x}$:

```
>> norm(A*x-b)
ans = 1.2247
```

The backward error is not zero, which means that not all equations can be satisfied simultaneously. This is expected for an inconsistent equation.

6.2.3 Least Square Regression With the Normal Equations

With the aid of the normal equations, conducting least square fitting for linear models transforms into the task of solving a set of linear equations. The main property to understand here is that minimizing the square error SE of a model for a given set of data points is equivalent to minimizing the backward error $\|A\underline{x}-\underline{b}\|$ of the set of inconsistent equations derived by applying the model to each data point individually. This statement comes from the fact that the square error SE is, in fact, equal to the backward error squared $\|A\underline{x}-\underline{b}\|^2$ Consequently, both minimization problems lead to the same solution.

Algorithm 6.1 summarizes the approach to follow.

Algorithm 6.1. Least Square Regression Optimal Model Parameters

Consider a linear model having n parameters $a_0, a_1, \ldots, a_{n-1}$ that can be freely adjusted and m points $\{(x_1, y_1), \ldots, (x_m, y_m)\}$, with $m > n$. The most optimal parameters $\hat{a}_0, \hat{a}_1, \ldots, \hat{a}_{n-1}$, in the sense of least square regression, can be found as follows:

1. Using the data set and the model, write down all equations that should be satisfied. These equations will, in general, form a set of inconsistent equations
2. Find the least square solution of these equations

Let us illustrate this method with an example.

Example 6.4. Linear Least Square Regression

Consider the following set of data points:

x_i	1	−1	1	2
y_i	2	1	3	3

and the model

$$f(x) = a_0 + a_1 x \tag{6.18}$$

This model has two parameters that have to be adjusted in the sense of least square regression, i.e., chosen in such a way that the square error is minimal.

In the first step, we write down all equations that result from applying model (6.18) to the set of data points:

$$\begin{cases} a_0 + a_1 & = & 2 \\ a_0 - a_1 & = & 1 \\ a_0 + a_1 & = & 3 \\ a_0 + 2a_1 & = & 3 \end{cases} \tag{6.19}$$

Or, in a more compact form as a matrix equation:

$$\begin{bmatrix} 1 & 1 \\ 1 & -1 \\ 1 & 1 \\ 1 & 2 \end{bmatrix} \begin{bmatrix} a_0 \\ a_1 \end{bmatrix} = \begin{bmatrix} 2 \\ 1 \\ 3 \\ 3 \end{bmatrix} \tag{6.20}$$

Consider the coefficient matrix A. The first column contains only entries equal to 1. This is because the unknown a_0 is always multiplied by 1, regardless of the data points. The second column is made out of the x values of the data points. This comes from the fact that the unknown a_1 is always multiplied by the x value of the data point. Finally, the right-hand side vector is made of the y values of the data points. Understanding this structure allows for the rapid obtaining of the coefficient matrix A.

Finding the two parameters $\hat{a}_0$ and $\hat{a}_1$ that minimize the square error is equivalent to minimizing the backward error of the system (6.20). This can be achieved by solving the normal equations. Using Octave:

```
>> A = [1 1; 1 -1; 1 1; 1 2];
>> b = [2; 1; 3; 3];
>> A'*A\A'*b
ans =
 1.7368
 0.6842
```

Algorithm 6.1 applies to general linear models such as a model like this:

$$y = a_0 + a_1 g_1(x) + a_2 g_2(x) + \ldots a_{n-1} g_{n-1}(x) \tag{6.21}$$

which is a model made out of a linear combination of $n-1$ functions g_i. Each of these functions can be a non-linear expression. They are often referred to as support or basis functions. This model has n parameters a_i that can be selected in the sense of least square regression. Following Algorithm 6.1, we determine the set of inconsistent equations that result from applying the model to the set of data points. If m points $\{(x_1, y_1), \ldots, (x_m, y_m)\}$, with $m > n$, are available, these m equations will, in matrix form, be written as:

$$\begin{bmatrix} 1 & g_1(x_1) & g_2(x_1) & \cdots & g_{n-1}(x_1) \\ 1 & g_1(x_2) & g_2(x_2) & \cdots & g_{n-1}(x_2) \\ 1 & \vdots & \vdots & \vdots & \vdots \\ 1 & g_1(x_m) & g_2(x_m) & \cdots & g_{n-1}(x_m) \end{bmatrix} \begin{bmatrix} a_0 \\ a_1 \\ \vdots \\ a_{n-1} \end{bmatrix} = \begin{bmatrix} y_1 \\ y_2 \\ \vdots \\ y_m \end{bmatrix} \tag{6.22}$$

From this point on, one uses the method of the normal equations to find the solution of the system (6.22) that has the smallest possible backward error.

Example 6.5. Polynomial Regression

In this example, we consider the following set of points:

x_i		−1	4	2
y_i	0	10	19	4

and the polynomial model

$$f(x) = a_0 + a_1 x + a_2 x^2 \tag{6.23}$$

This model has three parameters and is a model formed of linear combinations of the two functions $f_1(x) = x$ and $f_2(x) = x^2$. Using Octave, constructing and solving the normal equations becomes simple. We start by defining the data points:

```
>> x = [1; -1; 4; 2];
>> y = [0; 10; 19; 4];
```

The coefficient matrix A of the inconsistent equations can be computed like this:

```
>> A = [x.^0 x x.^2]
A =
    1  1  1
    1 -1  1
    1  4 16
    1  2  4
```

Consider matrix A a: each column (with the exception of the first column whose entries all equal 1) is built by applying f_i to the x_i. To construct the first column, which is made of only ones, we use a trick and implement it as `x.^0`. The resulting matrix belongs to the family of the Vandermond matrices, and as we noted in Chapter 4, these matrices often have a high conditioning number. As a result, solving the normal equations becomes challenging from a numerical point of view. We will come back to this problematic in the next section.

Finding the parameters $\hat{a}_i$ that minimize the backward error is achieved by solving the normal equations:

```
>> a = A'*A\A'*y
a =
   3.3397
  -4.3654
   2.0833
```

We can compute the residual errors like this:

```
>> r = A*a-y
r =
   1.0577
  -0.2115
   0.2115
  -1.0577
```

This shows that, as expected, the equations cannot be all satisfied at the same time. However, these are the smallest errors we can obtain in the sense of the least squares. For example, the root mean square error RMSE is equal to

```
>> RMSE = norm(r)/length(r)
RMSE = 0.3814
```

where we used the `norm()` function of Octave to compute the Euclidian norm of the residual error, which is, in fact, the square root of the square error SE. The `length()` function gives the number of elements in a vector, which, in our case, is the number of data points m.

6.2.4 QR Decomposition for Solving the Normal Equations

The method of the normal equations is a very elegant way to solve the problem of least square regression. From a practical point of view, despite its apparent simplicity in applying it, there is a hidden pitfall in the method. Unfortunately, the calculation of the matrix $A^T A$, often leads to a matrix with a high condition number. This leads to a solution of the normal equations that is very sensitive to round-off errors.

There exist algorithms that approach the least square regression while avoiding the calculation of the matrix $A^T A$. One such method is based on the *QR* decomposition (see Chapter 5) of the matrix A and is based on the following theorem:

Theorem 6.2. Least Square Linear Regression Using QR Decomposition

The vector $\hat{\underline{x}}$ that minimizes the backward error $\|A\underline{x}-\underline{b}\|$ of a linear system of equations $A\underline{x}=\underline{b}$, is given by the solution of the equations:

$$R\underline{a}=Q^{T}\underline{y} \tag{6.24}$$

where R and Q are the matrices found by QR decomposition of the matrix A.

We will give the proof of Theorem 6.2 later after we introduce least square regression in abstract vector spaces. The solution found using the approach of Theorem 6.2 is often called the *minimum norm solution* for reasons that will become apparent once we outline its proof.

Example 6.6. Linear Regression Using QR Decomposition

Consider the following set of data points:

x_i	1.0	2.0	3.0	4.0
y_i	3.2	4.3	4.7	5.9

and the model

$$y=a_1+a_2x \tag{6.25}$$

Let us adjust the model to the data points, in the sense of the least squares, by using QR decomposition.

We start by defining the data points and the coefficient matrix A of the set of inconsistent equations resulting from applying the model to the data points:

```
>> x = [1; 2; 3; 4];
>> y = [3.2; 4.3; 4.7; 5.9];
>> A = [x.^0 x];
```

Next, we compute the QR decomposition of the coefficient matrix A:

```
>> [Q, R] = qr(A, 0);
```

And, finally, solve Equation (6.24) to find the parameters $\hat{a}_i$ of the model in the sense of the least squares:

```
>> a = R\Q'*y
a =
     2.4000
     0.8500
```

Octave solves inconsistent equations while avoiding the formation of the matrix $A^T A$ with the left division operator. For our example, this is written:

```
>> A\y
ans =
     2.4000
     0.8500
```

6.3 LINEARIZATION OF MODELS

The method of the normal equations applies to linear models only. There are cases where, with an appropriate change of variable, it is possible to transform a non-linear model into a linear model. We present here as illustrations two examples.

6.3.1 Exponential Model

Let us consider the exponential model. The general form of an exponential model is given by:

$$y = a_0 e^{a_1 x} \tag{6.26}$$

It has two parameters a_0 and a_1. It is possible to convert this non-linear model into a linear one by applying the logarithmic function on both sides of the equation:

$$\log y = \log a_0 + a_1 x \tag{6.27}$$

The following example illustrates this approach.

Example 6.7. Exponential Model

Several attempts exist to estimate the number of websites present on the World Wide Web. The table that follows shows a summary according to Internet Live Stats.[3] Note that the definition of websites is not as simple as one may think. The interested reader is invited to read the methodology followed by Internet Live Stats to gather their data.

Figure 6.4 displays the data on a graph. We observe that, on a logarithmic scale, the trend is linear. This justifies the usage of an exponential model to fit the data.

Year	Number of websites (in millions)
2000	17
2001	29
2002	39
2003	41
2004	52
2005	65
2006	86
2007	122
2008	172
2009	238
2010	207
2011	346
2012	697
2013	673
2014	969
2015	863
2016	1046
2017	1767
2018	1630

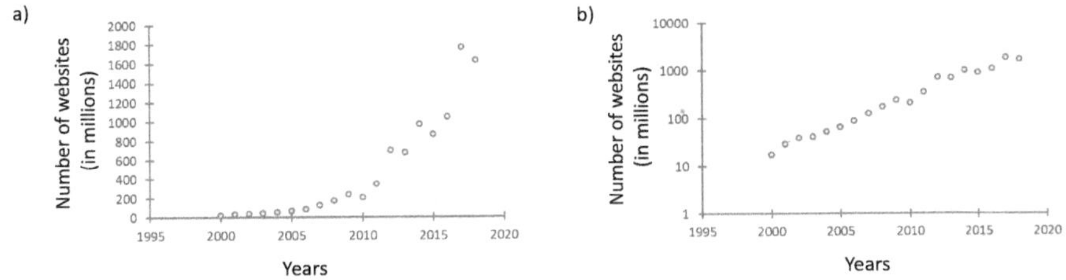

FIGURE 6.4 (a) Number of websites from the years 2000 to 2018 according to Internet Live Stats. (b) Same data on a logarithmic scale.

In order to find the model parameters, we start by providing the data and defining the coefficient matrix A of the set of inconsistent equations resulting from applying model (6.27) to the data points (for space reasons, we do not show the full code when defining x and y):

```
>> x = [2000; 2001; . . . ; 2017; 2018];
>> y = [17; 29; . . . ; 1767; 1630];
>> A = [(x-2000).^0 x-2000];
```

We chose to count the years since 2000 when building the coefficient matrix A. This has the advantage of providing a lower conditioning number for matrix $A^T A$ (480 instead of $5\,10^{11}$ when using x directly). It is, in general, a good idea to keep the numerical values in as similar ranges as possible to minimize the effect of round-off errors. Recall that the first column of A is made of ones. By using x-2000 instead of x, we obtain a second column that is more similar to the first one.

The model parameters are estimated by solving the normal equations. Note that we have to apply the logarithmic function (any base for the logarithmic function can be used; in our case, we used the natural logarithmic function) to the y values:

```
>> a = A\log(y)
a =
    2.9974
    0.2593
```

We can verify on a graph that model (6.27) reproduces well the data set (Figure 6.5a):

```
>> plot(x-2000, log(y), x-2000, A*a)
```

Similarly, the residuals can be displayed (Figure 6.5b):

```
>> r = A*a-log(y);
>> plot(r, 'o')
```

Note that the fitted model predicts $\log y$ and not y, which explains the way we computed the residuals.

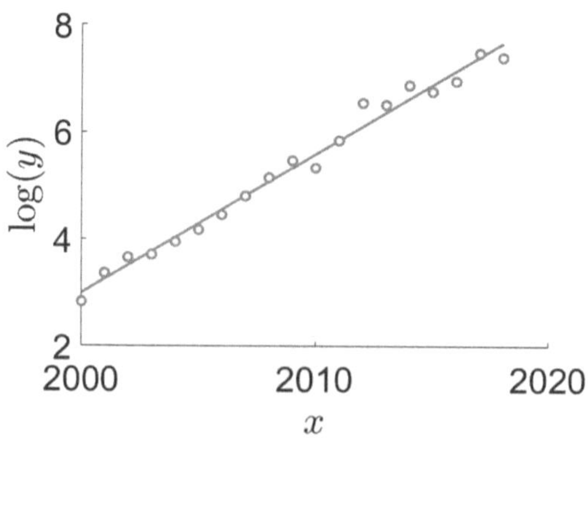

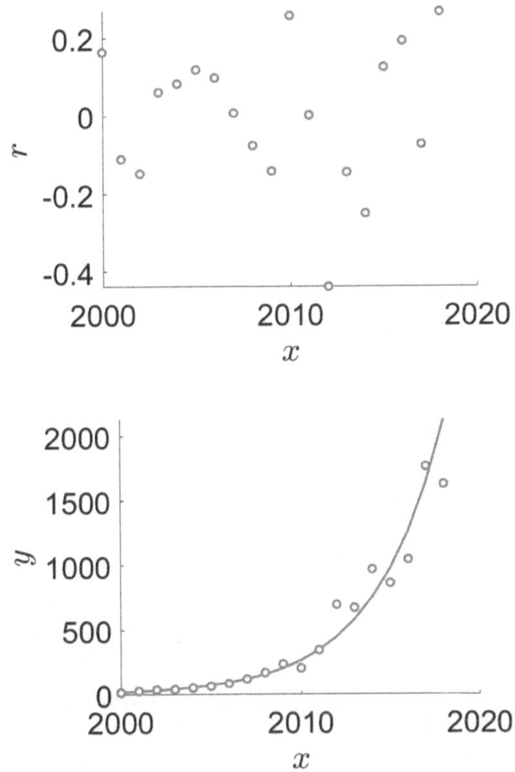

FIGURE 6.5 (a) Transformed data set and fitted model, (b) residuals computed using the transformed data, (c) original data with fitted model.

Plotting the original data and the fitted model (6.26) can be done as follows (Figure 6.5c):

```
>> plot(x, y, x, exp(A*a))
```

The residuals of the model fitted to the original data become

```
>> r = exp(A*a)-y;
```

With the root mean square error RMSE:

```
>> RMSE = norm(r)/length(r)
RMSE = 0.040516
```

To find the actual model parameters, we must convert back `a(1)` to a_0 by computing the inverse of the logarithmic transformation. In our case, as we used the natural logarithmic function, this requires the usage of the exponential function:

```
>> exp(a(1))
ans = 20.033
```

In summary, our model writes:

$$y = 20.033e^{0.2593(x-2000)} \tag{6.28}$$

where x is the calendar year.

From the parameter a_1, one can compute the doubling rate (i.e., the time it takes the number of websites to double):

```
>> log(2)/a(2)
ans = 2.6737
```

In average, about every 2.7 years, the number of websites doubles.

6.3.2 Power Law

Consider the power law model. Such a model takes the following general form:

$$y = a_0 x^{a_1} \tag{6.29}$$

It has two parameters a_0 and a_1. It is possible to convert this non-linear model into a linear one by applying the logarithmic function on both sides of the equation:

$$\log y = \log a_0 + a_1 \log x \tag{6.30}$$

The following example illustrates this approach.

Example 6.8. Power Law—Kepler's Third Law

A famous example of a power law model is Kepler's third law of planetary motion. The following table shows data that was available to Kepler[4] in 1618 (converted into modern units):

A log-log plot of the mean distance from the sun versus the period of revolution (Figure 6a) suggests a power-law type of relation between these two variables.

In order to find the parameters of a power-law model (6.29), we start by providing the data and defining the coefficient matrix A of the set of inconsistent equations resulting from applying the model (6.30) to the data points:

Planet	Mean Distance from the Sun (Astronomical Units AU)	Period of Revolution around the Sun (Days)
Mercury	0.389	87.77
Venus	0.724	224.7
Earth	1	365.25
Mars	1.524	686.95
Jupiter	5.2	4332.62
Saturn	9.51	10759.2

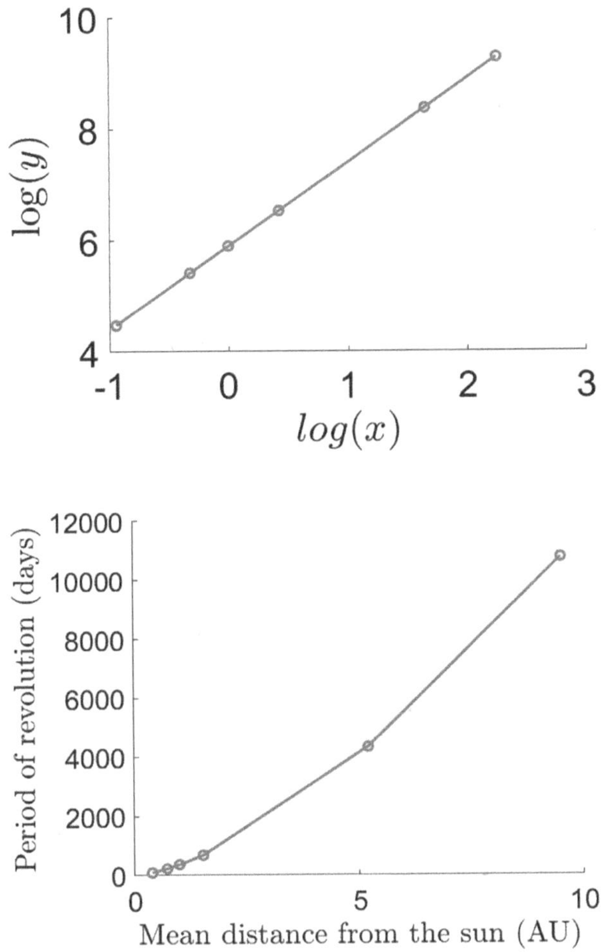

FIGURE 6.6 (a) Log-log plot of the mean distance x between the sun and a planet versus the period of revolution y of the planet (o) together with the fitted model (–). (b) Original data (o) and fitted model (–).

```
>> x = [0.389; 0.724; 1; 1.524; 5.2; 9.51];
>> y = [87.77; 224.7; 365.25; 686.95; 4332.62;
10759.2];
>> A = [x.^0 log(x)];
```

The model parameters are estimated by solving the normal equations. Note that we have to apply the logarithmic function to the y values as we did for the x values in the coefficient matrix A:

```
>> a = A\log(y)
a =
 5.8979
 1.5031
```

We can verify on a graph that the model (6.30) reproduces well the data set (Figure 6a):

```
>> plot(log(x), log(y), log(x), A*a)
```

The original data and the fitted model (6.29) can be plotted as follows (Figure 6b)

```
>> plot(x, y, x, exp(A*a))
```

In both cases, the fit is of excellent quality. This is confirmed by calculating the residuals of the model fitted to the original data:

```
>> r = exp(A*a)-y;
```

With the root mean square error RMSE:

```
>> RMSE = norm(r)/length(r)
RMSE = 1.5567
```

This means that the model estimates the period of revolution of a planet with an error in the order of 1.5 days.

To find the actual model parameters, we must convert back `a(1)` to a_0 by computing the inverse of the logarithmic transformation. In our case, as we used the natural logarithmic, this requires the usage of the exponential function:

```
>> exp(a(1))
ans = 364.27
```

In summary, the model writes:

$$y = 364.27x^{1.5031} \tag{6.31}$$

Where x is the mean distance from the sun in astronomical units and y is the period of revolution around the sun of the planet in days. This result is very close to the values expected from Newton's theory of gravity ($a_0 = 365.25$ and $a_1 = 1.5$).

6.4 ABSTRACT NORMAL EQUATIONS

In Section 6.2, we learned how to use the approach of the normal equations to solve linear regression problems. In this section, we take a further step of abstraction to approach this problem by solving it in a real-valued vector space V. These developments will serve as the starting point for function approximations discussed in Section 6.5. We will conclude this section by providing the proof of Theorem 6.2.

6.4.1 Normal Equations in an Abstract Vector Space

The presentation in this section closely mirrors that of Section 6.2.2, and readers are encouraged to compare the two to help the understanding of the abstract derivations.

Consider a given vector $|y\rangle$. of a real-valued vector space V of dimension m. Consider $n < m$ given vectors $\left|x^{(i)}\right\rangle$ of this vector space, which we number from 1 to n. We want to find real numbers a_i such that

$$|y\rangle = \sum_{i=1}^{n} a_i \left|x^{(i)}\right\rangle \tag{6.32}$$

Let us have a closer look at Equation (6.32). The vector $\sum_{i=1}^{n} a_i \left|x^{(i)}\right\rangle$ is a linear combination of the vectors $\left|x^{(i)}\right\rangle$. This vector is consequently in the sub-space spanned by n vectors $\left|x^{(i)}\right\rangle$. Unless the vector $|y\rangle$ is in that sub-space, Equation (6.32) has no solution. It represents a set of inconsistent equations.

Generally, Equation (6.32) will not have a solution. Therefore, we search for a set of numbers $\hat{a}_i$ that minimize the norm $\left\| |y\rangle - \sum_{i=1}^{n} \hat{a}_i \left|x^{(i)}\right\rangle \right\|$. A question immediately arises. Which norm should we use? In the following, we will use induced norms:

$$\|x\| = \sqrt{\langle x|x\rangle} \tag{6.33}$$

We will use the concept of orthogonality to solve this minimization problem. This remark is important as we will use the idea of orthogonality to find the solution to this minimization problem. The choice (6.33) as the norm leads to least square regression. Using other norms would lead to other problems, such as the min-max problem if one uses the infinity norm.

Intuitively, we want to find numbers $\hat{a}_i$ such that the vector $\sum_{i=1}^{n} \hat{a}_i \left| x^{(i)} \right\rangle$ is as close as possible to the vector $\left| y \right\rangle$. This will be the case if the vector $\left| y \right\rangle - \sum_{i=1}^{n} \hat{a}_i \left| x^{(i)} \right\rangle$ is as short as possible. Based on Theorem 5.3 in Chapter 5, we have to look for a vector that is orthogonal to any vector of the sub-space spanned by the n vectors $\left| x^{(i)} \right\rangle$. Note the importance of using an induced norm at this stage. It is because we use an induced norm that we can use the idea of orthogonality.

The condition of orthogonality can be translated by the equation stating that the inner product (the one that is used to define our induced norm) between the vector $\left| y \right\rangle - \sum_{i=1}^{n} \hat{a}_i \left| x^{(i)} \right\rangle$ and a general vector of the sub-space spanned by the vectors $\left| x^{(i)} \right\rangle$ is equal to zero:

$$\left(\sum_{k=1}^{n} a_k\, x^{(k)} \middle| \left(\left| y \right\rangle - \sum_{i=1}^{n} \hat{a}_i \left| x^{(i)} \right\rangle \right) \right) = 0 \quad \text{for any numbers } a_k \tag{6.34}$$

In order for Equation (6.34) to hold for any number a_k, indexes $k = 1, 2, \ldots, n$ must satisfy the following equations:

$$a_k \left\langle x^{(k)} \middle| \left(\left| y \right\rangle - \sum_{i=1}^{n} \hat{a}_i \left| x^{(i)} \right\rangle \right) \right) = 0 \quad \text{for any numbers } a_k \tag{6.35}$$

Equation (6.35) must hold for any number a_k. Consequently, it follows that:

$$\left\langle x^{(k)} \middle| y \right\rangle = \sum_{i=1}^{n} \hat{a}_i \left\langle x^{(k)} \middle| x^{(i)} \right\rangle \; k = 1, 2, \ldots, n \tag{6.36}$$

Equations (6.36) are the *abstract normal equations*. The name comes from the fact that the key idea behind them is the orthogonal (normal) conditions between vectors.

The n Equations (6.36) are actually the same equations as the normal Equations (6.16). As we will show shortly, the normal Equations (6.16) are the matrix representation of the Equations (6.36). Formulating the normal equations in the form (6.36) rather than in a matrix Equation (6.16) allows for function approximations. This will be discussed in Section 6.5.

To understand the equivalence between Equations (6.16) and (6.36), we use the representation of the various abstract vectors of V in $\mathbb{R}^m$. For this, we introduce $\mathcal{B} = \{ |e_1\rangle, |e_2\rangle, \ldots, |e_m\rangle \}$, an ordered orthonormal base of the vector space V. It is important to choose an orthonormal base; otherwise, there will be no match between the inner product of V and $\mathbb{R}^m$ (i.e. $\langle x | y \rangle = \underline{x}^T \underline{y}$). We use the standard representation of these base vectors. In this representation, the vector $|y\rangle$ is represented by the column vector

$$|y\rangle := \begin{bmatrix} y_1 \\ \vdots \\ y_k \\ \vdots \\ y_m \end{bmatrix} \tag{6.37}$$

where the numbers y_k are the components of $|y\rangle$ in the base $\mathcal{B}$. In the same way, we represent each of the n vectors $|x^{(i)}\rangle$ by a column vector $\underline{x}^{(i)}$ from $\mathbb{R}^m$:

$$|x^{(i)}\rangle := \underline{x}^{(i)} = \begin{bmatrix} x_1^{(i)} \\ \vdots \\ x_k^{(i)} \\ \vdots \\ x_m^{(i)} \end{bmatrix} \tag{6.38}$$

Note that we use a subscript to number these column vectors and an index to number the coordinates.

Using the representation of our vectors by elements of $\mathbb{R}^m$, Equation (6.32) is written as

$$\underline{y} = A\underline{a} \tag{6.39}$$

by introducing the matrix $A \in \mathbb{R}^{m \times n}$ constructed by combining all column vectors $\underline{x}^{(i)}$:

$$A = \left[\underline{x}^{(1)}, \underline{x}^{(2)}, \ldots, \underline{x}^{(n)}\right] \tag{6.40}$$

and the column vector $\underline{a} = \left[a_1, a_2, \ldots a_n\right]^T$.

Equations (6.39) are, as Equation (6.32), a set of inconsistent equations. The vector $\underline{\hat{a}}$ that minimizes the backward error $\left\|\underline{y} - A\underline{a}\right\|$, is given by the solution of the normal Equations (6.36). To find their matrix representation, we start by writing the n Equations (6.36) using the representations in $\mathbb{R}^m$ (note the importance that we did chose an orthonormal base for the representation of the vectors in $\mathbb{R}^m$, which allows us to match the inner products of V and $\mathbb{R}^m$):

$$\left(\underline{x}^{(k)}\right)^T \underline{y} = \sum_{i=1}^{n} \hat{a}_i \left(\underline{x}^{(k)}\right)^T \underline{x}^{(i)} \quad k = 1, 2, \ldots, n \tag{6.41}$$

Using definition (6.40) of matrix A, these equations can be further written as

$$A^T \underline{y} = A^T A \underline{a} \tag{6.42}$$

which is, indeed, Equation (6.16). One says that Equation (6.42) is the matrix representation of the abstract normal Equations (6.36).

An interesting consequence is that the matrix elements of the matrix $A^T A$ are given by

$$\left(A^T A\right)_{ij} = \left\langle x^{(i)} \middle| x^{(j)} \right\rangle = \left\langle x^{(j)} \middle| x^{(i)} \right\rangle = \left(\underline{x}^{(i)}\right)^T \underline{x}^{(j)} = \left(\underline{x}^{(j)}\right)^T \underline{x}^{(i)} \tag{6.43}$$

Equation (6.43) shows that if the vectors $\left|x^{(i)}\right\rangle$ are orthogonal to each other, then the matrix $A^T A$ is diagonal, and the normal equations are simple to solve. In this case, the solutions are:

$$\hat{a}_i = \frac{\left\langle x^{(i)} \middle| y \right\rangle}{\left\langle x^{(i)} \middle| x^{(i)} \right\rangle} \tag{6.44}$$

This property will be exploited in the field of function approximations in Section 6.5.

6.4.2 Linear Regression by Orthogonalization

In this section, a proof of Theorem 6.2 is given. The reader may skip this section in the first reading. In Section 6.4.1, we formulated the problem of linear regression as a geometrical problem in a real-valued vector space V. The problem consists in decomposing a vector $|y\rangle$ of V according to n linearly independent vectors $\left|x^{(i)}\right\rangle$ as "good as possible". "As well as possible" was defined as finding the decomposition, not of the vector $|y\rangle$, but of its orthogonal projection $P_U|y\rangle$ on the sub-space U spanned by the n linearly independent vectors $\left|x^{(i)}\right\rangle$. This resulted in a problem of solving the abstract normal Equations (6.36). One issue with the normal equations is that they often result in a system of linear equations with a high conditioning number. However, the solutions of the normal equations would be particularly simple to find if the n vectors $\left|x^{(i)}\right\rangle$ would form an orthonormal base. A way to avoid the high-condition number resulting from the formation of the matrix $A^T A$, consists in first decomposing the projection $P_U|y\rangle$ onto an orthonormal base $\left\{\left|z^{(1)}\right\rangle, \ldots, \left|z^{(n)}\right\rangle\right\}$ of the sub-space U:

$$|\hat{y}\rangle = P_U|y\rangle = \sum_{i=1}^{n} \hat{a}_i \left|x^{(i)}\right\rangle = \sum_{i=1}^{n} \hat{b}_i \left|z^{(i)}\right\rangle \tag{6.45}$$

Using the base $\left\{\left|z^{(1)}\right\rangle, \ldots, \left|z^{(n)}\right\rangle\right\}$ results in a system of equations that is diagonal, and their solutions are straightforward to write down:

$$\hat{b}_i = \langle z^{(i)} | y \rangle \tag{6.46}$$

We solved the problem of linear regression without passing by a system of equations with high conditioning numbers. However, we calculate the numbers $\hat{b}_i$ and not the numbers $\hat{a}_i$ that we actually aim to find. Luckily, it is possible to calculate the numbers $\hat{a}_i$ from the numbers $\hat{b}_i$. To find the relations existing between $\hat{a}_i$ and $\hat{b}_i$, we start by observing that for every vector $|x^{(i)}\rangle$, its orthogonal projection on the sub-space U is equal to itself. Using the fact that $\{|z^{(1)}\rangle, ..., |z^{(n)}\rangle\}$ is an orthonormal base of U, we can write down the orthogonal projection as:

$$|x^{(i)}\rangle = P_U |x^{(i)}\rangle = \sum_{j=1}^{n} \langle z^{(j)} | x^{(i)} \rangle |z^{(j)}\rangle \tag{6.47}$$

With this observation, we can rewrite Equation (6.45) as

$$|\hat{y}\rangle = \sum_{i=1}^{n} \hat{a}_i |x^{(i)}\rangle = \sum_{i=1}^{n} \hat{a}_i \sum_{j=1}^{n} \langle z^{(j)} | x^{(i)} \rangle |z^{(j)}\rangle = \sum_{j=1}^{n} \hat{b}_j |z^{(j)}\rangle \tag{6.48}$$

from which it follows that

$$\hat{b}_j = \sum_{i=1}^{n} \hat{a}_i \langle z^{(j)} | x^{(i)} \rangle \tag{6.49}$$

Solving this system of Equations (6.49) allows us to calculate the numbers $\hat{a}_i$ from the numbers $\hat{b}_i$. It turns out that the system of Equations (6.49) is simple to solve as it is an upper diagonal system. To understand why Equations (6.49) form an upper diagonal system, let us rewrite it in a matrix form:

$$\underline{\hat{b}} = Q^T A \underline{\hat{a}} \tag{6.50}$$

with the matrix A formed in the usual way as

$$A = \left[\underline{x}^{(1)}, \underline{x}^{(2)}, \dots, \underline{x}^{(n)}\right] \tag{6.51}$$

and the matrix Q formed in a similar way, but using the vectors $|z^{(i)}\rangle$:

$$Q = \left[\underline{z}^{(1)}, \underline{z}^{(2)}, \dots, \underline{z}^{(n)}\right] \tag{6.52}$$

For both matrices, we chose an orthonormal base $\mathcal{B} = \{|e_1\rangle, |e_2\rangle, ..., |e_m\rangle\}$ of the vector space V to create the representations in $\mathbb{R}^m$ of the various vectors. The vectors $\underline{\hat{a}}$ and $\underline{\hat{b}}$ are the column vectors of $\mathbb{R}^n$, regrouping the numbers $\hat{a}_i$ and $\hat{b}_i$.

With these notations, and using the QR decomposition of the matrix A, we realize that Equations (6.49) and Equation (6.50) are, in fact,

$$\underline{\hat{b}} = Q^T A\underline{\hat{u}} = R\underline{\hat{a}} \tag{6.53}$$

Since $A = QR$, it follows that $Q^T A = R$ as $Q^T Q = I_n$ (all vectors $\underline{z}^{(i)}$ being orthonormal). As R is an upper diagonal matrix, the system (6.49) is indeed an upper diagonal system that can be solved by backward substitution.

We can write Equation (6.46) in matrix notation as

$$\underline{\hat{b}} = Q^T \underline{y} \tag{6.54}$$

Combining Equations (6.53) and (6.54) outlines the steps to calculate the numbers $\hat{a}_i$ that give a decomposition of the vector $|y$ according to the n linearly independent vectors $|x^{(i)}$ in the best way possible (in the sense of the least squares). This result is summarized in Theorem 6.2.

6.5 APPROXIMATION OF FUNCTIONS

All examples of linear regression discussed so far were examples from a vector space of finite dimension and used the matrix formulation (6.16) of the normal equations. On the other hand, the abstract normal Equations (6.36) derived in Section 6.4.1 can be applied to any vector space, such as the vector space of real functions. This allows the development of the theory of function approximations. We will discuss two examples: approximating functions with polynomials and approximating functions with Fourier polynomials.

6.5.1 Polynomial Function Approximation

The problem consists of finding, for a given real function f, a polynomial P_{n-1} of degree $n-1$ such that the error E over an interval $[a, b]$ is minimal. We define the error E as

$$E = \int_a^b \left[f(x) - P_{n-1}(x)\right]^2 dx \tag{6.55}$$

This problem can be formulated in the framework of abstract normal equations. Therefore, we work in the vector space V of the real functions with the standard inner product:

$$\langle f | g \rangle = \int_a^b f(x) g(x) dx \tag{6.56}$$

Let us define the vector $|y\rangle$ as the function f, i.e., $|y\rangle = |f\rangle$, and the vectors $|x^{(i)}\rangle$ as the monomials x^{i-1}, i.e., $|x^{(i)}\rangle = |x^{i-1}\rangle$. The reader is invited to pay attention to the used notations and avoid confusion between the numbered vectors $|x^{(i)}\rangle$

and the vectors of the monomials $\left|x^{i-1}\right\rangle$. In the first case, the superscript (i) is the number of the abstract vector. In the second case, the superscript $i-1$ is the exponent of the variable x.

With these definitions, the problem of approximating the real function f with a polynomial P_{n-1} with a minimum error (6.55) can now be translated into the following geometric problem: find the numbers $\hat{a}_i$ such as the following equation is satisfied "as well as possible" in the sense of the least squares:

$$|f\rangle = \sum_{i=1}^{n} a_i \left|x^{i-1}\right\rangle \tag{6.57}$$

Equation (6.57) means that we try to satisfy the equation $f(x) = P_{n-1}(x)$ for all x in the interval $[a, b]$, which, of course, unless f is a polynomial of at maximum degree $n-1$, cannot be achieved. We can only minimize the norm $\left\| |f\rangle - \sum_{i=1}^{n} \hat{a}_i \left|x^{i-1}\right\rangle \right\|$, which, thanks to our choice of the inner product (6.56), is equivalent to minimizing the error E defined by Equation (6.55).

Based on this formulation of the problem of approximating a function by a polynomial, we can write down the abstract normal Equations (6.36):

$$\int_a^b x^{k-1} f(x)\,dx = \sum_{i=1}^{n} \hat{a}_i \int_a^b x^{i+k-2}\,dx \quad k = 1, 2, \ldots, n \tag{6.58}$$

The solutions $\hat{a}_i$ of these n linear equations give the coefficients of the approximating polynomial of $f(x)$ with the smallest possible error in the sense of the least squares. In deriving these equations, which is a straight re-formulation of Equation (6.36), the reader is invited to pay attention to the fact that $\left|x^{(k)}\right\rangle = \left|x^{k-1}\right\rangle$.

Example 6.9. Polynomial Function Approximation

Find a quadratic polynomial $P_2(x)$ that approximates with the smallest possible error, in the sense of the least squares, the function $y = \sin(\pi x)$ in the interval $[0, 1]$.

In this case, we have three parameters $\hat{a}_i$ to calculate. We obtain them by solving the three normal Equations (6.58):

$$\hat{a}_1 \int_0^1 1dx + \hat{a}_2 \int_0^1 xdx + \hat{a}_3 \int_0^1 x^2 dx = \int_0^1 \sin(\pi x)\,dx \tag{6.59}$$

$$\hat{a}_1 \int_0^1 xdx + \hat{a}_2 \int_0^1 x^2 dx + \hat{a}_3 \int_0^1 x^3 dx = \int_0^1 x \sin(\pi x)\,dx \tag{6.60}$$

$$\hat{a}_1 \int_0^1 x^2 dx + \hat{a}_2 \int_0^1 x^3 dx + \hat{a}_3 \int_0^1 x^4 dx = \int_0^1 x^2 \sin(\pi x)\,dx \tag{6.61}$$

After integration and solving the linear system of equations, we find:

$$\hat{a}_1 = \frac{12\pi^2 - 120}{\pi^3} \tag{6.62}$$

$$\hat{a}_2 = -\hat{a}_3 = \frac{720 - 60\pi^2}{\pi^3} \tag{6.63}$$

It is instructive to compare the quadratic polynomial obtained previously with the Taylor series expansion of $y = \sin(\pi x)$ around $x = 0$.

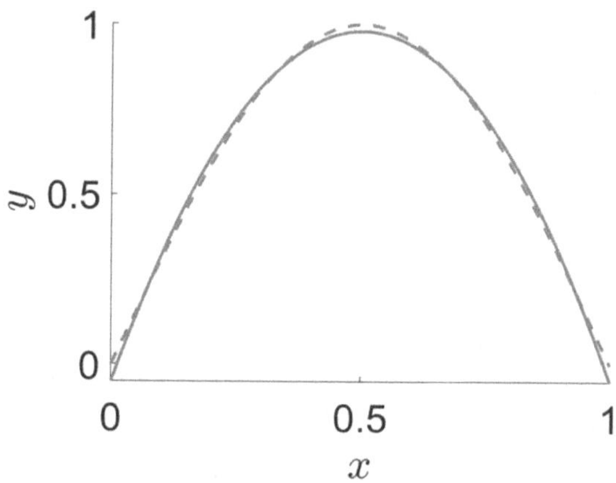

FIGURE 6.7 Quadratic polynomial (continuous line) approximating $y = \sin(\pi x)$ (dashed line) with the smallest possible error in the sense of the least squares.

Figure 6.7 shows the function $y = \sin(\pi x)$ together with the calculated quadratic polynomial $P_2(x) = \hat{a}_1 + \hat{a}_2 x + \hat{a}_3 x^2$ that minimizes the error in the sense of the least squares.

6.5.2 Fourier Polynomials Function Approximation

The normal Equations (6.36) are simple to solve when using orthogonal vector $\left|x^{(i)}\right\rangle$. In this case, the normal equations become a diagonal linear system of equations, and their solution is given by Equation (6.44). In Section 6.5.1, this was not the case, as the vectors $\left|x^{(i)}\right\rangle = \left|x^{i-1}\right\rangle$ are not orthogonal. In the present section, we use orthogonal base functions ϕ_i for the function approximation. There exist several sets of orthogonal functions, including orthogonal polynomials (for example, the Legendre polynomials discussed in Chapter 5). A popular set of orthogonal functions is the set of trigonometric functions. Trigonometric functions are particularly useful for approximating periodic functions.

We will not present the general theory of approximating functions by trigonometric functions (which would lead to the Fourier series) but give just an example to get a first feeling about it.

Consider the following set of trigonometric functions, where n is a positive integer number:

$$\phi_o(x) = \frac{1}{\sqrt{2\pi}} \tag{6.64}$$

$$\phi_k(x) = \frac{1}{\sqrt{\pi}} \cos kx \qquad \text{for all } k = 1, 2, \ldots, n \tag{6.65}$$

$$\phi_{n+k}(x) = \frac{1}{\sqrt{\pi}} \sin kx \qquad \text{for all } k = 1, 2, \ldots, n-1 \tag{6.66}$$

If we define the inner product as

$$\langle f | g \rangle = \int_{-\pi}^{+\pi} f(x) g(x) dx \tag{6.67}$$

then the set of functions $\{\phi_o, \phi_1, \ldots, \phi_{2n-1}\}$ is a set of orthogonal functions according to the inner product. The coefficients in front of the trigonometric functions are chosen such that this set is not only orthogonal but further orthonormal (i.e., the norm of each function ϕ_j is equal to one). Note that we numbered our functions starting with 0, a common practice when using the Fourier series. A Fourier polynomial is defined as

$$S_n(x) = \sum_{k=0}^{2n-1} a_k \phi_k(x) \tag{6.68}$$

Fourier polynomials $S_n(x)$ can be used similarly to polynomial functions to approximate a real continuous function $f(x)$ on the interval $[-\pi, +\pi]$ such that the error E defined as

$$E = \int_a^b \left[f(x) - S_n(x)\right]^2 dx \tag{6.69}$$

is as small as possible. The wording “Fourier polynomials” is common in the literature even though $S_n(x)$ isn’t a polynomial in the traditional sense.

As the set of functions $\{\phi_o, \phi_1, \ldots, \phi_{2n-1}\}$ are all orthogonal, the coefficients $\hat{a}_k$ that minimize the error E are given by Equation (6.44), which writes for our case in a particularly simple form as the functions ϕ_i are orthonormal (i.e., $\langle \phi_i | \phi_j \rangle$ is equal to zero if $i \neq j$ and equal to one if $i = j$):

$$\hat{a}_k = \int_{-\pi}^{+\pi} f(x) \phi_k(x) dx \tag{6.70}$$

Equation (6.70) may be familiar to the reader from the theory of the Fourier series.

Example 6.10. Fourier Polynomial Function Approximation

Approximate function $y = |x|$ with the smallest possible error, in the sense of the least squares, in the interval $[-\pi, +\pi]$ with a Fourier series $S_3(x) = a_o\phi_o(x) + a_1\phi_1(x) + a_2\phi_2(x) + a_3\phi_3(x)$.

To calculate the coefficients $\hat{a}_k$ that minimize the error in the sense of the least squares, we apply Equation (6.70):

$$\hat{a}_o = \int_{-\pi}^{+\pi} \frac{|x|}{\sqrt{2\pi}} dx = \frac{\sqrt{2}\pi^2}{2\sqrt{\pi}} \tag{6.71}$$

$$\hat{a}_1 = \int_{-\pi}^{+\pi} \frac{|x|}{\sqrt{\pi}} \cos x \, dx = -\frac{4}{\sqrt{\pi}} \tag{6.72}$$

$$\hat{a}_2 = \int_{-\pi}^{+\pi} \frac{|x|}{\sqrt{\pi}} \sin x \, dx = 0 \tag{6.73}$$

$$\hat{a}_3 = \int_{-\pi}^{+\pi} \frac{|x|}{\sqrt{\pi}} \cos 3x \, dx = -\frac{4}{9\sqrt{\pi}} \tag{6.74}$$

The Fourier polynomial becomes

$$S_3(x) = \frac{\pi}{2} - \frac{4}{\pi}\cos x - \frac{4}{9\pi}\cos 3x \tag{6.75}$$

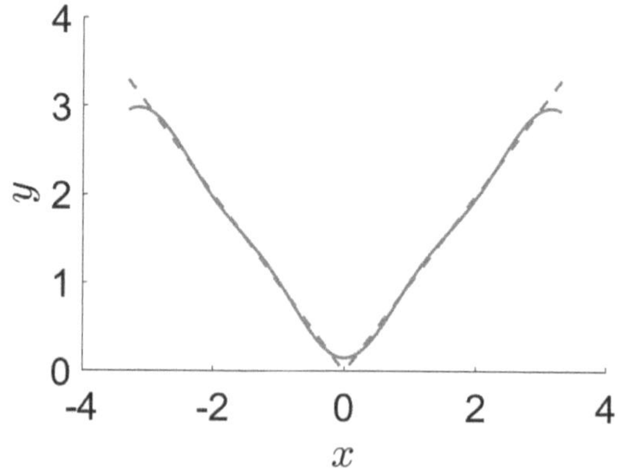

FIGURE 6.8 Fourier polynomial $S_3(x)$ (continuous line) approximating $y = |x|$ (dashed line) with the smallest possible error in the sense of the least squares.

Figure 6.8 shows on a graph the Fourier polynomial $S_3(x)$, which approximates $y=|x|$ with the smallest possible error in the sense of the least squares.

Note that the Fourier polynomial can be considered as a truncated polynomial from the exact Fourier series, which would be given by $S_\infty(x)$. In this case, $S_\infty(x)$ converges and exists. $S_\infty(x)$ would be exactly equal to $y=|x|$.

6.6 LAB: MODELING WEATHER DATA

As an illustration of the concepts of this chapter, we will model the temperature in Montreal, Canada. This table gives the hourly temperatures on May 7, 2013 in Montreal.

Time	Temperature [C]
12:00 AM	19
1:00 AM	15
2:00 AM	15
3:00 AM	14
4:00 AM	13
5:00 AM	13
6:00 AM	13
7:00 AM	14
8:00 AM	16
9:00 AM	17
10:00 AM	20
11:00 AM	21
12:00 PM	23
1:00 PM	25
2:00 PM	26
3:00 PM	27
4:00 PM	27
5:00 PM	28
6:00 PM	28
7:00 PM	26
8:00 PM	24
9:00 PM	23
10:00 PM	20
11:00 PM	19
12:00 AM	19

We will adjust the following linear model:

$$T = a_o + a_1\sin(2\pi t) + a_2\cos(2\pi t) \tag{6.76}$$

where T is temperature in Celsius and t the time since midnight in a fraction of days (i.e., 12PM is $t = 0.5$).

To fit our model, we start by defining the variables t for the time and T for the temperature (for clarity, we do not show all entries of the vector T):

```
>> t = linspace(0, 1, 24)';
>> T = [19; 15; 15; . . . ; 20];
```

The coefficient matrix A is built the usual way:

```
>> A=[t.^0 sin(2*pi*t) cos(2*pi*t)];
```

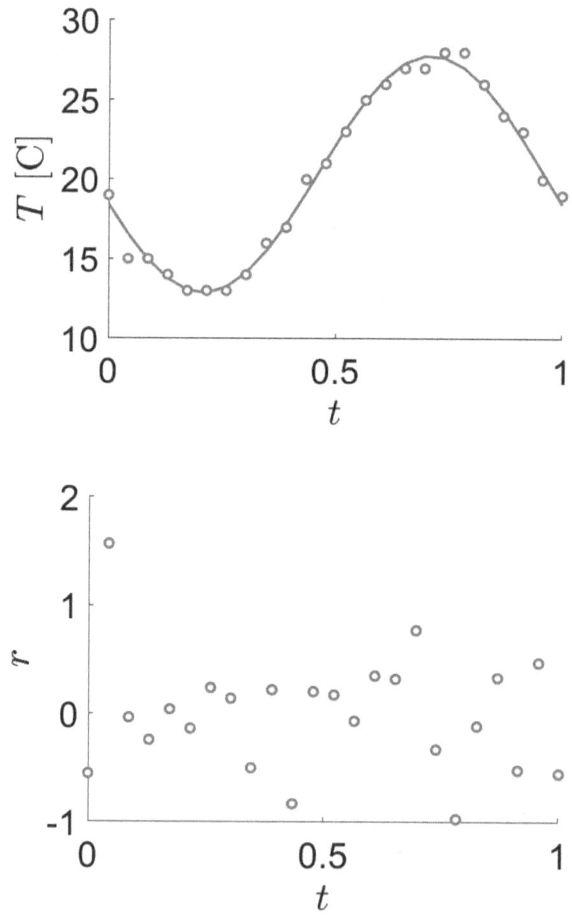

FIGURE 6.9 (a) Temperature T in Montreal on May 7, 2013 (o) together with the fitted model. $T = a_o + a_1 \sin(2\pi t) + a_2 \cos(2\pi t)$ (b) Residuals of the fitted model.

To find the model parameters a_i, we solve the normal equations:

```
>> a = A'*A\A'*T
ans =
 20.3283
 -7.2337
 -1.8786
```

The residuals can be computed, together with the root mean square error like so:

```
>> r = A*a-T;
>> RMSE = norm(r, 2) / sqrt(length(t))
RMSE =
 0.5312
```

In other words, our model fits the data with an error of about half degree Celsius.

Figure 6.9 shows the data points together with the fitted model as well as the residuals.

6.7 EXERCISES

6.7.1 Regression Problems

1. Consider the following dataset:

x_i	1	−1	0	2	1.5
y_i	0.5	1	−1	7	1

For each of the proposed models $y = f(x)$, compute the squared error SE, the mean squared error MSE, and the root mean squared error RMSE. Which model presents the smallest error?

a. $y = 2x - 3$
b. $y = 2x^2 - 1$
c. $y = 0.5x^3 + x^2 - 2$
d. $y = 2\sin(x)$

2. Consider the model $y = ax^2 - 1$ with the parameter a that can be freely chosen.

a. Compute a general expression $\mathrm{SE}(a)$ for the squared error when using the dataset of Exercise 1.
b. Find the parameter for which value a, the squared error $\mathrm{SE}(a)$ becomes minimal.

6.7.2 Normal Equations

3. Prove that the following set of inconsistent equations has no solution.

$$\begin{bmatrix} 1 & -1 \\ 1 & 1 \\ 1 & 3 \end{bmatrix} \begin{bmatrix} x_1 \\ x_2 \end{bmatrix} = \begin{bmatrix} 1 \\ 2 \\ 6 \end{bmatrix}$$

4. Verify that in the case of a consistent system of linear equations (i.e., the same number of equations and unknowns), the solutions of the normal equations are the same as the solution of the system of equations $A\underline{x} = \underline{b}$ for situations where $A\underline{x} = \underline{b}$ has a solution.

   Hint: as in this case, A is a square matrix, its inverse can be computed. Use this to simplify the normal equations. Explain why this cannot be done in case of inconsistent equations.

5. Solve the following systems of inconsistent equations in the sense of least squares, i.e., find a vector $\hat{\underline{x}}$ which minimizes the backward error $\|A\underline{x} - \underline{b}\|$. In each case, compute this minimal backward error.

   a. $$\begin{bmatrix} 2 & 2 \\ -3 & 5 \\ 4 & -1 \end{bmatrix} \begin{bmatrix} x_1 \\ x_2 \end{bmatrix} = \begin{bmatrix} 3 \\ 4 \\ 1 \end{bmatrix}$$

   b. $$\begin{bmatrix} 5 & 0 \\ -1 & 2 \\ 7 & 4 \end{bmatrix} \begin{bmatrix} x_1 \\ x_2 \end{bmatrix} = \begin{bmatrix} 10 \\ 5 \\ 6 \end{bmatrix}$$

   c. $$\begin{bmatrix} 1 & -5 & -3 \\ 1 & 15 & -7 \\ 3 & 11 & 8 \\ -7 & 8 & 1 \end{bmatrix} \begin{bmatrix} x_1 \\ x_2 \\ x_3 \end{bmatrix} = \begin{bmatrix} 21 \\ -7 \\ 13 \\ -4 \end{bmatrix}$$

   d. $$\begin{bmatrix} 1 & -5 \\ 1 & 15 \\ 3 & 11 \\ -7 & 8 \end{bmatrix} \begin{bmatrix} x_1 \\ x_2 \end{bmatrix} = \begin{bmatrix} 21 \\ -7 \\ 13 \\ -4 \end{bmatrix}$$

6. Using the dataset from Exercise 1, find the values of the parameters a_i of the following models, which minimize the squared errors by solving the normal equations. For each adjusted model, compute the squared error

SE on the dataset. Which model presents the lowest squared error and which the highest?

a. $y = a_0 x + a_1$
b. $y = a_0 x^2 + a_1$
c. $y = a_o x^3 + a_1 x^2 + a_2 x + a_3$
d. $y = a_0 x^2 + 1$
e. $y = a_o \sin(x) + a_1 \cos(x)$

7. The following table shows the measurement of time t an object took to fall from a given height h when dropped at zero velocity. Find the earth's gravitational constant g by adjusting the model $h = \frac{1}{2} g t^2$ to the dataset.

$h[m]$	0.1	0.2	0.5	1	1.5
t [s]	0.1427	0.2020	0.3194	0.4514	0.5532

8. Repeat Exercise 7 using the model $t = \sqrt{\frac{2h}{g}}$. Compare the two values you obtain for g from Exercise 7 and this exercise.

6.7.3 Linearization of Models

9. Using the dataset from Exercise 1, find the values of the parameters a_i of the following models, which minimize the squared errors by solving the normal equations after linearizing them. For each adjusted model, compute the squared error SE on the dataset. Which model presents the lowest squared error and which the highest?

a. $y = \dfrac{1}{a_o + a_1 x}$
b. $y = \dfrac{1}{a_o + a_1 x + a_2 x^2}$
c. $y = a_o e^{a_1 x^2}$
d. $y = a_0 + a_1 \sin(x) + a_2 x + a_3 x^2$

6.7.4 Approximation of Functions

10. Find a quadratic polynomial $P_2(x)$ that approximates with the smallest possible error, in the sense of the least squares, the function $y = \cos(\pi x)$ in the interval $[0, 1]$. Plot the approximating polynomial and the cosine function on the same graph.

11. Approximate the function $y=\pi(x-\pi)$ with the smallest possible error, in the sense of the least squares, in the interval $[-\pi,+\pi]$ with a Fourier series $S_2(x)=a_o\phi_o(x)+a_1\phi_1(x)+a_2\phi_2(x)$. Plot the approximating Fourier series $S_2(x)$ and the function $y=\pi(x-\pi)$ on the same graph.

12. Approximate the function $f(x)=-1$ if $-\pi\leq x\leq 0$ and $f(x)=1$ if $0<x\leq\pi$ with the smallest possible error, in the sense of the least squares, in the interval $[-\pi,+\pi]$ with a Fourier series $S_2(x)=a_o\phi_o(x)+a_1\phi_1(x)+a_2\phi_2(x)$. Plot the approximating Fourier series $S_2(x)$ and the function $y=f(x)$ on the same graph.

NOTES

1 Jean-Baptiste Joseph Fourier (1768–1830), French mathematician and physicist.
2 In case of consistent equations, $\underline{b}$ will be within the plan spanned by the vectors $\underline{v}_1$ and $\underline{v}_2$.
3 www.internetlivestats.com/total-number-of-websites/ (Retrieved February 2023).
4 Johannes Kepler (1571–1630), German astronomer, mathematician, astrologer, philosopher, and composer.

7 Polynomial Interpolation

Interpolation has a goal similar to that of regression. A mathematical model is adjusted in order to represent a given dataset. The difference between regression and interpolation is in the way the goodness of the model adjustment is measured. In the case of regression, the least square error is minimized. In the case of interpolation, we ensure the model passes exactly by all available data points. Interpolation is, hence, fundamentally different from the concept of regression. From a practical point of view, it imposes restrictions on the mathematical model. Except for very particular cases, the model must have the same number of freely adjustable parameters as data points. Furthermore, the predictive value of the adjusted model is very different from that of regression. So far, the concept of predictive value has not yet been introduced. This discussion will be conducted in the second part of this book. For now, without delving into detailed justifications, we will outline the two primary cases for using either the regression approach or the interpolation approach. If the goal is to predict values not present in the dataset, the regression approach is preferred, as will be explained in the second part of this book. On the other hand, interpolation is better suited for approximating values near existing data points (typically between points in the dataset). A typical example of the use of interpolation is to determine values of a tabulated function (e.g., the surface below the Gaussian distribution as used in statistics). In general, regression is to be used in case the dataset contains points with statistical errors (e.g., from a measurement process), and interpolation is to be used in case the dataset contains (at least in principle) error-free points.

In this chapter, we will discuss only polynomial interpolation. Three different approaches to adjusting a polynomial model to the dataset will be presented. The first approach, the so-called Vandermond interpolation, consists of solving the n equations resulting from imposing that the polynomial $P(x) = a_o + a_1 \mathrm{x} + \cdots + a_{n-1} x^{m-1}$ passes by the m points of the dataset. A second approach, the Newton interpolation, uses the same idea but writes the interpolating polynomial $P(x)$ in a different way, resulting in a solution process with a smaller condition number. The third approach, the Lagrange[1] interpolation, uses a different method by writing the polynomial $P(x)$ as a linear combination of carefully chosen base functions, the so-called Lagrange polynomials.

7.1 DEFINITIONS

Before outlining approaches to solve polynomial interpolation problems, let us first define the problem. The following definition states what a polynomial interpolation problem is and summarizes the notations used in this chapter.

DOI: 10.1201/9781003262121-8

Definition 7.1. Polynomial Interpolation Problem

Consider a dataset consisting of m ordered pairs (x_i, y_i), with $i = 1,\ldots,m$. The aim of polynomial interpolation is to find m real numbers $a_o, a_1, \ldots, a_{m-1}$, such as for all i, the equations $P_{m-1}(x_i) = y_i$ are satisfied with $P_{m-1}(x) = a_o + a_1 x + \ldots + a_{m-1} x^{m-1}$.

We would like to draw the attention of the reader of the way the different quantities are indexed. The coefficients a_i of the polynomial $P_{m-1}(x)$ are numbered from 0 to $m-1$, resulting in a polynomial of degree $m-1$. The m points (x_i, y_i) of the dataset are numbered from 1 to m.

There is a more abstract way to present the same problem, which takes advantage of the concept of real-valued vector spaces. Consider a dataset of m points (x_i, y_i) numbered from 1 to m. The aim is to interpolate these points with a polynomial. As we have m interpolation points, we need a polynomial of degree $m-1$. We know that the set of polynomials of degree $m-1$ forms a vector space of dimension m. We can choose a vector base $\mathcal{B} = \{e_1(x), e_1(x), \ldots, e_m(x)\}$ of this space and represent our polynomial in that base:

$$P_{m-1}(x) = p_1 e_1(x) + p_2 e_2(x) + \ldots + p_m e_m(x) \tag{7.1}$$

where the m numbers p_i are the coordinates of our interpolation polynomial in the base $\mathcal{B}$.

Interpolating the m points (x_i, y_i) with the polynomial $P_{m-1}(x)$, means that we have to satisfy the following m equations:

$$y_i = P_{m-1}(x_i) = p_1 e_1(x_i) + p_2 e_2(x_i) + \ldots + p_m e_m(x_i) \tag{7.2}$$

It is convenient to write these equations in matrix form as

$$\underline{y} = A\underline{p} \tag{7.3}$$

with the column vectors $\underline{y} = [y_1, y_2, \ldots, y_m]^T$ and $\underline{p} = [p_1, p_2, \ldots, p_m]^T$ and the matrix

$$A = \begin{bmatrix} e_1(x_1) & e_2(x_1) & \ldots & e_m(x_1) \\ e_1(x_2) & e_2(x_2) & \ldots & e_m(x_2) \\ \vdots & \vdots & & \vdots \\ e_1(x_m) & e_2(x_m) & \ldots & e_m(x_m) \end{bmatrix} \tag{7.4}$$

Solving Equations (7.2), or (7.3), for the m unknowns p_i will yield the interpolation polynomial that passes exactly through all m data points. Depending on the choice of our vector base $\mathcal{B}$ to represent the interpolation polynomial, the system of equations (7.2), respectively (7.3) will be more or less straightforward to solve. In this chapter, we will discuss three examples of the choice of the vector base $\mathcal{B}$.

Once the coefficients p_i are found, the interpolating polynomial can be evaluated at any point x according

$$P_{m-1}(x) = [e_1(x), e_2(x), \ldots, e_m(x)] \begin{bmatrix} p_1 \\ p_2 \\ \vdots \\ p_m \end{bmatrix} \quad (7.5)$$

This matrix notation is convenient for performing calculations in Octave, as illustrated in the following example.

Example 7.1. Evaluating Polynomials

Consider the following cubic polynomial:

$$P_3(x) = 2 - x + 2x^2 + x^3 \quad (7.6)$$

Let us evaluate this polynomial in the following points: –1, 0, 1, and 2.

Using Octave, the first method consists of using a function definition for the polynomial $P_3(x)$:

```
>> P = @(x) 2-x+2*x.^2+x.^3
```

And then evaluate this function in the desired points:

```
>> x = [-1, 0, 1, 2]';
>> P(x)
y =
    4
    2
    4
   16
```

Alternatively, we could also take advantage of Equation (7.5). We select a vector base $\mathcal{B}$. For this example, we choose the natural base $\mathcal{B} = \{1, x, x^2, x^3\}$. We will first evaluate the polynomial in a single point, for example $x = 2$. Following Equation (7.5), we define the column vector $\underline{p}$ with the coefficients of the polynomial:

```
>> p = [2 -1 2 1]';
```

To evaluate $P_3(x)$ in $x=2$, we need to evaluate the base polynomials $e_i(x)$ in $x=2$ to compute the matrix A (which in this case is a line vector) and then calculate the matrix product (7.5):

```
>> [1 2 2^2 2^3]*p
ans = 16
```

If we wish to evaluate $P_3(x)$ in several x values, we can achieve this by calculating:

```
>> x = [-1, 0, 1, 2]';
>> A = [x.^0 x x.^2 x.^3];
>> y = A*p
y =
    4
    2
    4
   16
```

which is the same answer as when we used the function definition for our polynomial.

7.2 VANDERMOND INTERPOLATION

In Vandermonde interpolation, the choice for the vector base $\mathcal{B}$ to represent the interpolation polynomial is the natural base $\mathcal{B}=\{1, x, x^2, \ldots, x^{m-1}\}$. In this base, we have

$$P_{m-1}(x)=a_0+a_1x+\ldots+a_{m-1}x^{m-1} \tag{7.7}$$

Note that we number the coefficients a_i from 0 to $m-1$.

Forcing this polynomial to pass by all the data points results in the following system of m linear equations:

$$y_i=a_0+a_1x_i+\ldots+a_{m-1}x_i^{m-1} \tag{7.8}$$

Using the matrix version for the equations (7.8) results in the following coefficient matrix

$$A=\begin{bmatrix} 1 & x_1 & \cdots & x_1^{m-1} \\ 1 & x_2 & \cdots & x_2^{m-1} \\ \vdots & \vdots & & \vdots \\ 1 & x_m & \cdots & x_m^{m-1} \end{bmatrix} \tag{7.9}$$

The matrix A has a very particular structure and is called a Vandermond matrix. Solving system (7.8) yields the coefficients a_i.

Once the solutions are found, it is simple to evaluate the interpolating polynomial in any point x according to:

$$P_{m-1}(x) = \left[1, x, x^2, \ldots, x^{m-1}\right] \begin{bmatrix} a_0 \\ a_1 \\ a_2 \\ \vdots \\ a_{m-1} \end{bmatrix} \tag{7.10}$$

One of the possible issues with Vandermond interpolation is the structure of the coefficient matrix A. A Vandermond matrix usually has a high condition number (see Chapter 4). Consequently, the solution of the system of equations (7.8) is sensitive to round-off errors.

Example 7.2. Vandermond Interpolation

Consider the following dataset:

x_i	-1	0	1	2
y_i	4	2	4	16

Let us find the Vandermond interpolation polynomial.

The previous dataset consists of four data points. Therefore, the interpolating polynomial is of degree three. To find this polynomial, we solve the system of equations (7.8). Using Octave, this can be achieved efficiently. We start by defining the dataset:

```
>> x = [-1, 0, 1, 2]';
>> y = [4 2 4 16]';
```

We then compute the coefficient matrix A of the system (7.8):

```
>> A = [x.^0 x x.^2 x.^3];
```

And solve it using Gauss elimination to find the coefficients a_i of our interpolating polynomial:

```
>> a = A\y
a =
    2
```

```
-1
2
1
```

Consequently, our interpolating polynomial is:

$$P_3(x) = 2 - x + 2x^2 + x^3 \tag{7.11}$$

The reader is invited to verify that this polynomial indeed interpolates the dataset.

If one wants to evaluate the interpolating polynomial at some point (e.g., $x = 1.5$), this can be achieved like this:

```
>> [1 1.5 1.5^2 1.5^3]*a
ans = 8.3750
```

We can plot on the same graph the interpolating points and the interpolating polynomial using the approach described in Example 7.1 to evaluate a polynomial in multiple points. For this, we construct a column vector `xint`, which holds points in which we want to evaluate our interpolation polynomial:

```
>> xint = linspace(-1.15, 2.15, 100)';
```

and then evaluate them with the help of the matrix `A`:

```
>> A = [xint.^0 xint xint.^2 xint.^3];
>> plot(xint, A*a, '-', x, y, 'o')
```

which results in Figure 7.1.

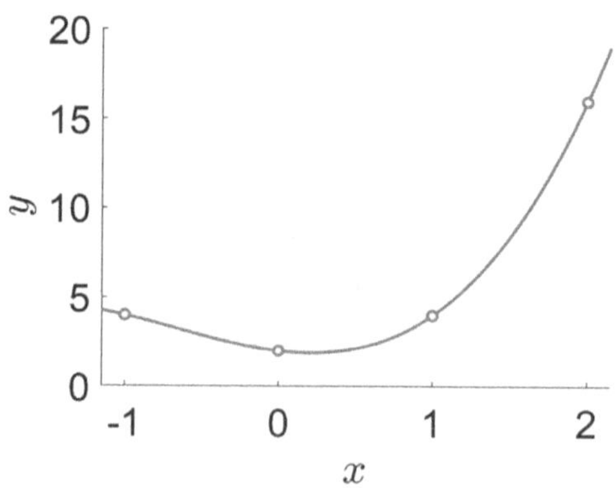

FIGURE 7.1 Interpolation points (o) together with the Vandermond interpolation polynomial (–).

7.3 NEWTON INTERPOLATION

Newton interpolation uses a vector base $\mathcal{B}$ such that the Equations (7.2) become a lower triangular system. The vector base is the following:

$$\mathcal{B} = \left\{1, x - x_1, (x - x_1)(x - x_2), \ldots, (x - x_1)(x - x_2)\ldots(x - x_{m-1})\right\} \quad (7.12)$$

where x_i are the first $m-1$ interpolating points. We will refer to this base as the Newton base. In this base, the interpolation polynomial writes

$$P_{m-1}(x) = b_0 + b_1(x - x_1) + \ldots + b_{m-1}(x - x_1)(x - x_2)\ldots(x - x_{m-1}) \quad (7.13)$$

Note that we number the coefficients b_i from 0 to $m-1$.

The system of equations (7.3) becomes a lower diagonal system with the coefficient matrix A:

$$A = \begin{bmatrix} 1 & 0 & 0 & 0 & \ldots & 0 \\ 1 & x_2 - x_1 & 0 & 0 & \ldots & 0 \\ 1 & x_3 - x_1 & (x_3 - x_1)(x_3 - x_2) & 0 & \ldots & 0 \\ \vdots & \vdots & \vdots & & & \vdots \\ 1 & x_m - x_1 & (x_m - x_1)(x_m - x_2) & \ldots & \ldots & (x_m - x_1)\ldots(x_m - x_{m-1}) \end{bmatrix} \quad (7.14)$$

The solution of this lower diagonal system can be expressed by the Newton finite differences:

$$b_k = y[x_1, \ldots, x_k] \quad (7.15)$$

which are defined recursively, starting with

$$y[x_i] = y_i \quad (7.16)$$

followed by

$$y[x_i, x_{i+1}, \ldots, x_{i+k}] = \frac{y[x_{i+1}, \ldots, x_{i+k}] - y[x_i, x_{i+1}, x_{i+k-1}]}{x_{i+k} - x_i} \quad (7.17)$$

One of the advantages of Newton interpolation is that the solutions (7.15) can be computed efficiently by hand calculations, as illustrated in the following example.

Example 7.3. Netwon Interpolation

Consider the following dataset:

x_i	-1	0	1	2
y_i	4	2	4	16

Let us find the Newton interpolating polynomial.

The Newton interpolating polynomial will take the following form:

$$P_3(x) = b_0 + b_1(x+1) + b_2(x+1)x + b_3(x+1)x(x-1) \quad (7.19)$$

We can verify that using the vector base $\mathcal{B} = \{1, x+1, (x+1)x, (x+1)x(x-1)\}$, the coefficient matrix A of system (7.3) is indeed a triangular matrix:

```
>> x = [-1, 0, 1, 2]';
>> A = [x.^0 x+1 (x+1).*x (x+1).*x.*(x-1)]
A =
     1     0     0     0
     1     1     0     0
     1     2     2     0
     1     3     6     6
```

Solving system (7.3) results in the coefficients b_i of the Newton polynomial:

```
>> y = [4 2 4 16]';
>> b = A\y
b =
     4
    -2
     2
     1
```

If one wants to evaluate the Newton interpolating polynomial in a particular point (e.g., $x_p = 1.5$), this can be achieved like this:

```
>> xp = 1.5
>> [1 xp+1 (xp+1)*xp (xp+1)*xp*(xp-1)]*b
ans = 8.3750
```

which yields the same numerical answer as the Vandermond interpolating polynomial evaluated at $x = 1.5$.

Similarly to Example 7.2, a graph displaying the interpolation points and the interpolation polynomial can be constructed:

```
>> xint = linspace(-1.15, 2.15, 100)';
>> A = [xint.^0 xint+1 (xint+1).* xint (xint+1).*
xint.*(xint-1)];
>> plot(xint, A*b, '-', x, y, 'o')
```

which results in the same graph as Figure 7.1.

One of the benefits of Newton interpolation is that it can be conducted efficiently by hand. We can organize our work in a table. In the first column, we write the x values and, in the second column, the corresponding y values:

-1	4
0	2
1	4
2	16

Note that in this table, the second column is the first level of the Newton differences (i.e., the $y[x_i]$). We now construct the second-level Newton differences $y[x_i, x_{i+1}]$. For example:

$$y[-1,0]=\frac{y[0]-y[1]}{0-(-1)}=\frac{2-4}{0-(-1)}=-2 \tag{7.20}$$

We add the results in the next column:

-1	4	
		$y[-1,0]=-2$
0	2	
		$y[0,1]=2$
1	4	
		$y[1,2]=12$
2	16	

We now compute the third level Newton differences $y[x_i, x_{i+1}, x_{i+2}]$. For example:

$$y[-1,0,1]=\frac{y[0,1]-y[-1,0]}{1-(-1)}=\frac{2-(-2)}{1-(-1)}=2 \tag{7.22}$$

and add the results to the table:

-1	4		
		$y[-1,0]=-2$	
0	2		$y[-1,0]=2$
		$y[0,1]=2$	
1	4		$y[0,1,2]=5$
		$y[1,2]=12$	
2	16		

And finally, we compute the fourth level Newton differences $y[x_i, x_{i+1}, x_{i+2}, x_{i+3}]$, where in our case, only one remains to be computed:

$$y[-1,0,1,2] = \frac{y[0,1,2] - y[-1,0,1]}{2-(-1)} = \frac{5-2}{2-(-1)} = 1 \tag{7.24}$$

We add the result to the table:

-1	4			
		$y[-1,0]=-2$		
0	2		$y[-1,0]=2$	
		$y[0,1]=2$		$y[-1,0,1,2]=1$
1	4		$y[0,1,2]=5$	
		$y[1,2]=12$		
2	16			

From this table, we can read the Newton interpolating polynomial (the first row of each column):

$$P_3(x) = 4 - 2(x+1) + 2(x+1)x + (x+1)x(x-1) \tag{7.26}$$

The reader is invited to compare this result with the previously computed solution using Octave and to verify that this polynomial indeed interpolates the dataset.

Even though the Vandermond interpolating polynomial (7.7) and the Newton interpolating polynomial (7.13) look, at first sight, very different, they are actually the same function. Indeed, both polynomials evaluate in m points exactly to the same value. They must consequently be identical. The reader can verify this fact by checking that, indeed, polynomials (7.11) and (7.26) are identical.

7.4 LAGRANGE INTERPOLATION

Lagrange interpolation uses a vector base $\mathcal{B}$ such that Equations (7.3) become a diagonal system. The vector base we will construct will be written as $\mathcal{B} = \{L_1(x), L_1(x), \ldots, L_m(x)\}$, and we will refer to it as the Lagrange base. To construct the Lagrange base, we start from Equation (7.4). A possible choice to make the coefficient matrix A diagonal is to impose that A is the identity matrix:

$$L_i(x_j) = \delta_{ij} \tag{7.27}$$

where δ_{ij} is the Kronecker symbol ($\delta_{ij} = 1$ if $i = j$ and 0 otherwise). Equation (7.27) allows constructing the m base vectors $L_i(x)$, which are called Lagrange polynomials. Two important facts follow from Equation (7.27). First, the roots of the polynomial L_i are the interpolating points x_j where $j \neq i$. Second, the polynomial $L_i(x)$ evaluates to 1 in x_i, i.e., $L_i(x_i) = 1$. Using these two facts, we write the Lagrange polynomials in the following way:

$$L_i(x) = \prod_{j=1,\, j\neq i}^{m} \frac{x - x_j}{x_i - x_j} \tag{7.28}$$

In the Lagrange base, the interpolation polynomial can be written directly without any need for calculations:

$$P_{m-1}(x) = L_1(x)y_1 + L_2(x)y_2 + \ldots + L_m(x)y_m \tag{7.29}$$

Example 7.4. Lagrange Interpolation

Consider the following dataset:

x_i	-1	0	1	2
y_i	4	2	4	16

Let us find the Lagrange interpolation polynomial. As the dataset has four points, the Lagrange base will have four Lagrange polynomials. They can be written directly by inspecting the dataset. For example, the Lagrange polynomial $L_1(x)$ has the three roots 0, 1, and 2 and must evaluate to one in $x = -1$. Consequently, we can write:

$$L_1(x) = \frac{(x-0)(x-1)(x-2)}{(-1-0)(-1-1)(-1-2)} = -\frac{1}{6}x(x-1)(x-2) \tag{7.30}$$

Proceeding the same way for the three remaining Lagrange polynomials results in:

$$L_2(x) = \frac{(x+1)(x-1)(x-2)}{(0+1)(0-1)(0-2)} = \frac{1}{2}(x+1)(x-1)(x-2) \tag{7.31}$$

$$L_3(x) = \frac{(x+1)(x-0)(x-2)}{(1+1)(1-0)(1-2)} = -\frac{1}{2}x(x+1)(x-2) \tag{7.32}$$

$$L_4(x) = \frac{(x+1)(x-0)(x-1)}{(2+1)(2-0)(2-1)} = \frac{1}{6}x(x+1)(x-1) \tag{7.33}$$

The interpolating polynomial $P_3(x)$ becomes:

$$P_3(x) = 4L_1(x) + 2L_2(x) + 4L_3(x) + 16L_4(x) \tag{7.34}$$

The reader is invited to verify that this polynomial indeed interpolates the dataset.

It is interesting to check that the coefficient matrix A of system (7.3) becomes the identity matrix when using the Lagrange base. Let us first define the four Lagrange polynomials of our dataset:

```
>> L1 = @(x) -1/6.*x.*(x-1).*(x-2);
>> L2 = @(x) 1/2.*(x+1).*(x-1).*(x-2);
>> L3 = @(x) -1/2.*x.*(x+1).*(x-2);
>> L4 = @(x) 1/6.*x.*(x+1).*(x-1);
```

The coefficient matrix A is indeed the identity matrix:

```
>> x = [-1, 0, 1, 2]';
>> A = [L1(x) L2(x) L3(x) L4(x)]
A =
     1     0     0     0
     0     1     0     0
     0     0     1     0
     0     0     0     1
```

If one wants to evaluate the Lagrange interpolating polynomial in a particular point (e.g., $x_p = 1.5$), this can be achieved as follows:

```
>> xp = 1.5
>> [L1(1.5) L2(1.5) L3(1.5) L4(1.5)]*y
ans = 8.3750
```

which is the same as in the case of Vandermond and Newton interpolation.

Similarly to Example 7.2, a graph displaying the interpolation points and the interpolation polynomial can be constructed:

```
>> xint = linspace(-1.15, 2.15, 100)';
>> A = [L1(xint) L2(xint) L3(xint) L4(xint)];
>> plot(xint, A*a, '-', x, y, 'o')
```

which results in the same graph as Figure 7.1.

7.5 INTERPOLATING FUNCTIONS

A common application of interpolation is to approximate functions. In this case, the dataset is replaced by m ordered pairs $(x_i, y_i = f(x_i))$, with $i = 1,\ldots,m$. In this situation, it becomes possible to define an interpolation error as being the difference between the interpolating polynomial and the actual function value:

$$E(x) = \left|f(x) - P_{m-1}(x)\right| \tag{7.35}$$

There exists a theorem that allows estimating the interpolation error, which we state without further proof:

Theorem 7.1. Interpolation Error

If $P_{m-1}(x)$ is interpolating the m points $(x_i, f(x_i))$ with $f(x)$ a sufficiently derivable function, then

$$f(x) - P_{m-1}(x) = \frac{(x-x_1)(x-x_2)\ldots(x-x_m)}{m!} f^{(m)}(c) \tag{7.36}$$

with c some number between the smallest and largest values of the interpolating points x_i.

Note that the number c will change if one evaluates the error in different points x. That is, c is a function of x.

Example 7.5. Interpolating the Sine Function

As an illustration, we interpolate the sine function between 0 and $\frac{\pi}{2}$, with four equally spaced points. From trigonometry, we know the exact values of the sine function in those points.

x_i	0	$\frac{\pi}{6}$	$\frac{2\pi}{6}$	$\frac{\pi}{2}$
$y_i = \sin x_i$	0	$\frac{1}{2}$	$\frac{\sqrt{3}}{2}$	1

To determine the interpolation polynomial, let us use Vandermond interpolation:

```
>> x = [0 pi/6 2*pi/6 pi/2]';
>> y = [0 0.5 sqrt(3)/2 1]';
```

```
>> A = [x.^0 x x.^2 x.^3];
>> a = A\y
a =

         0
    1.0204
   -0.0655
   -0.1139
```

The interpolating polynomial is:

$$P_3(x) = 1.0204x - 0.0655x^2 - 0.1139x^3 \tag{7.37}$$

The interpolation polynomial, together with the interpolation points, can be plotted the usual way:

```
>> xint = linspace(0, pi/2, 100)';
>> A = [xint.^0 xint xint.^2 xint.^3];
>> plot(xint, A*a, '-', x, y, 'o')
```

which results in Figure 7.2a.
Using Theorem 7.1, it is possible to estimate the error when computing the sine function using the interpolation polynomial $P_3(x)$.

$$\left|\sin(x) - P_3(x)\right| = \frac{(x-0)\left(x-\frac{\pi}{6}\right)\left(x-\frac{2\pi}{6}\right)\left(x-\frac{\pi}{2}\right)}{4!}\left|\frac{d^4}{dx^4}\sin(c)\right| \tag{7.38}$$

As $\left|\frac{d^4}{dx^4}\sin(c)\right|$ is, at maximum, equal to one, we can further write the following:

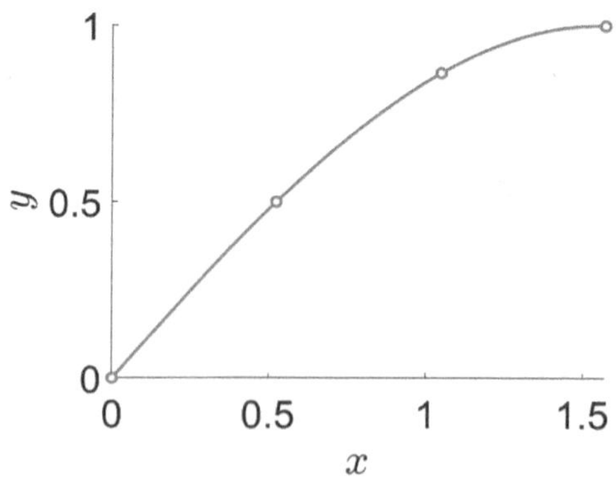

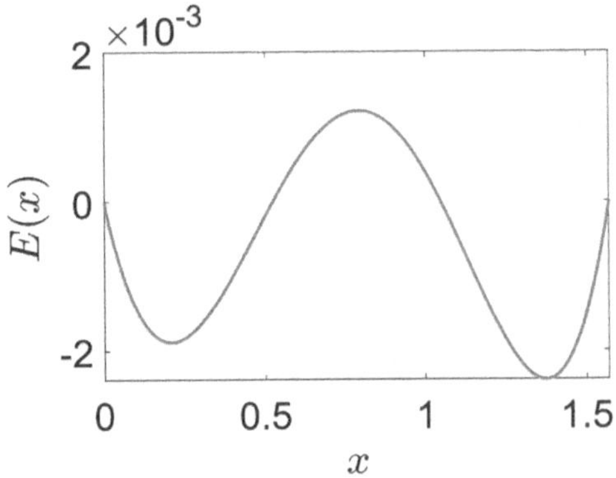

FIGURE 7.2 (a) Interpolation polynomial (–) of four points (o) of the sine functions. (b) Interpolation error $E(x)$ (difference between the interpolation polynomial and the sine function).

$$\left|\sin(x) - P_3(x)\right| \le \frac{(x-0)\left(x-\frac{\pi}{6}\right)\left(x-\frac{2\pi}{6}\right)\left(x-\frac{\pi}{2}\right)}{4!} \tag{7.39}$$

For example, the error in computing $\sin(1)$ using $P_3(1)$ would be:

$$\left|\sin(1) - P_3(x1)\right| \le \frac{(1-0)\left(1-\frac{\pi}{6}\right)\left(1-\frac{2\pi}{6}\right)\left(1-\frac{\pi}{2}\right)}{4!} \cong 5 \cdot 10^{-4} \tag{7.40}$$

which slightly over-estimates the true error:

```
>> sin(1)-[1 1 1 1]*a
ans = 3.8497e-04
```

The true interpolation error can be computed and plotted over the full interval:

```
>> plot(xint, sin(xint)-A*a
```

which results in Figure 7.2b.

One may wonder how many interpolation points should be used to approximate a function. Theorem 7.1 allows us to estimate the interpolation error. It is, however, not obvious from Equation (7.36) how the error will behave as we increase the number of interpolation points. At first sight, we could intuit that this error must decrease as we increase the number of interpolation points. This is, however, not the case, as illustrated in the following example.

Example 7.6. Runge Phenomenon

Consider the following function:

$$f(x) = \frac{1}{1+12x^2} \tag{7.41}$$

Let us build an interpolation polynomial using 11 interpolation points uniformly distributed over the interval $[-1, 1]$. Using Octave, this can be done as follows.

First, let us define the 11 interpolation points x_i together with their corresponding values $f(x_i)$:

```
>> x = linspace(-1., 1., 11)';
>> y = 1./(1+12*x.^2);
```

Next, we use Vandermond interpolation to construct the interpolating polynomial $P_{10}(x)$:

```
>> A = [x.^0 x.^1 x.^2 x.^3 x.^4 x.^5 x.^6 x.^7
x.^8 x.^9 x.^10];
>> a = A\y;
```

The column vector a contains the coefficients of the interpolating polynomial. Let us now plot the interpolation points together with the polynomial $P_{10}(x)$:

```
>> xf = linspace(-1.025, 1.025, 100)';
>> Af = [xf.^0 xf.^1 xf.^2 xf.^3 xf.^4 xf.^5 xf.^6
xf.^7 xf.^8 xf.^9 xf.^10];
>> plot(x, y, 'o', xf, Af*a, '-');
```

Figure 7.3 shows that the interpolating polynomial $P_{10}(x)$ is fluctuating around the interpolating points. This effect, known as the Runge[2] phenomenon, is typical for situations using many interpolating points.

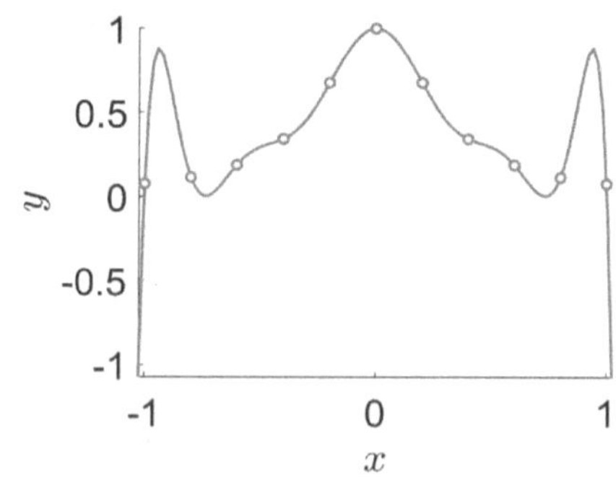

FIGURE 7.3 Interpolating some function with a large number of interpolating points. Often, this results in the so-called Runge phenomenon, where the interpolating polynomial is wiggling around the interpolating points.

This example shows that using an excessive number of interpolation points to approximate functions is generally not a good idea.

7.6 LAB: INTERPOLATING SURFACE BELOW THE STANDARD NORMAL DISTRIBUTION

The standard normal distribution plays an important role in statistics and is the surface below its curve. For example, quantities like these need to be evaluated:

$$I(Z)=\frac{2}{\sqrt{\pi}}\int_{-\infty}^{z} e^{-x^2/2}dx \tag{7.42}$$

which are known in the literature as the area left of the Z score. The definite integral (7.42) cannot be computed analytically, and numerical approximations are required. In Chapter 8, we will learn how to compute numerically definite integrals. For the moment, we will use tabulated values such as in this table:

Z_i	-0.4	-0.3	-0.2	-0.1	0.0
$I(z_i)$	0.34458	0.38209	0. 42074	0. 46017	0.5

If other quantities than the ones listed need to be known, for example, the quantity $I(Z=-0.26)$, interpolation can be used.

As an example, let us use Newton interpolation. We start by defining our dataset:

```
>> x = [-0.4, -0.3, -0.2, -0.1, 0]';
>> y = [0.34458, 0.38209, 0.42074, 0.46017, 0.5]';
```

We then build the coefficient matrix A (7.14) of system (7.3):

```
>> A = [x.^0 x+0.4 (x+0.4).*(x+0.3)
(x+0.4).*(x+0.3).*(x+0.2) (x+0.4).*(x+0.3).*(x+0.2)
.*(x+0.1)]
A =
    1.0000         0         0         0         0
    1.0000    0.1000         0         0         0
    1.0000    0.2000    0.0200         0         0
    1.0000    0.3000    0.0600    0.0060         0
    1.0000    0.4000    0.1200    0.0240    0.0024
```

We can verify that, indeed, the matrix A is a lower diagonal matrix as it should be. The coefficients b_i of the Newton interpolating polynomial (7.13) are computed by solving system (7.3):

```
>> A\b
b =
       0.3446
       0.3751
       0.0570
      -0.0600
      -0.0083
```

To find the value $I(Z = -0.26)$, we compute:

```
>> xp = -0.26;
>> [1 xp+0.4 (xp+0.4).*(xp+0.3)
(xp+0.4).*(xp+0.3).*(xp+0.2) (xp+0.4).*(xp+0.3).*
(xp+0.2).*(xp+0.1)]*b
ans =
     0.3594
```

This value is close to the value we would find by numerically computing integral (7.42). As we don't know that value, a simple way to check if the computed interpolated value is sound is to plot the dataset together with the interpolation polynomial (Figure 7.4).

As shown in Figure 7.4, the interpolating polynomial does not present the Runge phenomenon, and it is reasonable to assume it represents well the desired quantity over the considered range. The reader will note that we are unable to attribute any quantitative error estimation to the interpolated value of $I(Z = -0.26)$.

7.7 EXERCISES

7.7.1 DEFINITIONS

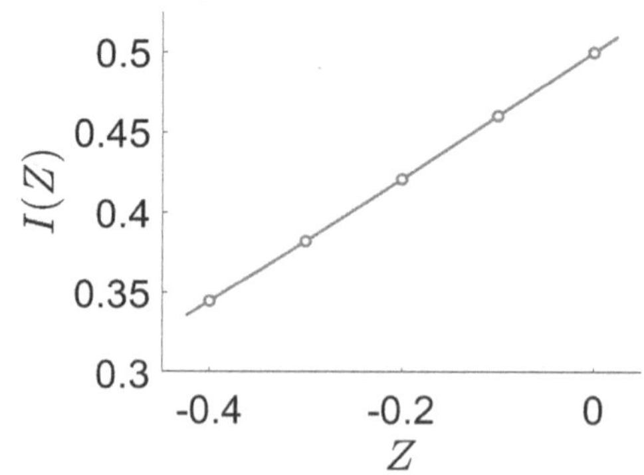

FIGURE 7.4 Area left to the Z score of the standard normal distribution (o) together with a Newton interpolation polynomial (–).

1. For each of the following polynomials $P_n(x)$, represent them in the natural vector base $\mathcal{B}=\{1,x,x^2,\ldots,x^n\}$ (where n is the degree of the polynomial) and afterward evaluate them in the points -1, 0, 1, and 2 following the method explained in Example 7.1.

   a. $P_4(x)=2+3x-4x^2+5x^4$

   b. $P_2(x)=-3+x-4x^2$

   c. $P_5(x)=4-x+3x^3-4x^4+x^5$

   d. $P_3(x)=2x-4x^2-3x^3$

7.7.2 Vandermond Interpolation

2. For each of the following, provided datasets determine the Vandermond interpolation polynomial

a.

x_i	3	5	6
y_i	3	-2	5

b.

x_i	1	-1	0	2	1.5
y_i	0.5	1	-1	7	1

c.

x_i	1	-1	0
y_i	-2	3	1

d.

x_i	-4	-2	2	5
y_i	2	1	-3	-8

7.7.3 Newton Interpolation

3. For each dataset from Exercise 2, determine the Newton interpolation polynomial. Use two methods. First, determine the coefficient matrix (7.4), and afterward, solve the system (7.3) (check the coefficient matrix is indeed a triangular matrix). Second, use hand calculations as outlined in Example 7.3.

7.7.4 Lagrange Interpolation

4. For each dataset from Exercise 2, determine the Lagrange interpolation polynomial. First, determine the individual Lagrange polynomials, and second, build the coefficient matrix (7.4). Verify that the coefficient matrix is indeed the unity matrix.

7.7.5 Interpolation Error

5. Determine an interpolation polynomial for the cosine function using the points from the table. Estimate an upper boundary for the error using Theorem 7.1 and plot it over the interval 0 to $\frac{\pi}{2}$.

x_i	0	$\frac{\pi}{6}$	$\frac{2\pi}{6}$	$\frac{\pi}{2}$
$y_i = \cos x_i$	1	$\frac{\sqrt{3}}{2}$	$\frac{1}{2}$	0

NOTES

1 Joseph-Louis Lagrange (1736–1813), Italian mathematician, physicist, and astronomer; later naturalized French.

2 Carl David Tolmé Runge (1856–1927), German mathematician, physicist, and spectroscopist.

8 Numerical Integration

Computing the definite integral of a function is a challenging problem. Except for a few exceptions, no closed-form solution can be provided. However, from a practical point of view, the calculation of the definite integral is a very common problem in engineering and science that needs to be addressed frequently. A large number of important quantities are defined through the calculations of definite integrals, such as the center of mass, the RMS value of signals, probabilities of continuous random variables, the mass flow of fluids, or the calculation of the position from the signal measured by an accelerometer are only a very few of them. Luckily, from a numerical point of view, the problem can be solved, and, as discussed in the present chapter, a large number of very accurate algorithms exist to estimate the numerical values of definite integrals.

8.1 NUMERICAL INTEGRATION BASED ON POLYNOMIAL INTERPOLATION

Several strategies have been developed to estimate numerical definite integrals. A popular approach, which we will explore in this chapter, involves two main steps. First, the function $f(x)$ to be integrated is approximated by a polynomial $P_n(x)$. Second, the definite integral is approximated using that polynomial accordingly:

$$\int_a^b f(x)\,dx \cong \int_a^b P_n(x)\,dx \tag{8.1}$$

To calculate the polynomial $P_n(x)$, a polynomial interpolation can be used by choosing $n+1$ distinct interpolation points $x_o, x_1, \ldots, x_n$ in the integration interval $[a, b]$ as explained in Section 7.5. Figure 8.1 illustrates the concept of quadrature formulas developed using polynomial interpolation. The figure demonstrates a case where three interpolation points are selected to define the quadratic interpolating polynomial $P_2(x)$. Instead of integrating the function $f(x)$, the quadratic polynomial is integrated.

An important characteristic of interest is the error of a quadrature formula. It is defined as the difference between the true value of the definite integral and its estimation using the quadrature formula.

DOI: 10.1201/9781003262121-9

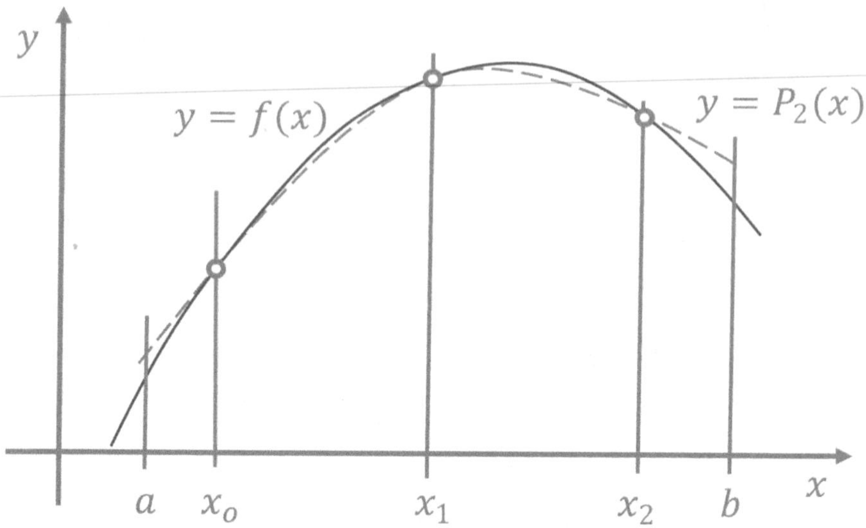

FIGURE 8.1 Replacing a function $f(x)$ (–) to be integrated by an interpolating polynomial $P_n(x)$ (––) computed based on $n+1$ distinct interpolation points $x_o, x_1, \ldots, x_n$ in the integration interval $[a, b]$.

Definition 8.1. Error of a Quadrature Formula

The error E of a quadrature formula is the difference between the true value of the definite integral and its estimation using the quadrature formula. In the case of quadrature formulas developed based on polynomial interpolation, the error is obtained as follows:

$$E = \left| \int_a^b f(x)\,dx - \int_a^b P_n(x)\,dx \right| = \int_a^b \left| f(x) - P_n(x) \right| dx \qquad (8.2)$$

where $P_n(x)$ is an interpolation polynomial computed based on $n+1$ distinct interpolation points $x_o, x_1, \ldots, x_n$ in the integration interval $[a, b]$.

Figure 8.2 provides a visual representation of the error of a quadrature formula based on a polynomial interpolation using two interpolating points $x_o = a$ and $x_1 = b$. In Section 8.3.2, we will derive approaches to estimate the truncation errors of quadrature formulas.

Let us illustrate the introduced concepts with an example.

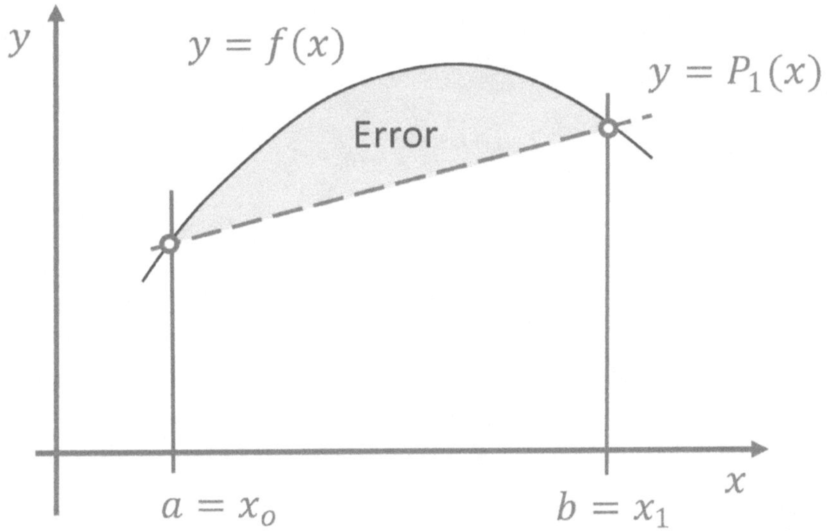

FIGURE 8.2 Error of a quadrature formula developed based on polynomial interpolation using two interpolating points $x_o = a$ and $x_1 = b$. The function $y = f(x)$ (–) is approximated by the interpolation polynomial $y = P_1(x)$ (––). The error of the quadrature formula is the surface between f and P_1.

Example 8.1. Trapezoidal Rule

In this example, we develop the quadrature formula that uses two interpolation points $x_o = a$ and $x_1 = b$. Figure 8.2 demonstrates why this formula is called the trapezoidal rule. The definite integral, which is the surface below the function $f(x)$, is replaced by the surface of a trapezoid formed by the interpolation polynomial $P_1(x)$.

The linear interpolation polynomial $P_1(x)$ can be written using Lagrange interpolation:

$$P_1(x) = f(a)\frac{x-b}{a-b} + f(b)\frac{x-a}{b-a} \tag{8.3}$$

To obtain the quadrature formula, we compute the definite integral of $P_1(x)$:

$$\int_a^b P_1(x)dx = f(a)\int_a^b \frac{x-b}{a-b}dx + f(b)\int_a^b \frac{x-a}{b-a}dx \tag{8.4}$$

To compute the quadrature formula, we have to integrate the two Lagrange polynomials L_1 and L_2:

$$\int_a^b \frac{x-b}{a-b}dx = \frac{1}{2}(b-a) \quad (8.5)$$

and

$$\int_a^b \frac{x-a}{b-a}dx = \frac{1}{2}(b-a) \quad (8.6)$$

Inserting the previous results into Equation (8.4) yields the trapezoidal rule:

$$\int_a^b f(x)dx \cong \int_a^b P_1(x)dx = (b-a)\frac{f(a)+f(b)}{2} \quad (8.7)$$

Let us now apply this quadrature formula to estimate the following definite integral:

$$I = \int_1^2 \text{In}(x)dx \quad (8.8)$$

From calculus, we know that the exact value is:

$$\int_1^2 \ln(x)dx = 2\ln(2) - \ln(1) - 1 \cong 0.3863 \quad (8.9)$$

Using our trapezoidal rule, we obtain:

$$\int_1^2 \ln(x)dx \cong (2-1)\frac{\ln(2)-\ln(1)}{2} \cong 0.3466 \quad (8.10)$$

In this example, the error is about:

$$E \cong |0.3863 - 0.3466| \cong 0.04 \quad (8.11)$$

Quadrature formulas can be characterized according to their degree of precision defined as follows:

Definition 8.2. Degree of Precision of a Quadrature Formula

The degree of precision of a quadrature formula is the highest degree of a polynomial, which will be integrated exactly with the quadrature formula.

Based on Definition 8.2, it follows that any quadrature formula constructed using an interpolation polynomial $P_n(x)$ of degree n, is at least of degree of precision n. A quadrature formula could be of a higher degree of precision, but only under certain circumstances we will discuss later.

8.2 NUMERICAL INTEGRATION BASED ON LAGRANGE INTERPOLATION

In Example 8.1, we used Lagrange interpolation to determine the interpolating polynomial that approximates the function $f(x)$ to be integrated. To find the corresponding quadrature formula, we had to integrate the individual Lagrange polynomials. This approach offers various advantages. In fact, the Lagrange interpolation polynomial $P_n(x)$ is generally written as follows (see Chapter 7):

$$P_n(x) = \sum_{i=0}^{n} f(x_i) L_i(x) \tag{8.12}$$

with $L_i(x)$ the $n+1$ Lagrange polynomials associated to the $n+1$ interpolation points. Using Lagrange interpolation is an excellent choice, as the quadrature formula becomes:

$$\int_a^b f(x)\,dx \cong \int_a^b P_n(x)\,dx = \sum_{i=0}^{n} \omega_i f(x_i) \tag{8.13}$$

with the coefficients ω_i calculated as

$$\omega_i = \int_a^b L_i(x)\,dx \tag{8.14}$$

The coefficients ω_i are often termed the weights of the quadrature formula. The practical value of this approach lies in the fact that the Lagrange polynomials $L_i(x)$ do not depend on the function $f(x)$, but only on the choice of the interpolation points. In other words, once we have chosen our interpolation points, we can compute the weights ω_i independently of $f(x)$. It becomes possible to establish tables for these weights that can then be used later with Equation (8.13) to estimate the definite integral of a function $f(x)$. In Example 8.1, we computed the weights ω_o and ω_1 for the trapezoidal rule when evaluating Equations (8.5) and (8.6).

The weights ω_i of a quadrature formula will always sum to $b-a$.

Theorem 8.1. Sum of Quadrature Weights

The sum of the weights of a Quadrature Formula (8.13) based on Lagrange interpolation is always equal to $b-a$.

To prove this theorem, consider the function $f(x)=1$. The interpolating polynomial of this function will be $P_n(x)=\sum_{i=0}^{n} L_i(x)$. The quadrature formula

(8.13) is at least a degree of precision 1. Consequently, we can calculate the exact value of the definite integral of $f(x)=1$:

$$b-a=\int_a^b 1\cdot dx=\int_a^b\left[\sum_{i=0}^{n}L_i(x)\right]dx=\sum_{i=0}^{n}\left[\int_a^b L_i(x)\,dx\right]=\sum_{i=0}^{n}\omega_i \tag{8.15}$$

which concludes the proof.

In practice, Theorem 8.1 can be used to verify if the computed weights of a quadrature formula seem to be correct or not. For example, the reader can verify that the weights ω_o and ω_1 of the trapezoidal rule computed in Equations (8.5) and (8.6) indeed sum to $b-a$.

Example 8.2. Open Trapezoidal Rule

In this example, we develop a quadrature formula using the two interpolation points $x_o=a+\frac{1}{3}(b-a)$ and $x_1=a+\frac{2}{3}(b-a)$. This results in a linear interpolating polynomial, as was the case in the trapezoidal rule, but this time, the polynomial will be slightly different as we change the interpolation points. This quadrature formula is often referred to as the open trapezoidal rule.

The linear interpolation polynomial $P_1(x)$ is obtained using Lagrange interpolation:

$$P_1(x)=f(x_o)\frac{x-x_1}{x_o-x_1}+f(x_1)\frac{x-x_o}{x_1-x_o} \tag{8.16}$$

The two quadrature weights are calculated as

$$\omega_o=\int_a^b\frac{x-x_1}{x_o-x_1}dx=\frac{b-a}{2} \tag{8.17}$$

and

$$\omega_1=\int_a^b\frac{x-x_o}{x_1-x_o}dx=\frac{b-a}{2} \tag{8.18}$$

As stated by Theorem 8.1, $\omega_0+\omega_1=b-a$.

The quadrature formula writes:

$$\int_a^b f(x)\,dx\cong\int_a^b P_n(x)\,dx=\omega_0 f(x_0)+\omega_1 f(x_1)=\frac{\mathrm{b}-\mathrm{a}}{2}\left[f(x_0)+f(x_1)\right] \tag{8.19}$$

8.3 NEWTON-COTES QUADRATURE

So far, we have discussed how to build a quadrature formula using polynomial interpolation to approximate the function $f(x)$ of the definite integral. This requires choosing some interpolation points $x_o, x_1, \ldots, x_n$ in the interval $[a, b]$. A straightforward choice consists in distributing these points uniformly over the interval $[a, b]$. This is exactly how Newton-Cotes[1] quadrature formulas proceed.

8.3.1 Newton-Cotes Formulas

Two families of Newton-Cotes formula exist depending on whether or not the endpoints a and b of the definite integral are used as interpolation points.

Definition 8.3. Definition Newton-Cotes Formulas

Newton-Cotes formulas are quadrature rules based on polynomial interpolation, built using uniformly distributed interpolation points $x_o, x_1, \ldots, x_n$, to approximate the definite integral of a function over the integration interval $[a, b]$.

- Closed Newton-Cotes formulas use the interpolating points $x_i = a + ih$, where $h = \frac{b-a}{n}$.
- Open Newton-Cotes formulas use the interpolating points $x_i = a + (i+1)h$, where $h = \frac{b-a}{n+2}$.

Using the methodology presented in Section 8.2, closed and open Newton-Cotes formulas can be derived for various values of n.

Table 8.1 lists four closed Newton-Cotes formulas. Note that n is, in agreement with the notation used in the present chapter, the degree of the interpolating

TABLE 8.1
Weights of Some Closed Newton-Cotes Formulas

n	A	w_o	w_1	w_2	w_3	w_4	Degree of precision
1	1/2	1	1				1
2	1/3	1	4	1			3
3	3/8	1	3	3	1		3
4	2/45	7	32	12	32	7	5

polynomial. Consequently, $n+1$ is the number of interpolation points and the number of weights ω_i. The weights ω_i of the quadrature formula, computed using Equation (8.14), are given by $\omega_i = hAW_i$ where $h = \frac{b-a}{n}$, the distance between the interpolating points. The numbers W_i in Table 8.1 are sometimes called the Cotes weights or Cotes numbers. Note that for each quadrature formula, the sum of the weights ω_i is equal to $b-a$, in agreement with Theorem 8.1. One can prove that if n is odd, the closed Newton-Cotes formulas have a degree of precision equal to n, but if n is even, one has a higher degree of precision equal to $n+1$.

The listed quadrature formulas are the trapezoidal rule ($n=1$), Simpson's-1/3 rule[2] ($n=2$), Simpson's-3/8 rule ($n=3$) and Boole's rule[3] ($n=4$). Example 8.1 shows, for the case of the trapezoidal rule, how the quadrature formulas are derived.

Example 8.3. Simpson's-1/3 Rule

Let us apply Simpson's-1/3 rule to the definite integral (8.8) already used to illustrate the trapezoidal rule in Example 8.1. From Table 8.1, we can write down Simpson's-1/3 rule as:

$$\int_a^b f(x)\,dx \cong \frac{h}{3}\left[f(a) + 4f\left(\frac{a+b}{2}\right) + f(b)\right] \tag{8.20}$$

with $h = \frac{b-a}{2}$.

Applying Simpson's-1/3 rule to the definite integral (8.8) results in:

$$\int_1^2 \ln(x)\,dx \cong \frac{2-1}{6}\left[\ln(1) + 4\ln(1.5) + \ln(2)\right] \cong 0.3858 \tag{8.21}$$

Comparing the previous numerical approximation with the value obtained using calculus via Equation (8.9), we can evaluate the error of Simpson's-1/3 rule in this case as follows:

$$E \cong \left|0.3863 - 0.3858\right| \cong 0.0005 \tag{8.22}$$

which is significantly lower than the error of the trapezoidal rule computed in Example 8.1.

The reader may have noted that the numbers W_i are symmetric in Table 8.1. This is not an accident but a consequence of the following theorem.

Theorem 8.2. Symmetry of Weights of Newton-Cotes Formulas

The weights ω_i of a Newton-Cotes quadrature formula are symmetric, meaning that $\omega_o = \omega_n$, $\omega_1 = \omega_{n-1}$, etc.

To prove this theorem, we start by noting that the weights ω_i are given by Equation (8.14), which represents the definite integral of the Lagrange polynomial $L_i(x)$. In the case of a Newton-Cotes formula, where the interpolation points are uniformly distributed across the integration interval, this polynomial is an odd function relative to the midpoint $(a+b)/2$ of the integration interval. Consequently, the definite integrals of, for example, $L_o(x)$ and $L_n(x)$, will be equal. The same holds for the definite integrals of $L_1(x)$ and $L_{n-1}(x)$ and so on, which concludes the proof.

There are closed Newton-Cotes quadrature formulas as well as open ones. Table 8.2 lists the first five open Newton-Cotes formulas. Here, as for Table 8.1, n is the degree of the interpolating polynomial and, consequently, $n+1$ is the number of interpolation points and weights ω_i. The weights ω_i of the quadrature formula are given by $\omega_i = hAW_i$ where $h = \dfrac{b-a}{n+2}$ is the distance between the interpolating points. Again, for each quadrature formula, the sum of the weights ω_i is equal to $b-a$, in agreement with Theorem 8.1. Similar to what we saw in Table 8.1, and for the same reasons, the weights are symmetric. Example 8.2 shows, for the case of the open trapezoidal rule, how to derive the values from Table 8.2.

We can prove that if n is odd, the open Newton-Cotes formulas have a degree of precision equal to n. When n is even, the open Newton-Cotes formulas have a degree of precision equal to $n+1$.

Some of the listed quadrature formulas have names such as the mid-point (or rectangle) rule ($n=0$), the open trapezoidal rule ($n=1$), and the Milne's rule[4] ($n=2$).

TABLE 8.2
Weights of Some Open Newton-Cotes Formulas

n	A	W_o	W_1	W_2	W_3	W_4	Degree of precision
0	2	1					1
1	3/2	1	1				1
2	4/3	2	-1	2			3
3	5/24	11	1	1	11		3
4	3/10	11	-14	26	-14	11	5

Example 8.4. Midpoint Rule

Derive the open 1-point Newton-Cotes formula.

For the open 1-point Newton-Cotes formula, the only interpolation point is $x_o = (a+b)/2$. Using Equation (8.14), we can compute the weight ω_o:

$$\omega_o = \int_a^b L_o(x)\,dx = \int_a^b 1 \cdot dx = b - a$$

Equation (8.13) yields the open 1-point Newton-Cotes formula:

$$\int_a^b f(x)\,dx \cong (b-a)\cdot f\left(\frac{a+b}{2}\right)$$

which is known as the midpoint rule.

In principle, we could use any number of interpolation points. However, Runge's phenomenon (wiggling of the interpolation polynomial as described in Section 7.5) could negatively affect quadrature rules for excessively large values of n. Furthermore, some coefficients ω_i may become negative. It can be shown that this makes the formulas sensitive to round-off errors. For the aforementioned reasons, we generally use a limited number of interpolation points. To further increase the accuracy associated with approximating definite integrals, other methods must be used. One of them is composite methods, as discussed in Section 8.3.3.

8.3.2 Truncation Errors of Newton-Cotes Formulas

In this section, we turn our attention to the important question of errors associated with using quadrature formulas. Definition 8.1 states what is the error of a quadrature formula. However, the error cannot be estimated without knowledge of the true value of the definite integral. In this section, we derive expressions for error estimations that do not require the knowledge of the true value of the definite integral.

Let us start by giving the general expression of the truncation error E of a Newton-Cotes formula using the expression of the error from polynomial interpolation derived in Theorem 7.1 of Chapter 7:

$$E = \int_a^b f(x)\,dx - \int_a^b P_n(x)\,dx = \int_a^b \frac{(x-x_o)\ldots(x-x_n)}{n+1!} f^{(n+1)}(c)\,dx \quad (8.23)$$

Let us draw the reader's attention to two common mistakes in relation to Equation (8.23).

First, note that the unknown point c is some number between a and b. It is important to realize that this number is, in fact, a function of x, as for every x, the error $f(x)-P_n(x)$ will be different, and, consequently, the number c will be different as well. It is wrong to believe that $f^{(n+1)}(c)$ is a constant that could be factorized out of the integral.

Furthermore, we note that generally, the function $q_n(x)=(x-x_o)\ldots(x-x_n)$ is not a purely positive or purely negative function on the interval $[a, b]$. Therefore, the second mean value theorem of definite integrals can not be used to simplify the expression of the error E.

In fact, deriving correct expressions for the truncation error of Newton-Cotes formulas is not a straightforward task. It can, however, be achieved (see, for example, the book *A first course in numerical analysis* by Ralston and Rabinowitz). In the case of closed Newton-Cotes formulas, it can be shown that there exists a number η in the interval $[a, b]$ such that

$$E=\frac{f^{(n+1)}(\eta)}{(n+1)!}\int_a^b(x-x_o)\ldots(x-x_n)\,dx \qquad \text{if } n \text{ odd} \tag{8.24}$$

$$E=\frac{f^{(n+2)}(\eta)}{(n+2)!}\int_a^b x\cdot(x-x_o)\ldots(x-x_n)\,dx \qquad \text{if } n \text{ even} \tag{8.25}$$

Equations (8.24) and (8.25) allow deriving expressions for the truncation errors for some Newton-Cotes formulas by computing the value of the definite integral over the function $(x-x_o)\ldots(x-x_n)$, respectively $x\cdot(x-x_o)\ldots(x-x_n)$.

Table 8.3 lists the truncation errors for some Newton-Cotes formulas together with the truncation errors for the open Newton-Cotes formulas. The definitions of n and h are the same as in Table 8.1 and Table 8.2. The quantity h is the distance between two interpolation points, as stated in Definition 8.3.

TABLE 8.3
Truncation Errors of Some Newton-Cotes Formulas. Refer to Definition 8.3 for the Values of *h*

n	Truncation Error for	
	Closed Newton-Cotes Formulas	**Open Newton-Cotes Formulas**
0	N.A.	$-\frac{1}{3}h^3 f^{(2)}(\eta)$
1	$-\frac{1}{12}h^3 f^{(2)}(\eta)$	$-\frac{1}{4}h^3 f^{(2)}(\eta)$
2	$-\frac{1}{90}h^5 f^{(4)}(\eta)$	$-\frac{28}{90}h^5 f^{(4)}(\eta)$
3	$-\frac{3}{80}h^5 f^{(4)}(\eta)$	$-\frac{95}{144}h^5 f^{(4)}(\eta)$
4	$-\frac{8}{945}h^7 f^{(6)}(\eta)$	$-\frac{41}{140}h^7 f^{(6)}(\eta)$

From Table 8.3, we can derive the degree of precision (Table 8.1 and Table 8.2) of the various quadrature formulas. Consider, for example, Simpon's-1/3 rule (the three points Newton-Cotes formula). Its truncation error is equal to a term involving the fourth derivative of $f(x)$. Consequently, applying that quadrature rule to any polynomial of degree three or less will result in a truncation error equal to zero. This means, according to Definition 8.2, that the degree of precision of Simpon's-1/3 rule is equal to three, as stated in Table 8.1.

As previously mentioned, the derivation of Equations (8.24) and (8.25) is not straightforward. In the case of the trapezoidal rule, the calculations are simple, and the second mean value theorem of definite integrals can be used as demonstrated in Example 8.5. However, for the Simpson's-1/3 rule and higher order methods, the calculations are difficult because the second mean value theorem of definite integrals can no longer be used. For open Newton-Cotes formulas, the difficulties begin at the first open Newton-Cotes rule, the mid-point rule.

Example 8.5. Truncation Error of the Trapezoidal Rule

Derive an expression for the truncation error of the trapezoidal rule.

For the trapezoidal rule, which uses the two interpolation points $x_o = a$ and $x_1 = b$, the truncation error is written using Equation (8.23):

$$E = \int_a^b f(x)dx - \int_a^b P_1(x)dx = \int_a^b \frac{(x-a)(x-b)}{2!} f^{(2)}[c(x)]dx \quad (8.26)$$

As $(x-a)(x-b) \leq 0$ in the integration interval $[a, b]$, we can use the second mean value theorem of definite integrals to simplify the expression of the truncation error to

$$E = \frac{f^{(2)}(\eta)}{2} \int_a^b (x-a)(x-b)dx = \frac{(a-b)^3}{12} f^{(2)}(\eta) = -\frac{h^3}{12} f^{(2)}(\eta) \quad (8.27)$$

with η some number in the interval $[a, b]$. This expression agrees with the general formula given by Equation (8.24).

The Newton-Cotes truncation errors of Table 8.3 can be employed to estimate the errors of practical situations, as illustrated in the following example.

Example 8.6. Estimating Truncation Errors of Newton-Cotes Formulas

In Example 8.1 and Example 8.3, we applied the trapezoidal and Simpon's-1/3 rules to numerically estimate the definite integral (8.8). Let us now illustrate how the expressions of the truncation errors of Newton-Cotes formulas from Table 8.3 can be used to estimate the truncation errors.

We first have to compute the second and fourth derivatives of $f(x)$, the natural logarithmic function in our example:

$$\frac{d^2}{dx^2}\ln(x) = -\frac{1}{x^2} \tag{8.28}$$

$$\frac{d^4}{dx^4}\ln(x) = -\frac{6}{x^4} \tag{8.29}$$

We don't know the value of the number η in the interval $[1, 2]$ needed to compute the truncation errors. But we can still find the maximal values of these truncation errors:

$$\max_{[1,2]}\left|\frac{(2-1)^3}{12}\cdot\frac{1}{x^2}\right| = \frac{1}{12} \cong 0.08 \tag{8.30}$$

$$\max_{[1,2]}\left|\frac{\left(\frac{2-1}{2}\right)^5}{90}\cdot\frac{6}{x^4}\right| = \frac{1}{480} \cong 0.002 \tag{8.31}$$

Comparing these error estimations with the true errors computed in Example 8.1 and Example 8.3, we observe that our error estimations are indeed conservative estimations and can consequently be used to give an upper bound to the truncation error.

8.3.3 Composite Newton-Cotes Formulas

Table 8.3, which lists the truncation errors of Newton-Cotes formulas, shows that as the number of interpolation points used to construct the quadrature method increases, the truncation error decreases. This is because the distance h between two interpolation points reduces, and its power increases with the number of interpolation points used. However, as discussed in Section 8.3, the Runge phenomenon and other effects can lead to numerically unstable methods as the number of interpolation points increases. If higher precision quadrature formulas are required, an alternative approach must be developed. One such approach is to use the additive property of definite integrals. For example, if we add a point, $c \in [a, b]$, we can write:

$$\int_a^b f(x)\,dx = \int_a^c f(x)\,dx + \int_c^b f(x)\,dx \tag{8.32}$$

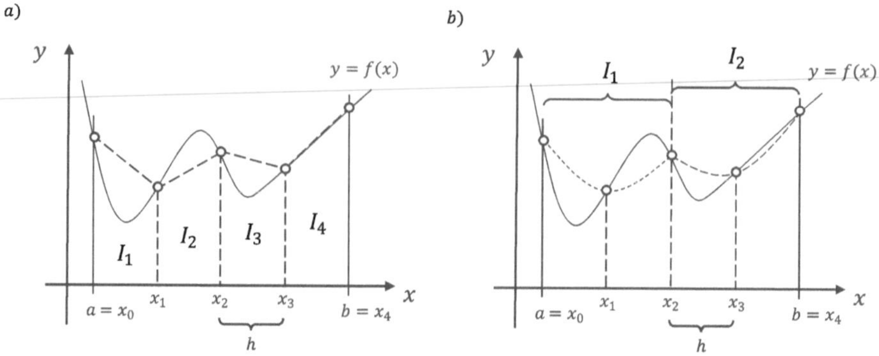

FIGURE 8.3 Two examples of composite Newton-Cotes formulas: (a) composite trapezoidal rule and (b) composite Simpson's-1/3 rule. For both examples, five interpolation points are used. This results into four panels for the composite trapezoidal rule and two panels for the composite Simpson's-1/3 rule.

This leads to the idea of composite Newton-Cotes formulas. The integration interval $[a, b]$ is subdivided into m equal-sized sub-intervals called panels. In each panel, a Newton-Cotes formula is used to approximate the definite integral. Depending on the chosen Newton-Cotes formula, each panel is subdivided into the required interpolation points. In this chapter, all the interpolation points within the integration interval $[a, b]$ are numbered consecutively as $x_o, x_1, \ldots, x_n$ where x_o is the first interpolation point in the first panel and x_n is the last interpolation point of the last panel.

Figure 8.3 provides two examples (the composite trapezoidal and composite Simpson's-1/3 rule). The reader is invited to carefully observe how the interpolation points are numbered on the figure. Different authors often use different conventions in numbering these points. This results in different composite Newton-Cotes formulas and is often a source of confusion when learning these methods.

For example, for the composite Simpson's-1/3 rule (Figure 8.3b), we used two panels. The first panel is the interval $[x_o = a, x_2]$ and the second panel is the interval $[x_2, x_4 = b]$. Each panel contains three interpolation points, as needed by the Simpson's-1/3 rule. In all cases, we keep our definition of h as being the distance between two interpolation points. The reader should be careful not to confuse h with the size of a panel. In the case of the composite trapezoidal rule, it happens that h also represents the size of a panel. However, this isn't universally true; for instance, in the composite Simpson's 1/3 rule, h is half the size of a panel.

Example 8.7. Composite Trapezoidal Rule

Let us show how to use the composite trapezoidal rule to estimate numerically the definite integral (8.8) from Example 8.1. We will use four panels:

$$\int_{1}^{2} \ln(x)\,dx = \int_{1}^{1.25} \ln(x)\,dx + \int_{1.25}^{1.5} \ln(x)\,dx + \int_{1.5}^{1.75} \ln(x)\,dx + \int_{1.75}^{2} \ln(x)\,dx \quad (8.33)$$

In each panel, we apply the trapezoidal rule. For example, in the second panel:

$$\int_{1.25}^{1.5} \ln(x)\,dx \cong (1.5-1.25)\frac{\ln(1.5)-\ln(1.25)}{2} \cong 0.0228 \quad (8.34)$$

Summing the results on all panels yields:

$$\int_{1}^{2} \ln(x)\,dx \cong 0.3837 \quad (8.35)$$

Comparing with the exact solution (8.9) obtained from calculus gives us the following error:

$$E \cong |0.3863-0.3837| \cong 0.003 \quad (8.36)$$

Note that this error is significantly lower than the error from Example 8.1, where we applied the trapezoidal rule.
As an exercise, the reader is invited to repeat the same calculations but using Simposn's-1/3 rule, which will lead to the numerical estimation of the definite integral to be equal to 0.3863, identical with the answer from calculus (the difference will start to appear after the fifth significant digit).

The approach (explicitly computing in each panel the value of the numerical estimation of the definite integral using a given quadrature rule) used in Example 8.7 is well suited for hand calculations. It is, however, not very efficient. For example, for every start and endpoint of the panels, the function $f(x)$ will be evaluated twice (once for the left panel and once for the right panel). It is possible to develop compact formulas for composite methods that avoid evaluating the function f multiple times. As an example, let us show it for the composite trapezoidal method.

We start by dividing the integration interval into m panels. Defining $h = \frac{b-a}{m}$, then the interpolation points are $x_i = a + ih$ with $i = 1, 2, \ldots, m$. With these notations, we obtain for composite trapezoidal method:

$$\int_{a}^{b} f(x)\,dx \cong h\left[\frac{1}{2} f(x_o) + f(x_1) + \ldots f(x_{m-1}) + \frac{1}{2} f(x_m)\right] \quad (8.37)$$

which can be written as:

$$\int_a^b f(x)\,dx \cong \frac{h}{2}\left[f(a)+f(b)+2\sum_{i=1}^{m-1} f(x_i)\right] \tag{8.38}$$

Equation (8.38) is known as the composite trapezoidal rule.

Similarly, we can derive expressions for other composite Newton-Cotes formulas. For example, the composite Simpson's-1/3 rule writes:

$$\int_a^b f(x)\,dx \cong \frac{h}{3}\left[f(a)+f(b)+4\sum_{i=1}^{m} f(x_{2i-1})+2\sum_{i=1}^{m-1} f(x_{2i})\right] \tag{8.39}$$

with $h=\dfrac{b-a}{2m}$ and $x_i = a+ih$.

The composite midpoint method writes as:

$$\int_a^b f(x)\,dx \cong h\sum_{i=0}^{m-1} f(x_i) \tag{8.40}$$

with $h=\dfrac{b-a}{m}$ and $x_i = a+h\left(\dfrac{1}{2}+i\right)$.

Equations such as (8.38), (8.39), and (8.40) can be used to code efficiently composite Newton-Cotes quadrature formulas. The following example gives a possible implementation in Octave.

Example 8.8. Composite Newton-Cotes Formulas in Octave

Let us demonstrate how to employ the composite trapezoidal rule (8.38) to approximate the definite integral (8.8) of Example 8.1. We will use $m=4$ panels.

We start by defining the function that needs integrating and the interpolation points:

```
>> f = @(x) log(x);
>> a = 1; b = 2;
>> m = 4;
>> x = linspace(a, b, m+1);
```

The parameter h, the distance between two interpolation points, is given by

```
>> h = (b-a)/m;
```

Equation (8.38) becomes:

```
>> I = h/2*(2*sum(f(x)) - f(a)-f(b))
I = 0.38370
```

Note that, as the Octave function call `sum(f(x))` computes the sum of all $f(x_i)$, including for $x_o = a$ and $x_m = b$, we need to subtract $f(a)$ and $f(b)$ to reproduce Equation (8.38).

Similarly, we can implement the composite Simpon's-1/3 rule (8.39):

```
>> f = @(x) log(x);
>> a = 1; b = 2;
>> m = 4;
>> h = (b-a)/(2*m);
>> x1 = linspace(a+h, b-h, m);
>> x2 = linspace(a+2*h, b-2*h, m-1);
>> I = h/3*(f(a)+f(b)+4*sum(f(x1))+2*sum(f(x2)))
I = 0.38629
```

Note that h, the distance between two interpolation points, needs to be computed differently than in the case of the trapezoidal rule. Furthermore, two sets of interpolation points, stored in the vectors `x1` and `x2`, need to be computed.

8.3.4 Truncation Errors of Composite Newton-Cotes Formulas

An expression for the truncation errors of composite formulas can be derived by leveraging the truncation errors of the corresponding Newton-Cotes formulas from Table 8.3. To carry out these calculations, we need to utilize the generalized intermediate value theorem.

Theorem 8.3. Generalized Intermediate Value Theorem

Let f be a continuous function on the interval $[a, b]$ and let $x_1, x_2, \ldots, x_n$ be points in $[a, b]$. Given some strictly positive numbers $a_1, a_2, \ldots, a_n > 0$, there exists a number c such that

$$(a_1 + a_2 + \ldots + a_n) f(c) = a_1 f(x_1) + a_2 f(x_2) + \ldots + a_n f(x_n) \quad (8.41)$$

Theorem 8.3 is a generalization of the intermediate value theorem. If $f(x_i)$ is the smallest and $f(x_j)$ the largest value among the function values $f(x_1), f(x_2), \ldots, f(x_n)$, then we have

$$\begin{aligned} a_1 f(x_i) + \ldots + a_n f(x_i) &\leq a_1 f(x_1) + \ldots + a_n f(x_n) \\ &\leq a_1 f(x_j) + \ldots + a_n f(x_j) \end{aligned} \quad (8.42)$$

from which it follows that:

$$f(x_i) \le \frac{a_1 f(x_1) + \ldots + a_n f(x_n)}{a_1 + \ldots + a_n} \le f(x_j) \tag{8.43}$$

Using the intermediate value theorem, we can conclude that there exists a number c such that

$$f(c) = \frac{a_1 f(x_1) + \ldots + a_n f(x_n)}{a_1 + \ldots + a_n} \tag{8.44}$$

which concludes the proof of Theorem 8.3.

Example 8.9. Truncation Error of the Composite Trapezoidal Rule

In this example, we derive an expression for the truncation error of the composite trapezoidal rule based on the truncation error of the trapezoidal rule and Theorem 8.3.

In order to apply the composite trapezoidal rule, we divide the integration interval $[a, b]$ into m equal-sized panels. The interpolation points become $x_i = a + ih$ with $h = (b-a)/m$. In each panel, we apply the trapezoidal rule, which, according to Table 8.3, is:

$$\int_{x_i}^{x_{i+1}} f(x)dx = \frac{h}{2}\left[f(x_i) + f(x_{i+1})\right] - \frac{h^3}{12} f''(c_i) \tag{8.45}$$

with c_i an unknown number in the interval $[x_i, x_{i+1}]$. Summing all these approximations over all panels results in

$$\int_a^b f(x)dx = \frac{h}{2}\left[f(a) + f(b) + 2\sum_{i=1}^{m-1} f(x_i)\right] - \sum_{i=0}^{m-1} \frac{h^3}{12} f''(c_i) \tag{8.46}$$

The error term can be simplified using Theorem 8.3:

$$\sum_{i=0}^{m-1} \frac{h^3}{12} f''(c_i) = m\frac{h^3}{12} f''(c) = \frac{b-a}{12} h^2 f''(c) \tag{8.47}$$

with c some point in $[a, b]$.

Similarly to Example 8.9, one can derive the expressions for the truncation error of other composite Newton-Cotes formulas. Table 8.4 lists three examples.

The composite Trapezoidal and midpoint rules are both of second order in h. Consequently, they will produce approximations of similar accuracy when using

TABLE 8.4
Truncation Errors of Some Composite Newton-Cotes Formulas

Composite Newton-Cotes Formulas	Truncation Error
Trapezoid	$-\frac{b-a}{12}h^2 f''(c)$
Simpson's-1/3	$-\frac{b-a}{180}h^4 f^{(4)}(c)$
Midpoint	$-\frac{b-a}{24}h^2 f''(c)$

interpolation points separated by the same distance h. The composite Simpson 1/3 rule is significantly better, as it is in order four. The following example illustrates this fact.

Example 8.10. Order of Convergence of Composite Quadrature Methods

In this example, we will verify by evaluating the definite integral

$$\int_0^{\pi} \sin(x)\, dx = 2 \tag{8.48}$$

with the composite trapezoidal and Simpson's-1/3 rules, the orders of the composite quadrature formulas are two and four, respectively.

We perform successive applications of the composite trapezoidal (8.38) and Simpson's-1/3 rule (8.39) with increasing numbers of panels. Next, we calculate the true error by comparing the numerical approximations to the true value. This allows for the construction of a graph of the true error as a function of h.

In Octave, we can achieve this as follows. First, we define, similarly to Example 8.6, a function to compute the numerical approximation of a definite integral. The following is the code for the trapezoidal composite rule:

```
>> function I = trapezoidal(f, a, b, m)
>> x = linspace(a, b, m+1);
>> h = (b-a)/m;
>> I = h/2*(2*sum(f(x)) - f(a)-f(b));
>> endfunction
```

Then, we compute the numerical approximations of the definite integral (8.48) for different numbers of panels and store them in the column vector `I`. For the number of panels m, we start with one panel and then successively double them. At the same time, we compute h, which we store in the column vector h. Recall that h is the difference between two interpolating points x_i and x_{i+1}. If m is the number of panels used, then h can be calculated as $h = \pi/m$ for the composite trapezoidal rule, and $h = \pi/2m$ for Simpson's-1/3 composite rule.

```
>> f = @(x) sin(x);
>> m = 1;
>> I = []; h = [];
>> for i = 1:13
>> I = [I; trapezoidal(f, 0, pi, m)];
>> h = [h; pi/m];
>> m = m*2;
>> endfor
```

Finally, we compute the true errors, which we store in the column vector err. We can then plot the results in a log-log plot:

```
>> err = abs(I-2);
>> loglog(h, err, 'o')
```

To find the order of convergence of our calculations, we fit a power law $E = Kh^q$ over the errors (see Example 6.8 of Chapter 6 for the details):

```
>> A = [h.^0 log(h)];
>> a = A\log(err)
```

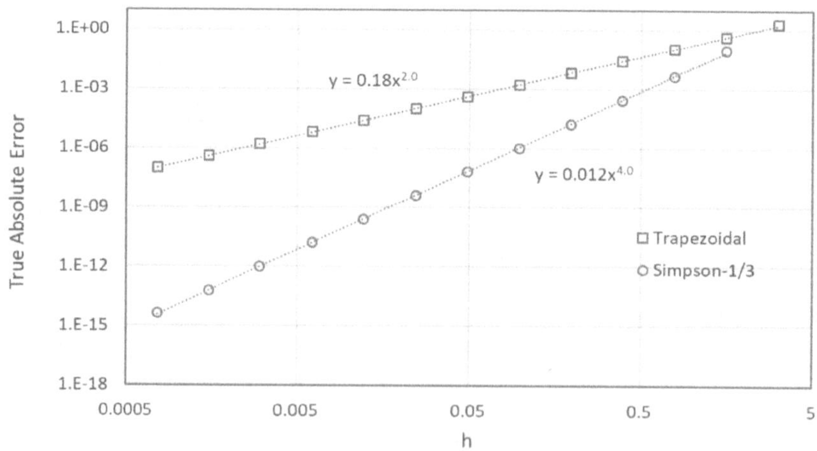

FIGURE 8.4 True absolute error of the trapezoidal and Simpons-1/3 rule when approximating the definite integral $\int_0^{\pi} \sin(x)\,dx$.

```
a =
    -1.7380
     2.0114
```

This confirms that the order of convergence of the error is quadratic ($q \cong 2.0114$).

Similarly, calculations can be repeated for Simpon's-1/3 composite method. Figure 8.4 summarizes the results and confirms that the composite trapezoidal method is in order two, whereas the composite Simpson 1/3 rule is in order four.

Note that to derive the expressions of Table 8.4, the function f must be sufficiently derivable and continuous. If that hypothesis is not satisfied, the truncation error can be of a different order in h than shown in Table 8.4. This is illustrated in Example 8.11.

Example 8.11. Reduced Order of Convergence of Composite Methods

Let us study the order of composite Newton-Cotes formulas applied to the following definite integral:

$$\int_0^1 \sqrt{x}\, dx = \frac{2}{3} \tag{8.49}$$

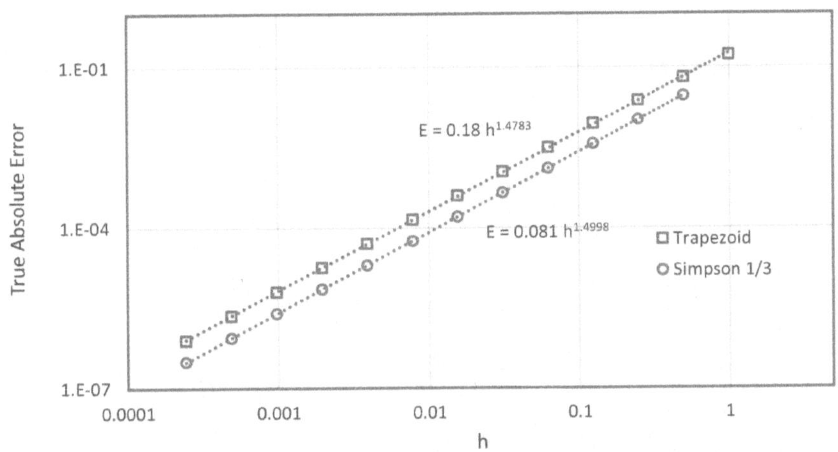

FIGURE 8.5 True absolute error of the trapezoidal and Simpsons-1/3 rule when approximating the definite integral $\int_0^1 \sqrt{x}\, dx$.

We start by noting that both the second and fourth derivatives of $f(x) = \sqrt{x}$ diverge in $x = 0$. Because of this fact, the truncation errors of Table 8.3 and Table 8.4 are no longer valid.

Successive application of the composite trapezoidal (8.38) and Simpson's-1/3 rule (8.39) followed by calculating the true error by comparing the approximations with the true value, allows constructing Figure 8.5 similarly to Example 8.10.

Fitting a power law $E = Kh^q$ over the errors (Figure 8.5) shows that the composite trapezoidal method and the composite Simpson's-1/3 rule are, in this case, about 1.5, significantly lower than in the case where the results of Table 8.4 can be applied.

As demonstrated in Example 8.11, if the function to be integrated isn't sufficiently derivable, then the results from Table 8.4 are, in general, not valid. As will be demonstrated in Example 8.12, the results from Table 8.4 give the minimal order of convergence (for the case where the function to be integrated is sufficiently derivable). It can happen that the order of convergence is higher.

Example 8.12. Increased Order of Convergence of Composite Methods

Let us study the order of the composite trapezoidal and Simpson's-1/3 formulas applied to the definite integral:

$$\int_0^1 e^x \cdot [1 - \cos 2\pi x] \, dx = \frac{4(e-1)\pi^2}{1+4\pi^2} \tag{8.50}$$

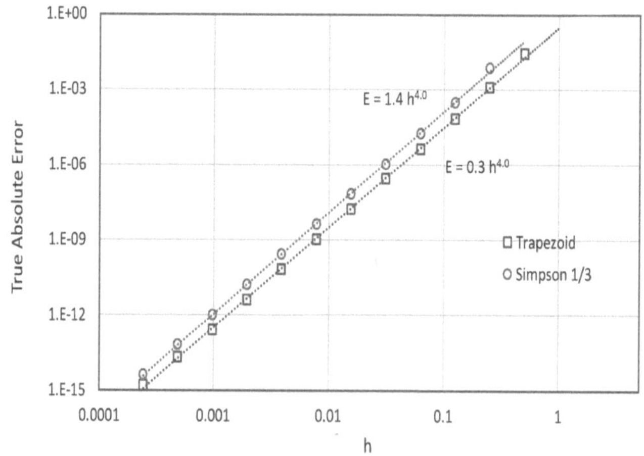

FIGURE 8.6 True absolute error of the trapezoidal and Simpons-1/3 rule when approximating the definite integral $\int_0^1 e^x \cdot [1 - \cos 2\pi x] dx$.

In the present case, all conditions required for the results of Table 8.4 to be valid are met. Successive applications of the composite trapezoidal (8.38) and composite Simpson 1/3 rule (8.39) followed by calculating the true error by comparing the approximations with the true value, allow constructing Figure 8.6 in a similar way as in Example 8.10.

We note that in this case, both methods, the composite trapezoidal and Simpson's-1/3 rule, are of order four. The trapezoidal method is, for this case, as precise as Simpson's-1/3 rule.

8.3.5 Estimating Truncation Errors of Composite Quadrature Formulas

Example 8.11 and Example 8.12 showed that the truncation error of a composite method may behave differently than the expected expressions listed in Table 8.4. This happens, for example, if the function f to be integrated is not sufficiently derivable. But, as shown in Example 8.12, sometimes a method behaves differently, even if the function f is sufficiently derivable (we will discuss the reason for this in Section 8.3.6). Such cases are rather difficult to detect, and it would be desirable to have a method that allows estimating the truncation error in a more practical way, without having to rely on the results of Table 8.4. This can be achieved and is the topic of the present section.

We start by noting that a given definite integral can be evaluated with the same composite quadrature formula but using two different numbers of panels, m_1 and m_2, resulting in two different values, h_1 and h_2. If the composite method is of order q, then we can write:

$$\int_a^b f(x)\,dx = I(h_1) + E(h_1) = I(h_2) + E(h_2) \tag{8.51}$$

where $I(h)$ stands for the numerical approximation computed by the composite method, with h being the distance between two interpolation points and $E(h)$ the corresponding truncation error. Assuming h_1 and h_2 are sufficiently small so the truncation error can be expressed as a power law $E = Kh^q$, then we can write:

$$\frac{E(h_1)}{E(h_2)} \cong \frac{h_1^q}{h_2^q} \tag{8.52}$$

Combining Equations (8.51) and (8.52) and solving for $E(h_2)$ results in Richardson's error estimation formula.[5]

Theorem 8.4. Richardson's Truncation Error Estimation

If $I(h_1)$ and $I(h_2)$ are two approximations produced with two-step sizes h_1 and h_2 by a composite quadrature formula in order $O(h^q)$ of a definite integral, the truncation error $E(h_2)$ of $I(h_2)$ is given by

$$E(h_2) \cong \left| \frac{I(h_2) - I(h_1)}{\left(\frac{h_1}{h_2}\right)^q - 1} \right| \tag{8.53}$$

as long as h_1 and h_2 are such that the truncation errors can be expressed reasonably as $E(h) = Kh^q$.

Richardson's error formula is of immense practical value as it allows a simple and precise estimation of the truncation error of a composite quadrature rule.

Example 8.13. Estimating Truncation Errors With Richardson's Method

Verify the validity of Richardson's error formula (8.53) by applying it to the numerical approximation of the definite integral using the trapezoidal composite rule:

$$\int_0^{\pi} \sin(x)\,dx = 2 \tag{8.54}$$

Using the numerical approximations of the definite integral, we can compute the error estimation using Richardson's error formula (8.53) and compare it with the true error. For example, applying the trapezoidal rule for $h_1 = \frac{\pi}{10}$ (i.e., $m = 10$ panels) and $h_2 = \frac{\pi}{20}$ (i.e., $m = 20$ panels), we obtain (the reader can use the codes from Example 8.8 to conduct the calculations):

$$E\left(h_2 = \frac{\pi}{20}\right) \cong \left| \frac{I(h_2) - I(h_1)}{2^2 - 1} \right| \cong \left| \frac{1.9959 - 1.9835}{3} \right| \cong 0.04 \tag{8.55}$$

The true error is:

$$\left| I(h_2) - 2 \right| \cong \left| 1.9959 - 2 \right| \cong 0.04 \tag{8.56}$$

This is in excellent agreement with the error estimation. Proceeding similarly with the other values of h shows an excellent match between both.

A key hypothesis behind the Richardson error formula (8.53) is that the values of h must be small enough to be able to represent the error by the power low

$E(h) = Kh^q$. Practically, this means one first has to find the range of values of h for which the hypothesis $E(h) = Kh^q$ is valid. As illustrated in Example 8.14, this task is simple to execute.

Example 8.14. Valid Range Of H in the Richardson Error Formula

Let us find the range for h where Richardson's error formula (8.53) can be used to estimate the truncation error of the definite integral:

$$I = \frac{2}{\sqrt{\pi}} \int_0^3 e^{-x^2} dx \tag{8.57}$$

when using the composite trapezoidal and Simpson's-1/3 rule. Furthermore, we will provide an approximation of the definite integral I with an absolute error below 10^{-9}.

To solve this problem, we start by applying the composite trapezoidal and Simpson's-1/3 rules for different values h to produce approximations of the definite integral I. Using the same function `trapezoidal` we defined in Example 8.10, we can achieve this in Octave:

```
>> f = @(x) 2/sqrt(pi)*exp(-x.^2);
>> m = 1;
>> I = []; h = [];
>> for i = 1:25
>> I = [I; trapezoidal(f, 0, 3, m)];
>> h = [h; 3/m];
>> m = m*2;
>> endfor
```

Note how the function `f` is defined. As this function has to handle vectors, the `.^` operator must be used. The `for` loop calculates the definite integral approximations $I(h)$ and the used values h to produce them in the two-column vectors `I` and `h`.

Next, we apply Richardson's error formula (8.53) to produce estimations of the truncation error in function of the used value h. As we double our number of panels in each step, the ratios between h_1 and h_2 will be equal to two. In Octave, we can compute the error estimations (stored in the column vector `err`) using Richardson's error formula (8.53) like this:

```
>> err = [];
>> for i = 2:length(I)
>> err = [err; abs(I(i)-I(i-1))/(2^2-1)];
>> endfor
```

The reader is invited to pay attention to the indexes used in column vector `I`. The same calculation can be achieved in a single call taking advantage of the vector notations of Octave (the vector index `end` is an Octave shortcut for the last element in the vector):

```
>> err = abs(I(2:end) - I(1:end-1)) / (2^2-1);
```

We can now present our results in a log-log plot:

```
>> loglog(h(2:end), err, 'o')
```

The reader should pay attention to how the correct values of h are matched with the corresponding error estimations.

Repeating the same calculations with Simpson's-1/3 composite method allows for the production of Figure 8.7, which shows the estimated truncation error versus h in a log-log scale. Inspecting the graphs, we identify values of h for which the hypothesis $E(h) = Kh^q$ is valid. This corresponds to values of h resulting in a straight line for the dependence of the error in the log-log scale. For these values of h, Richardson's error formula can be used to estimate accurately the truncation error. Let us explain why, for the other values of h, Richardson's formula will fail.

For too large values h (composite trapezoidal rule for values larger than about 0.5 and composite Simpson's 1/3 rule for values larger than about 0.2), the truncation errors do not yet follow the power law $E(h) = Kh^q$.

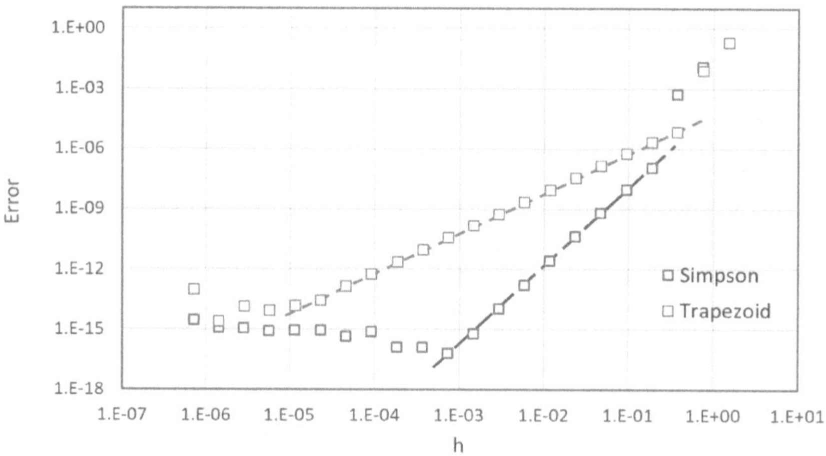

FIGURE 8.7 Truncation error estimations using Richardson's error formula for the composite trapezoidal and Simpson's-1/3 methods.

On the other side, for too-small values h (in the case of the composite trapezoidal rule for values smaller than about $1 \cdot 10^{-3}$ and for the composite Simpson's 1/3 rule for values smaller than about $2 \cdot 10^{-5}$), the total error isn't only determined by the truncation errors, but round-off errors become significant too. Richardson's formula can't estimate properly the total error (which is the sum of the truncation and round-off errors), as it will estimate only the contribution from the truncation errors.

Now that we know the range of values h where Richardon's formula is valid, we can select a particular h to compute an approximation of the definite integral I with an absolute error below 10^{-9}. According to Figure 8.7, we can, for example, choose $h = 10^{-3}$, which corresponds to $m = 3000$ panels in the case of the composite trapezoidal rule and to $m = 1500$ panels for the composite Simpson rule. For both cases, the approximations are below the requested precision of 10^{-9} and in the range where Richardson's error formula is valid. Now, using Simpson's-1/3 composite rule, we find:

$$I\left(h = 10^{-3}\right) \cong 0.999977909503001 \tag{8.58}$$

This approximation has an absolute error below 10^{-9}. In fact, according to Figure 8.7 Truncation error estimations using Richardson's error formula for the composite trapezoidal and Simpson's-1/3 methods, the absolute error is much smaller than 10^{-9} and is about 10^{-16}, indicating that almost all 16 digits of our approximation are correct. Note that we would not be able to achieve this accurate result with the composite trapezoidal method, which, at best, is able to give an approximation with an absolute error of about 10^{-14}.

The methodology presented in Example 8.14 not only allows identifying regions where the total error is dominated by the truncation errors (and consequently the total error follows a power law $E(h) = Kh^q$), but can also determine the order of the composite method used. Often, the order of the method is known. But, as discussed previously, if the function to be integrated isn't sufficiently derivable, the results from Table 8.4 are, in general, not valid. In such cases, the order of the method needs to be estimated.

This estimation can be obtained by analyzing the quantity $\Delta I(h) = |I(h) - I(2h)|$. From Richardson's formula (8.53) and $E(h) = Kh^q$ it follows:

$$\Delta I(h) = |I(h) - I(2h)| \cong \left[2^q - 1\right] K h_2^q = \tilde{K} h_2^q \tag{8.59}$$

Equation (8.59) shows that fitting a power low to $\Delta I(h)$ allows the estimation of the order q of the quadrature formula. Once the order q is known, the correct error estimation of $I(h)$ can be obtained from Richardson's formula. This approach is illustrated in Example 8.15.

Example 8.15. Estimating the Order of Convergence of a Composite Method

Let us estimate the order of the truncation error of the composite Simpson's-1/3 rule applied to:

$$I = \int_0^2 \sqrt{x + \sin x}\, dx \tag{8.60}$$

Start by noting that the first derivative of the integrant is:

$$\frac{d}{dx}\sqrt{x + \sin x} = \frac{\cos x + 1}{2\sqrt{x + \sin x}} \tag{8.61}$$

The first derivate diverges in $x = 0$ and, consequently, we know that the results of Table 8.4 are not valid. This implies that the order of convergence of the Simpson'-1/3 rule does not necessarily have to be equal to four.

To estimate the correct order of convergence, we start by applying the composite Simpson's-1/3 rule for different values of h to produce approximations $I(h)$ of the definite integral I. We choose the values $h = 1, 0.5, 0.25, \ldots$ that correspond to $m = 1, 2, 4, \ldots$ panels.[6] In Octave:

```
>> function I = simp(f, a, b, m)
>> h = (b-a)/(2*m);
>> x1 = linspace(a+h, b-h, m);
>> x2 = linspace(a+2*h, b-2*h, m-1);
>> I = h/3*(f(a)+f(b)+4*sum(f(x1))+2*sum(f(x2)));
>> endfunction
>> f = @(x) sqrt(x+sin(x));
>> m = 1;
>> I = []; h = [];
>> for i = 1:25
>> I = [I; simp(f, 0, 2, m)];
>> h = [h; 2/(2*m)];
>> m = m*2;
>> endfor
```

Next, we compute $\Delta I(h) = |I(h) - I(2h)|$ in function of h and fit a power law $\Delta I(h) = \tilde{K}h^q$.

```
>> deltaI = abs(I(1:end-1) - I(2:end));
>> A = [h(1:end-1).^0 log(h(1:end-1))];
>> a = A\log(deltaI)
a =
  -2.6055
   1.4997
```

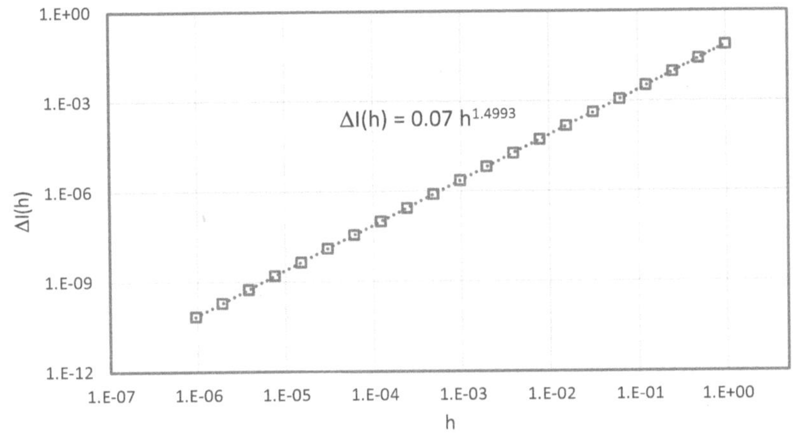

FIGURE 8.8 Truncation error estimations using Richardson's error formula for the composite Simpson's-1/3 rule applied to the definite integral $I = \int_0^2 \sqrt{x + \sin x}\, dx$.

From the fit, we estimate the order of the composite Simpson's-1/3 rule to be about 1.5. We further note (Figure 8.8) that in our case, the quantity $\Delta I(h)$ follows a power law over the full range of values h. Consequently, the truncation error will also follow a power law $E(h) = Kh^{1.5}$.

Knowing the order of convergence, we can use Richardson's formula (8.53) the usual way. For example, we can use the two approximations $I(0.02)$ and $I(0.01)$ to estimate the truncation error of $I(0.01)$:

$$E(0.01) \cong \left| \frac{I(0.01) - I(0.02)}{(0.02/0.01)^{1.5} - 1} \right| \cong \left| \frac{2.4902485 - 2.4900386}{2^{1.5} - 1} \right| \cong 1 \cdot 10^{-4} \tag{8.62}$$

We can conclude that:

$$I = \int_0^2 \sqrt{x + \sin x}\, dx = 2.4902 \pm 0.0001 \tag{8.63}$$

Note that if we had wrongly used an order of convergence equal to four, which corresponds to the value indicated by Table 8.4, we would have estimated the truncation error to be about $1 \cdot 10^{-5}$, which is a factor 10 too small.

8.3.6 Truncation Error of the Composite Trapezoidal Method

Before leaving the discussion about Newton-Cotes quadrature methods, let us discuss in more detail the truncation error of the trapezoidal method. This discussion will be used to develop another type of quadrature formula discussed in the next section, 8.4.

Example 8.12 demonstrated that there must exist cases where the truncation error of the composite trapezoidal method is higher than two. This comes from the very particular form the truncation error of the composite trapezoidal method takes. We can demonstrate the following theorem.

Theorem 8.5. Truncation Error of Composite Trapezoidal Method

For a function f sufficiently derivable, one has

$$\int_a^b f(x)dx = \frac{h}{2}\left[f(a)+f(b)+2\sum_{i=1}^{m-1} f(x_i)\right] - \frac{h^2}{12}\left[f'(b)-f'(a)\right] + O\left(h^4\right) \quad (8.64)$$

with $h = \dfrac{b-a}{m}$ and m an integer number.

In fact, the truncation error E of the composite trapezoidal rule can be written as

$$E = -\left[f'(b)-f'(a)\right]\frac{h^2}{12} + \left[f^{(3)}(b)-f^{(3)}(a)\right]\frac{h^4}{720} - \dots \quad (8.65)$$

Only even orders of h appear. This result follows from the Euler-Maclaurin formula, and the interested reader can find more details in *A first course in numerical analysis* by Ralston and Rabinowitz.

Theorem 8.5 explains why, in Example 8.12, the composite trapezoidal rule is of fourth order. Indeed, we have in this particular case that:

$$f'(x) = \frac{d}{dx}\left\{e^x \cdot \left[1-\cos 2\pi x\right]\right\} = e^x \cdot \left[2\pi \sin 2\pi x - \cos 2\pi x + 1\right] \quad (8.66)$$

and we realize that $f'(0) = f'(1) = 0$ such that according to Equation (8.64), the trapezoidal composite method will be in order four.

Theorem 8.5 can be used to increase dramatically the precision of the composite trapezoidal method if the derivatives of the function f to be integrated are known. For a function given explicatively, this task is possible. If the function isn't available (e.g., tabulated or the output of another algorithm), then the derivatives of f can be estimated using numerical derivation. This way of proceeding would result in Gregory's algorithm. The next example illustrates the first approach.

Example 8.16. Increasing the Precision of the Composite Trapezoidal Rule

Let us use Theorem 8.5 to improve the composite trapezoidal rule to an order four method for the following definite integral:

$$\int_0^{\pi} \sin(x)\,dx = 2 \tag{8.67}$$

As the first derivate of $\sin x$ is $\cos x$, we obtain, based on Equation (8.64) and after simplification, a first corrected trapezoidal rule:

$$\int_0^{\pi} \sin(x)\,dx = T(h) - \frac{h^2}{6} + O(h^4) \tag{8.68}$$

where $T(h)$ stands for the composite trapezoidal method with step size h calculated according to Equation (8.38).

We can correct further the composite trapezoidal method using Equation (8.65) to obtain a second corrected trapezoidal rule:

$$\int_0^{\pi} \sin(x)\,dx = T(h) - \frac{h^2}{6} - \frac{h^4}{360} + O(h^6) \tag{8.69}$$

Successive applications of the composite trapezoidal (8.38) and corrected trapezoidal rules (8.68) and (8.69), followed by calculating the true error by comparing the numerical approximations with the true value, allow the construction of the following Figure 8.9.

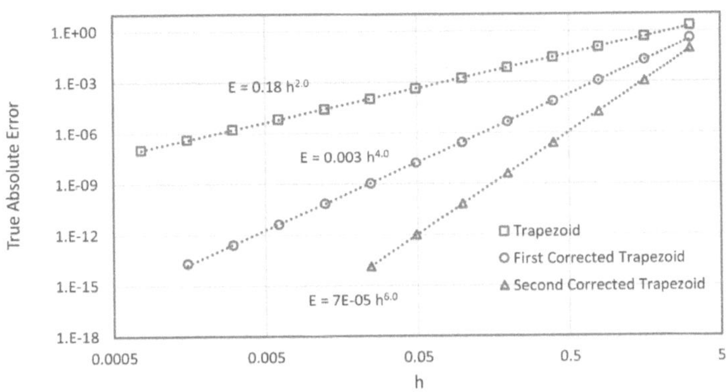

FIGURE 8.9 True errors for three quadrature methods applied to the definite integral $I = \int_0^{\pi} \sin(x)\,dx$.

As expected, the first and second corrected trapezoidal rules are now of order four and six, respectively, as show the fits of a power law over the errors.

Correcting the composite trapezoidal rule using the explicit expression of the truncation error according to Equation (8.65) allows deriving, at relatively low cost, higher order methods. However, we must compute the derivatives of the function to be integrated. The integration method presented in the next section, 8.4, will achieve, at a very similar cost, higher order methods too, but without the need to compute the derivatives of the function to be integrated.

8.4 ROMBERG INTEGRATION

Richardson's error formula (8.53) provides a precise estimation of the truncation error of a composite Newton-Cotes formula. Considering its precision, it is tempting to use that estimation to correct the numerical approximation of the definite integral with this error term. This idea turns out to be valid, and its application is the base of Romberg's integration method.[7]

8.4.1 Richardson Extrapolation

Let us consider a Newton-Cotes composite quadrature formula to approximate a definite integral. In general, such a formula will have a truncation error that can be expanded into a power series in h, the distance between two interpolation points, as

$$\int_a^b f(x)\,dx = I(h) + Kh^q + K'h^{q+1} + \ldots \tag{8.70}$$

where $I(h)$ stands for the composite quadrature formula applied with h and $K, K', \ldots$ are constants, depending on the integration interval and derivatives of f evaluated in some points between a and b, but no on h. The number q is the order of the method (for example, $q = 2$ for the composite trapezoidal method or $q = 4$ for the composite Simpson's-1/3 rule).

We can apply our composite method with twice as many panels, i.e., with $h/2$, to obtain

$$\int_a^b f(x)\,dx = I\left(\frac{h}{2}\right) + K\left(\frac{h}{2}\right)^q + K'\left(\frac{h}{2}\right)^{q+1} + \ldots \tag{8.71}$$

Multiplying Equation (8.71) by 2^q and subtracting it from Equation (8.70) yields

$$\left(2^q - 1\right)\int_a^b f(x)\,dx = 2^q \cdot I\left(\frac{h}{2}\right) - I(h) + O(h^{q+1}) \tag{8.72}$$

Note that we eliminated the constant K by performing this operation. We can rearrange this expression as

$$\int_a^b f(x)\,dx = \frac{2^q \cdot I(h/2) - I(h)}{2^q - 1} + O(h^{q+1}) \tag{8.73}$$

We obtain a new numerical approximation of our definite integral, but with a higher order of the truncation error by one. This result is summarized in Richardson's theorem:

Theorem 8.6. Richardson Extrapolation

Let $I(h)$ be a composite Newton-Cotes quadrature method of order $O(h^q)$ in h, the distance between two interpolation points for approximating a definite integral $\int_a^b f(x)\,dx$. If $I(h)$ and $I\left(\frac{h}{2}\right)$ are two approximations obtained using h and $\frac{h}{2}$, then

$$R(h) = \frac{2^q \cdot I(h/2) - I(h)}{2^q - 1} \tag{8.74}$$

is an approximation of $\int_a^b f(x)\,dx$ in order $O(h^{q+1})$.

In short, Richardson extrapolation allows computing an improved approximation based on the calculations of two approximations done with two different values of h (obtained by doubling the number of panels). The order of the truncation error of this improved approximation is one more than the original method.

8.4.2 Romberg Quadrature Method

Romberg integration consists of creating successive improved approximations based on Richardson extrapolation. If one starts with a Newton-Cotes formula in order q, one can get approximations in order $q+1, q+2, q+3, \ldots$.

It turns out that if one starts the process with the trapezoidal method, then one can even obtain a gain of two orders in each step. This comes from the very particular form of the truncation error of the trapezoidal method. As discussed in Section 8.3.6, the numerical approximation of a definite integral can be written as

$$\int_a^b f(x)\,dx = T(h) + Kh^2 + K'h^4 + K''h^6 + \ldots \tag{8.75}$$

where $T(h)$ is the composite trapezoidal rule. Only even powers of h are present in the series expansion of the truncation error. Recall that this is a property specific to the trapezoidal rule.

Richardson extrapolation applied to the trapezoidal rule gives

$$R(h)=\frac{2^2\cdot T\left(\frac{h}{2}\right)-T(h)}{2^2-1}=\frac{4\cdot T\left(\frac{h}{2}\right)-T(h)}{3} \tag{8.76}$$

but the truncation error will now be in order $O\left(h^{2+2}\right)=O\left(h^4\right)$, one order more than if we had used another second order quadrature composite formula. The reader is invited to verify this statement by repeating the steps that led to Equation (8.74) but using Equation (8.75) instead of Equation (8.70). In fact, this new approximation will again yield only even powers of h for the truncation error:

$$\int_a^b f(x)dx=\frac{4\cdot T\left(\frac{h}{2}\right)-T(h)}{3}+\tilde{K}h^4+\tilde{K}h^6+\tilde{K}h^8+\ldots \tag{8.77}$$

Applying Richardson extrapolation again gives an improved approximation, in order $4+2=6$ (again involving only even powers of h for the remaining terms).

Consequently, we can produce successive approximations that improve the order of the truncation error by two orders in each step:

$$T_1(h)=\frac{4\cdot T(h/2)-T(h)}{3} \qquad \text{error in order } O\left(h^4\right)$$

$$T_2(h)=\frac{16\cdot T_1(h/2)-T_1(h)}{15} \qquad \text{error in order } O\left(h^6\right)$$

$$T_3(h)=\frac{64\cdot T_2(h/2)-T_2(h)}{63} \qquad \text{error in order } O\left(h^8\right).$$

$$\vdots$$

or in general

$$\begin{aligned} T_{i+1}(h)&=\frac{2^{2i+2}\cdot T_i(h/2)-T_i(h)}{2^{2i+2}-1}\\ &=\frac{4^{i+1}\cdot T_i(h/2)-T_i(h)}{4^{i+1}-1} \qquad \text{Error in } O\left(h^{2(i+2)}\right) \end{aligned} \tag{8.78}$$

Equation (8.78) is the basis of Romberg integration. We organize the calculation in a table as shown:

R_{11}			
R_{21}	R_{22}		
R_{31}	R_{32}	R_{33}	
R_{41}	R_{42}	R_{43}	R_{44}

The first column is calculated with the trapezoidal rule:

$$R_{j1} = T\left(\frac{b-a}{2^{j-1}}\right) \tag{8.79}$$

Each line in the previous table corresponds to a certain value h. The first line is for $h = b-a$, the second is for $h = (b-a)/2$, and so on. The approximations listed in the first column are of order $O(h^2)$ since they are computed using the composite trapezoidal rule.

The second column contains approximations of order $O(h^4)$ computed using Richardson extrapolation based on the first column:

$$R_{j,2} = \frac{4 \cdot R_{j,1} - R_{j-1,1}}{4-1} \tag{8.80}$$

Similarly, the third column, of order $O(h^6)$, is computed using Richardson extrapolation using the results of the second column as:

$$R_{j,3} = \frac{4^2 \cdot R_{j,2} - R_{j-1,2}}{4^2 - 1} \tag{8.81}$$

The fourth column, of order $O(h^8)$, is computed using Richardson extrapolation performed on the third column as:

$$R_{j,4} = \frac{4^3 \cdot R_{j,3} - R_{j-1,3}}{4^3 - 1} \tag{8.82}$$

In general, a column k is computed using Richardson extrapolation using the previous column $k-1$ as

$$R_{j,k} = \frac{4^{k-1} \cdot R_{j,k-1} - R_{j-1,k-1}}{4^{k-1} - 1} \qquad j > 1 \tag{8.83}$$

These R_{jk} are approximations of order $O\left(h^{2(k-1)}\right)$. The following example illustrates how to use Romberg integration.

Example 8.17. Romberg Integration

Let us estimate the definite integral:

$$I = \int_2^3 \frac{x^3}{\sqrt{x^4 - 1}} dx \tag{8.84}$$

with the Romberg integration approximation R_{44}.

We start by computing the approximations of the definite integral I using the composite trapezoidal method using 1, 2, 3, and 4 panels and then produce a table according to Equation (8.83).

	$k = 1$	$k = 2$	$k = 3$	$k = 4$
$j = 1$	2.542141444			
$j = 2$	2.537384626	2.53579902		
$j = 3$	2.536088707	2.535656733	2.535647248	
$j = 4$	2.535756024	2.53564513	2.535644357	2.535644311

The approximation R_{44} of I is consequently:

$$I \cong R_{44} = 2.535644311 \tag{8.85}$$

To give an estimation of the error of R_{44}, we can use Richardson's error formula using the approximations R_{33} and R_{43} to give an error estimation of R_{43} (recall that column $k = 3$ is of order six):

$$E\left(R_{43}\right) \cong \left|\frac{R_{33} - R_{43}}{2^6 - 1}\right| \cong 5 \cdot 10^{-8} \tag{8.86}$$

We know that R_{44} is of order eight, so its error is smaller than the error of R_{43}. Since we don't have two estimations of order eight, we have no choice but to take the error estimation of R_{43}, even though we know the actual error of R_{44} is smaller:

$$I \cong 2.53564431 \pm 5 \cdot 10^{-8} \tag{8.87}$$

In fact, comparing with the true value of I, which is $2\sqrt{5} - \sqrt{15}/2$, we get a true error of R_{44}, about $3 \cdot 10^{-8}$.

8.5 GAUSS QUADRATURE

In this chapter, we discussed numerical integration formulas based on polynomial interpolation. The idea is to replace the function $f(x)$ we want to integrate by a polynomial $P_n(x)$ that interpolates the function. We introduced Newton-Cotes formulas in which $n+1$ interpolating points $x_o, x_1, \ldots, x_n$ are uniformly distributed in the integration interval $[a, b]$. We could think that by using appropriate interpolation points, we can obtain quadrature formulas of higher degree of precision than Newton-Cotes formulas. Figure 8.10 suggests that using different interpolation points to construct a quadrature formula can indeed reduce the error. In Figure 8.10, two interpolation points are used, resulting in the linear polynomial $P_1(x)$. Figure 8.10a shows the error induced when the two interpolation points correspond to the closed Newton-Cotes formula. When the two interpolating points are moved, as shown in Figure 8.10b, the error can be significantly reduced (almost zero in the example of the figure), as in some parts, the quadrature formula will over, and in others under, estimate the actual definite integral.

Once the locations of the interpolation points x_i are known, Equation (8.14) can be used to compute the weights of the quadrature method, and Equation (8.13) can be used to approximate the definite integral. In this section, a method based on orthogonal polynomials is presented that can identify how to place the interpolation points x_i to reduce the error corresponding to the quadrature method.

8.5.1 Legendre Polynomials

In Chapter 5, we discussed the concept of orthogonality and how polynomials of degree n can be considered as vectors from a real-valued vector space of dimension $n+1$. Orthogonality is defined using an inner product. In this chapter, we

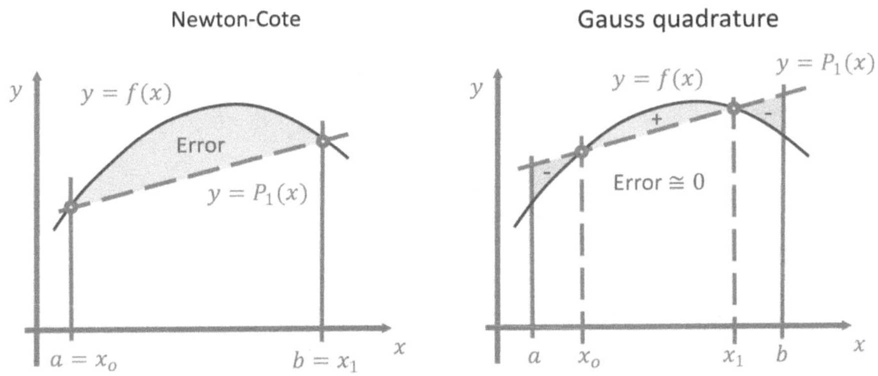

FIGURE 8.10 Reducing the error of a quadrature formula by placing smartly interpolation points. (a) Error of the closed two-points Newton-Cotes Formula. (b) By placing the two interpolation points differently, the error of the quadrature formula can be reduced.

employ the standard inner product between two polynomials $P_n(x)$ and $Q_n(x)$ defined as follows:

$$\langle P_n(x) | Q_n(x) \rangle = \int_{-1}^{+1} P_n(x) Q_n(x)\, dx \tag{8.88}$$

In Chapter 5, we discussed how, using Gram-Schmidt's algorithm, a set of orthogonal polynomials, the so-called Legendre polynomials, can be constructed. The following definition gives a practical way to compute these polynomials.

Definition 8.4. Legendre Polynomials

The set of Legendre polynomials $p_i(x)$ is defined by:

$$p_i(x) = \frac{1}{2^i i!} \frac{d^i}{dx^i}\left[(x^2 - 1)^i \right] \tag{8.89}$$

with $i = 0, 1, 2, \ldots$.

Legendre polynomial $p_i(x)$ are of degree i. Equation (8.89) is known as Rodrigues' formula.[8] Using this equation, one can immediately compute these polynomials. Note that the Legendre polynomials defined in Rodrigues' formula (8.89) are such that $p_i(x = 1) = 1$. Different definitions of Legendre polynomials that follow other conventions exist (for example, the Legendre polynomials we constructed in Chapter 5 using Gram-Schmidt's algorithm differ slightly).

Example 8.18. Legendre Polynomials

Let us determine the first four Legendre polynomials.

From Equation (8.89) follows:

$$p_o(x) = 1$$

$$p_1(x) = \frac{1}{2} \frac{d}{dx}(x^2 - 1) = x$$

$$p_2(x) = \frac{1}{2^2 2!} \frac{d^2}{dx^2}\left[(x^2 - 1)^2 \right] = \frac{1}{2}(3x^2 - 1)$$

$$p_3(x) = \frac{1}{2^3 3!} \frac{d^2}{dx^2}\left[(x^2 - 1)^3 \right] = \frac{1}{2} x (5x^2 - 3)$$

The reader is encouraged to compute some more examples.

Legendre polynomials have several interesting properties. The ones we will use in this section are summarized in the following theorem.

Theorem 8.7. Legendre Polynomials

The Legendre polynomials, as defined by Equation (8.89), have the following properties:

1. $p_i(x)$ is of degree i.
2. $p_i(x)$ has i distinct roots in $[-1, +1]$ for $i > 0$.
3. The set $\{p_o(x), p_1(x), \dots p_n(x)\}$ forms an orthogonal vector base of the real-valued vector space of real polynomials of degree n or less, that is

$$\langle p_i(x) | p_j(x) \rangle = \int_{-1}^{+1} p_i(x) p_j(x) dx = \begin{cases} 0, & i \neq j \\ \neq 0, & i = j \end{cases} \tag{8.90}$$

The reader is encouraged to verify this theorem for the particular case of $n = 2$ (the first three Legendre polynomials of Example 8.18).

8.5.2 Gauss Quadrature Formulas

Gauss quadrature takes advantage of the orthogonality of Legendre polynomials to get an approximation of a definite integral with a higher degree of precision than the Newton-Cotes formula for the same number of interpolating points.

Let us give an intuitive understanding behind the key ideas. The problem we have at hand is to develop an approximation of the definite integral

$$I = \int_{-1}^{+1} f(x) dx \cong \int_{-1}^{+1} P_n(x) dx \tag{8.91}$$

by using polynomial interpolation to approximate the function $f(x)$ in $n+1$ interpolating points. Note that in the first step, we restrict ourselves to the integration interval of $[-1, +1]$.

Contrary to the case of Newton-Cotes formulas, we do not impose any condition on the choice of the $n+1$ interpolating points, except that they must all be distinct and within the integration interval $[-1, +1]$. But regardless of the choice of the interpolation points, such a quadrature formula will be at least a degree of precision n. Can't we now choose these interpolation points in a smart way, such that the degree of precision of the quadrature formula will be higher than n?

Let us first note that the definite integral over the interpolating polynomial in Equation (8.91) can be seen as an inner product between the polynomials $P_n(x)$ and $Q(x) = 1$:

$$\int_{-1}^{+1} P_n(x)dx = \langle 1|P_n(x)\rangle \tag{8.92}$$

Equation (8.92) can lead us to the idea behind Gauss quadrature. If we use the orthogonal Legendre polynomials, we can, for example, write:

$$\int_{-1}^{+1} P_n(x)dx = \langle 1|P_n(x)\rangle + \langle p_1(x)|p_{n+1}(x)\rangle \tag{8.93}$$

because $\langle p_1(x)|p_{n+1}(x)\rangle = 0$ as the Legendre polynomials $p_1(x)$ and $p_{n+1}(x)$ are orthogonal. But we can, in fact, add a more general term $\langle S(x)|p_{n+1}(x)\rangle$ where:

$$S(x) = \sum_{i=0}^{n} a_i p_i(x) \tag{8.94}$$

is a polynomial made of a linear combination out of the first $n+1$ Legendre polynomials. The polynomial $S(x)$ will be of degree n. The term $\langle S(x)|p_{n+1}(x)\rangle$ still evaluates to zero as all polynomials $p_i(x), i = 0, 1, \ldots, n$ are orthogonal to $p_{n+1}(x)$. Consequently, we can rewrite Equation (8.92) as:

$$\int_{-1}^{+1} P_n(x)dx = \langle 1|P_n(x)\rangle + \langle S(x)|p_{n+1}(x)\rangle = \int_{-1}^{+1} [P_n(x) + S(x)p_{n+1}(x)]dx \tag{8.95}$$

The polynomial $Q(x) = P_n(x) + S(x)p_{n+1}(x)$ is of degree $2n+1$. If we choose the $n+1$ interpolating points that serve to build the interpolating polynomial $P_n(x)$ equal to the $n+1$ distinct roots x_i of the Legendre polynomial $p_{n+1}(x)$ then we obtain:

$$Q(x_i) = P_n(x_i) + S(x_i)p_{n+1}(x_i) = P_n(x_i) \text{ for all } i = 0, 1, \ldots, n. \tag{8.96}$$

Equation (8.96) shows that the polynomial $Q(x)$, of degree $2n+1$, is interpolated by $P_n(x)$ if we interpolate it with the interpolating points x_i. Furthermore, Equation (8.95) shows that the definite integral from -1 to +1 of this polynomial $Q(x)$ is evaluated without any truncation error using the quadrature formula (8.91) constructed based on the interpolation points x_i.

This result can be summarized by the Gauss-Legendre quadrature formula theorem.

Theorem 8.8. Gauss-Legendre Quadrature

The quadrature formula of the definite integral $I = \int_{-1}^{+1} f(x)dx$ developed based on the $n+1$ interpolating points x_i that are the $n+1$ distinct roots of the Legendre polynomial $p_{n+1}(x)$ has a degree of precision of $2n+1$.

The following example illustrates how Theorem 8.8 can be used to construct Gauss quadrature formulas.

Example 8.19. Two Point Gauss Quadrature Formula

Let us develop the Gauss quadrature formula based on two interpolation points.

According to Theorem 8.8, the two points Gauss quadrature formula is obtained using the roots of the $p_2(x)$ Legendre polynomial as interpolating points.

From Example 8.18, we know $p_2(x)$. The two roots of $p_2(x)$ are $x_0 = -\sqrt{\frac{1}{3}}$ and $x_1 = +\sqrt{\frac{1}{3}}$. Using them as interpolating points, we obtain the interpolating polynomial $P_2(x)$ as

$$P_2(x) = f(x_0)L_0(x) + f(x_1)L_1(x) \tag{8.97}$$

With the two Lagrange polynomials

$$L_0(x) = \frac{x - x_1}{x_0 - x_1} \quad \text{and} \quad L_1(x) = \frac{x - x_0}{x_1 - x_0} \tag{8.98}$$

The Gauss quadrature formula writes, using Equation (8.13)

$$I = \int_{-1}^{+1} f(x)\,dx \cong \omega_0 f(x_0) + \omega_1 f(x_1) \tag{8.99}$$

with the two weights

$$\omega_0 = \int_{-1}^{+1} L_0(x)\,dx = 1 \quad \text{and} \quad \omega_1 = \int_{-1}^{+1} L_1(x)\,dx = 1 \tag{8.100}$$

Combining everything results in the two points Gauss quadrature formula

$$I = \int_{-1}^{+1} f(x)\,dx \cong f\left(-\sqrt{\frac{1}{3}}\right) + f\left(+\sqrt{\frac{1}{3}}\right) \tag{8.101}$$

According to Theorem 8.8, this quadrature formula has a degree of precision of 3 (here $n = 1$), which is the same precision as Simpson's-1/3 rule, but we need only two function evaluations.

It is possible, similarly to Example 8.19, to compute the values of the interpolation points x_i and their corresponding weights ω_i for cases using more than two interpolation points. Table 8.5 lists these parameters for the first five Gauss quadrature methods.

TABLE 8.5
Weights and Interpolation Points of Gauss Quadrature Formulas

n	Weights ω_i	Gauss Interpolation Points x_i
2	1.0000000000000000	−0.5773502691896257
	1.0000000000000000	+0.5773502691896257
3	0.5555555555555556	−0.7745966692414834
	0.8888888888888888	0.0000000000000000
	0.5555555555555556	+0.7745966692414834
4	0.3478548451374538	−0.8611363115940526
	0.6521451548625461	−0.3399810435848563
	0.6521451548625461	+0.3399810435848563
	0.3478548451374538	+0.8611363115940526
5	0.2369268850561891	−0.9061798459386640
	0.4786286704993665	−0.5384693101056831
	0.5688888888888889	0.0000000000000000
	0.4786286704993665	+0.5384693101056831
	0.2369268850561891	+0.9061798459386640

Table 8.5, together with Equation (8.13), allows computing approximations of definite integrals over an interval $[-1, +1]$. The following example illustrates this.

Example 8.20. Gauss Quadrature for Definite Integrals Over an Interval $[-1, +1]$

Let us use the two points Gauss quadrature formula to approximate the following definite integral:

$$I = \int_{-1}^{1} e^{-\frac{x^2}{2}} dx \tag{8.102}$$

From Equation (8.13) and Table 8.5 it follows:

$$I = \int_{-1}^{1} e^{-\frac{x^2}{2}} dx \cong \omega_o f(x_1) + \omega_1 f(x_2) \cong 1.692963 \tag{8.103}$$

with $\omega_o = 1, \omega_1 = 1$ and $x_1 = -0.5773502691896257, x_2 = +0.5773502691896257$.

In case the definite integral is over a general interval $[a, b]$, a change of variable needs to be done first:

$$\int_a^b f(x) dx = \int_{-1}^{+1} f\left(\frac{(b-a)t + b + a}{2}\right) \frac{b-a}{2} dt \tag{8.104}$$

The following example illustrates this.

Example 8.21. Gauss Quadrature for Definite Integrals Over an Interval $[a, b]$

Let us use the four points Gauss quadrature formula to approximate the following definite integral:

$$I = \int_1^2 \ln(x)\,dx \tag{8.105}$$

In the first step, we must proceed with a change of variable to bring the integration interval to the interval $[-1, +1]$:

$$\int_1^2 \ln(x)\,dx = \int_{-1}^{+1} \ln\left(\frac{t+3}{2}\right)\frac{1}{2}\,dt \tag{8.106}$$

Next, we use Equation (8.13) and Table 8.5 to approximate our definite integral:

$$\int_1^2 \ln(x)\,dx = \int_{-1}^{+1} \ln\left(\frac{t+3}{2}\right)\frac{1}{2}\,dt = \sum_{i=1}^{4} \omega_i f(x_i) \cong 0.3862944969 \tag{8.107}$$

with $f(x_i) = \frac{1}{2}\ln\left(\frac{t+3}{2}\right)$ and the coefficients ω_i and interpolation points x_i taken from Table 8.5 (the row for $n = 4$).

8.6 LAB: COMPUTING A NUMERICAL APPROXIMATION OF THE ERROR FUNCTION

In statistics, an important integral that needs often to be evaluated is the following one:

$$\text{Erf}(z) = \frac{2}{\sqrt{\pi}} \int_0^z e^{-x^2}\,dx \tag{8.108}$$

which is known as the error function.[9] Our aim is to find an approximation of the value $\text{Erf}(z = 3)$ with an absolute error lower than 10^{-12}. We will use the composite Simpson 1/3 rule.

We start by defining the function we need to integrate:

```
>> f = @(x) 2*exp(-x.^2)/sqrt(pi);
```

To compute numerically the integral (8.108) using Simpson's 1/3 composite rule, we use the function `simp` defined in Example 8.15. The following code snippet approximates integral (8.108) using Simpson's 1/3 composite rule (Recall that, for Simpon's-1/3 rule, h is related to m by $h = \dfrac{b-a}{2m}$, which in our case is $h = \dfrac{3}{2m}$):

```
>> m = 1;
>> I = []; h = [];
>> for i = 1:25
>>    I = [I; simp(f, 0, 3, m)];
>>    h = [h; 3/(2*m)];
>>    m = m*2;
>> endfor
```

The errors of the approximations are computed using the Richardson's formula (8.53):

```
>> err = abs(I(2:end) - I(1:end-1)) / (2^4-1);
```

considering that Simpson's 1/3 composite rule is of order four. The results are best displayed in a log-log plot:

```
>> loglog(h(2:end), err)
```

Inspection of Figure 8.11 allows the conclusion that Richardson's formula is valid for values of h ranging from about 10^{-5} to 0.5. For comparison purposes, Figure 8.11 plots the estimated errors of the numerical approximations produced

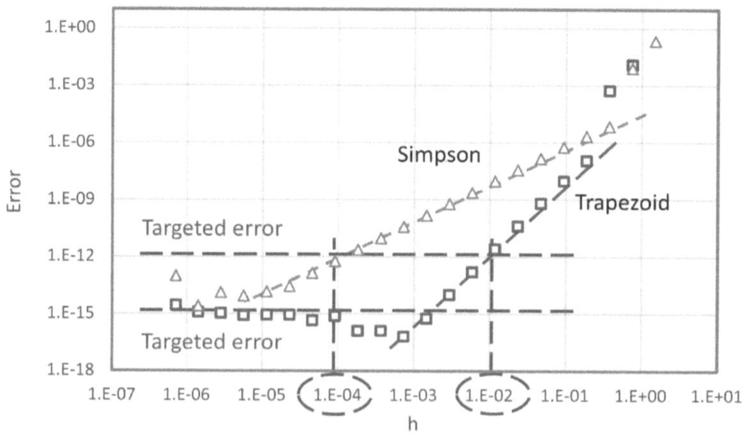

FIGURE 8.11 Estimated truncation errors, using Richardson's formula, of the numerical approximations of $Erf(z = 3)$ produced by Simpson's composite 1/3 rule and the trapezoidal composite rule.

by the composite trapezoidal rule. We observe that Richardson's formula is valid for values of h ranging from about 10^{-3} to 10^{-1}.

If an approximation with an error lower than 10^{-12} is desired, we must choose a value of h between 10^{-2} and 10^{-3} if we use Simpson's rule and between 10^{-4} and 10^{-5} if we use the trapezoidal rule. But, if we target an approximation with an error lower than 10^{-15}, only Simpson's 1/3 rule can yield such an approximation. The trapezoidal rule cannot provide such an approximation because the round-off errors take over before this accuracy can be reached.

Let us compute an approximation of $\mathrm{Erf}(z=3)$ with an absolute error lower than 10^{-12} with Simpson's composite rule. We choose $h=10^{-3}$. This corresponds to:

```
>> m = 3/(2*1E-3)
m = 1500
```

and consequently, an approximation of $\mathrm{Erf}(z=3)$, with an error lower than 10^{-12}, is:

```
>> format long
>> simp(f, 0, 3, 1500)
ans =
0.999977909503001
```

8.7 EXERCISES

8.7.1 Numerical Integration Based on Polynomial Interpolation

1. Following Example 8.1, develop quadrature rules for the definite integral $\int_a^b f(x)dx$ using the following interpolation points:

   a. $x_o = \dfrac{a+b}{2}$

   b. $x_o = a$ and $x_1 = \dfrac{a+b}{2}$

   c. $x_o = \dfrac{1}{3}(a+b)$ and $x_1 = \dfrac{2}{3}(a+b)$

   d. $x_o = a$, $x_1 = \dfrac{a+b}{3}$ and $x_2 = b$

2. Using the quadrature rules you developed in Exercise 1, approximate the following definite integrals and, for each of them, evaluate their errors by comparing the produced approximation with the value obtained by calculus. Are there cases where this error is zero? If yes, explain why.

   a. $\int_1^2 \ln(x)dx$

b. $\int_{-3}^{4} \left(1 + 2x + 3x^2\right) dx$

c. $\int_{0}^{2\pi} \sin(x)\, dx$

d. $\int_{1}^{2} e^x (x)\, dx$

8.7.2 Numerical Integration Based on Lagrange Interpolation

3. Repeat Exercise 1 using Lagrange interpolation and compute explicitly the quadrature weights as defined by Equation 8.14. Verify that, as stated by Theorem 8.1, the sum of the weights is equal to $b-a$.

4. Write an Octave function that approximates a definite integral using Equation (8.13). The function takes as arguments a vector containing the interpolation points, a vector containing the quadrature weights, the function to be integrated, and the start and end points a and b.

   Test your function using your results from Exercise 2.

8.7.3 Newton-Cotes Formulas

5. Evaluate the integrals of Exercise 2 using the Trapezoid and Simpson's 1/3 rule.

   Are there cases where this error is zero? If yes, explain why.

   For the cases where the error is not zero, give an estimation of the error by following the method outlined in Example 8.6.

6. Evaluate the definite integrals of Exercise 2 using the composite Trapezoid and Simpson's 1/3 rule. For each composite rule, perform the calculations using two panels the first time and three panels the second time.

7. Using an Octave code, plot the true error E when approximating the definite integrals of Exercise 2 using the composite Trapezoid and Simpson's 1/3 rule in function of h, the distance between two interpolation points. Verify by fitting the power law $E = Kh^q$ that the order of convergence of the composite Trapezoid rule is two and the order of convergence of the composite Simpson's 1/3 rule is four.

8. Using an Octave code, plot the true error E when approximating the definite integrals a and b using the composite Trapezoid and Simpson's 1/3 rule in function of h, the distance between two interpolation points.

By fitting the power law $E = Kh^q$, determine the order of convergence of the methods. If they differ from the expected value based on Table 8.4, explain why this is the case.

a. $\int_1^5 (x-1)^{2/3} dx$

b. $\int_2^3 \sqrt{x-2} dx$

9. Repeat Exercise 7, but instead of computing the true error E, estimate the error using Richardson's method as in Example 8.14. For which range of h, the distance between two interpolation points, is the Richardson formula valid? Compare the estimated errors with the true errors.

10. Repeat Example 8.16 but using the definite integral $\int_0^\pi \cos(x) dx$.

8.7.4 Romberg Integration

11. Evaluate the integrals of Exercise 2 using Romberg integration. Compute the values up to $k = 4$. Are there cases where this error is zero? If yes, explain why. Compare your answers with the ones from Exercise 2 and the true values.

8.7.5 Gauss Quadrature Formulas

12. Compute the roots of the Legendre polynomial $p_3(x)$ and verify explicitly that all properties of Theorem 8.7 are satisfied for $p_3(x)$.

13. Following Example 8.19, determine the weights and interpolation points of the three points Gauss quadrature formula.

14. Evaluate the integrals of Exercise 2 using 2, 3, and 4 point Gauss quadrature formulas. Are there cases where this error is zero? If yes, explain why. Compare your answers with the ones from Exercise 2 and the true values.

NOTES

1 Roger Cotes (1682–1716), English mathematician.
2 Thomas Simpson (1710–1761), British mathematician and inventor.
3 George Boole (1815–1864), British mathematician, philosopher, and logician.
4 James S. Milne (born October 10, 1942), New Zealand mathematician.
5 Lewis Fry Richardson (1881–1953), English mathematician, physicist, and meteorologist.

6 Recall that for Simpon's-1/3 rule, h is related to m by $h = \frac{b - a}{2m}$
7 Werner Romberg (1909–2003), German mathematician and physicist.
8 Benjamin Olinde Rodrigues (1795–1851), French banker, mathematician, and social reformer.
9 The word "error" in "error function" is not related in anyway to numerical errors.

9 Initial Value Problems

Initial value problems play a key role in engineering and science. They are, for example, the fundamental tool to describe the evolution of systems. However, solving initial value problems is a difficult task. Very few examples exist that can be solved analytically. Unfortunately, it turns out that the calculation of numerical approximations is a challenging problem as well. As will be described in this chapter, estimating the error of numerical approximations is not straightforward, and often, it is time-consuming. We will learn approaches used to address this problematic.

In this chapter, the numerical approximations of initial value problems involving ordinary differential equations of order n and systems of ordinary differential equations will be presented. We will learn that numerical initial value problem solvers can be stable or unstable, depending on some parameters controlling the solution process. Furthermore, we will discover that truncation errors will have to be classified into local and global truncation errors, which add further difficulty in the error estimation of the produced approximations.

9.1 INTRODUCTION

Let us start by recalling some concepts and definitions from calculus in relation to initial value problems. An equation that is composed of an unknown function and its derivatives is called a differential equation. The highest derivative that appears in the equation determines the order of the equation. For example, a differential equation in which the highest derivative is a second derivative of the unknown function f is said to be a second order differential equation. An important class of differential equations, the main object of the present chapter, is the class of ordinary differential equations defined as follows:

Definition 9.1. Ordinary Differential Equation

An ordinary differential equation (ODE) for the unknown function y is a differential equation dependent on only a single independent variable t. Given f a function of t, y and its derivatives, we distinguish between

- Explicit ordinary differential equations of order n:

$$\frac{d^n}{dt^n} y(t) = f\left(t, y, y', \ldots, y^{(n-1)}\right) \tag{9.1}$$

DOI: 10.1201/9781003262121-10

- Implicit ordinary differential equations of order n:

$$f\left(t, y, y', \ldots, y^{(n)}\right) = 0 \tag{9.2}$$

The majority of the present chapter is focused on explicit first order ODEs:

$$\frac{d}{dt} y(t) = f(t, y,)] \tag{9.3}$$

As the reader knowns from calculus, ODEs do not have a unique solution, but rather a family of solutions. For numerical algorithms, this is a serious issue. This is why algorithms focus on solving initial value problems (IVPs) rather than ordinary differential equations (ODEs). In an IVP, additional conditions, called initial values, are specified for the solution $y(t)$ that must be satisfied alongside the differential equation. These initial values ensure a unique solution to the problem. In the case of a first order ODE, an IvP is defined as follows:

Definition 9.2. First Order Initial Value Problem

Given a function $f(t, y)$, a first order initial value problem (IvP) consists in finding a function $y(t)$ for all $t \in [t_o, t_f]$ such that

$$\begin{cases} \dfrac{dy}{dt} = f(t, y) \\ y(t_o) = y_o \\ t \in [t_o, t_f] \end{cases} \tag{9.4}$$

where $y(t_o) = y_o$ is called the initial condition.

The next definition introduces the concept of numerical IvP solver and notations related to it.

Definition 9.3. Numerical Initial Value Problem Solver

A numerical initial value problem solver is an algorithm that produces approximations of the solution of an initial value problem. The exact solution of the initial value problem in $t = t_i$ will be written as $y_i = y(t_i)$ and the corresponding approximation by the numerical solver as w_i.

9.2 EULER'S ALGORITHM

In this section we present a simple initial value problem solver known as Euler's algorithm.[1] Despite its simplicity, the study of this algorithm will allow us to understand several important properties of numerical initial value problem solvers.

9.2.1 Basics of the Algorithm

Euler's algorithm aims to approximate the solution of a first order initial value problem as defined in Definition 9.1. Let $y(t)$ be the solution to this IvP. We can write, using the Taylor series expansion of y in some point t_i:

$$y(t_i + h) = y(t_i) + h\frac{dy(t_i)}{dt} + \mathcal{O}(h^2) \tag{9.5}$$

where h is a number we choose, and which we will call the step size. From the initial value problem we aim to solve, we know that:

$$\frac{dy(t_i)}{dt} = f(t_i, y(t_i)) \tag{9.6}$$

And, consequently, Equation (9.5) can be rewritten as

$$y(t_i + h) = y(t_i) + hf(t_i, y(t_i)) + \mathcal{O}(h^2) \tag{9.7}$$

This relation inspires the definition of Euler's algorithm.

Algorithm 9.1. Euler's Algorithm to Approximate First Order IVPS

Given a first order initial value Problem (9.4), Euler's initial value problem solver produces successive approximations w_i of the exact solution $y_i = y(t_i)$ as follows:

$$w_0 = y_o$$

$$w_{i+1} = w_i + hf(t_i, w_i) \tag{9.8}$$

with h the step size of the algorithm.

The following example illustrates the application of Euler's algorithm:

Example 9.1. Euler's Method

Let us approximate the solution to the following initial value problem using Euler's method:

$$\begin{cases} \dfrac{dy}{dt} = 1 + \dfrac{y}{t} \\ y(1) = 1 \\ t \in [1, 4] \end{cases} \tag{9.9}$$

To apply Euler's method, we need first to choose a step size. At this stage we have no idea how to chose it. We will discuss this aspect later in this chapter. For the moment we choose arbitrarily the value $h = 0.5$. Having chosen h, we know that the algorithm will produce approximations w_i of $y(t_i)$ for $t_i = 1, 1.5, 2, 2.5, 3, 3.5,$ and 4. Following Algorithm 9.1, we obtain successive approximations. First, we have the approximation w_o for $t_0 = 1.5$ from the initial conditions:

$$w_o = 1 \tag{9.10}$$

The approximation w_1 for $t_1 = 1.5$ is then computed with Equation (9.8):

$$w_1 = 1 + 0.5f\left(t_o = 1, w_o = 1\right) = 1 + 0.5\left(1 + \frac{1}{1}\right) = 2 \tag{9.11}$$

In the same way, we compute w_2 for $t_2 = 2.0$:

$$w_2 = 2 + 0.5f\left(t_1 = 1.5, w_1 = 2\right) = 2 + 0.5\left(1 + \frac{2}{1.5}\right) = 3.166666667 \tag{9.12}$$

Continuing the calculations until w_6, the approximation of $y_6 = y\left(t_6 = 4\right)$, allows us to build Table 9.1.

TABLE 9.1
Approximations w_i of the Initial Value Problem (9.9) According to Euler's Method

i	t_i	w_i
0	1	1
1	1.5	2
2	2	3.166667
3	2.5	4.458333
4	3	5.85
5	3.5	7.325
6	4	8.871429

The same calculations can be achieved efficiently in Octave. We start by defining a new function `Euler`

```
function [t, w] = Euler(f, to, yo, h, n)
 w=[yo];
 t=[to];
 for i = 1:n
 w=[w ; w(i)+h*f(t(i), w(i))];
 t=[t ; t(i)+h];
 endfor
endfunction
```

This function implements Euler's method. It returns two column vectors, `t` and `w`, that contain the values of t_i and w_i. With this new Octave function, it becomes straightforward to compute the Euler approximations of our initial value Problem (9.9):

```
>> f = @(t,y) 1+y/t;
>> [t w]=Euler(f, to=1, yo=1, h=0.5, n=6);
>> [t, w]
ans =
   1.0000   1.0000
   1.5000   2.0000
   2.0000   3.1667
   2.5000   4.4583
   3.0000   5.8500
   3.5000   7.3250
   4.0000   8.8714
```

As the function `Euler` has several arguments, it becomes easy to mix up some of them. Because of this, we used a function call where we explicitly named the arguments. This adds clarity to the code and avoids mistakes resulting from providing arguments in the wrong order.

The exact solution of our initial value Problem (9.9) is

$$y(t) = t + t \cdot \ln(t) \tag{9.13}$$

We can plot this solution together with the Euler approximations (note the. `*` operator as our function `y` must be able to handle vectors):

```
>> y = @(t) t+t.*log(t);
>> plot(t, y(t), t, w, 'o')
```

As shown in Figure 9.1, Euler's approximations reproduce the general trend of the exact solution. But the errors are rather large and increase with t.

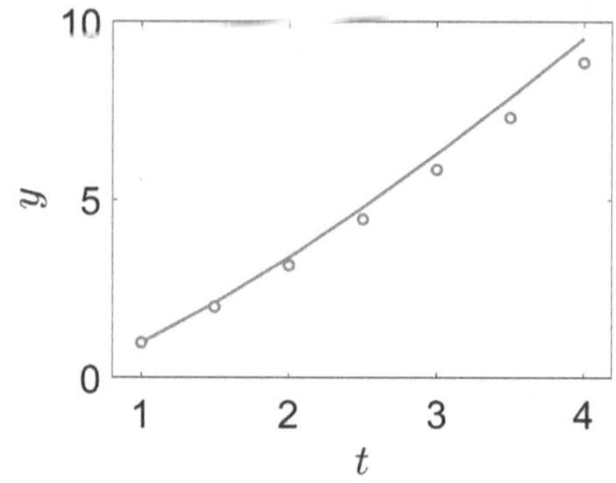

FIGURE 9.1 Exact solution (–) together with the approximations (o) produced by Euler's method of the initial value Problem (9.9).

9.2.2 Stability of Euler's Method

As discussed in Section 9.2.1, Euler's method produces a succession of approximations w_i of the exact solution $y(t)$ in points t_i distant by a step size h. In the present section, we want to discuss in more detail how the choice of step size h influences these approximations. For this, we use the following initial value problem

$$\begin{cases} \dfrac{dy}{dt} = -\beta y \\ y(0) = y_o \end{cases} \tag{9.14}$$

with β some strictly positive numbers. Problem (9.14) has the following solution:

$$y(t) = y_o e^{-\beta t} \tag{9.15}$$

As the number β is strictly positive, the solution (9.15) will be a strictly decreasing function.

Let us now examine how Euler's method will produce approximations of the solution (9.15). Euler's method applied to Problem (9.14) writes:

$$w_{i+1} = w_i - h\beta w_i = w_i(1 - h\beta) \tag{9.16}$$

where h is the step size we choose to use. Let us calculate a few approximations:

$$\begin{aligned} w_1 &= w_o(1-h\beta) = y_0(1-h\beta) \\ w_2 &= w_1(1-h\beta) = y_0(1-h\beta)^2 \\ w_3 &= w_2(1-h\beta) = y_0(1-h\beta)^3 \end{aligned} \tag{9.17}$$

In fact, the general expression for approximation w_i is:

$$w_i = y_0 \left(1 - h\beta\right)^i \tag{9.18}$$

Let us analyze the approximation (9.18) deeper. Depending on the value of the quantity $|1 - h\beta|$, the approximations w_i will decrease or increase with i. Indeed, if $|1 - h\beta| < 1$, then w_i will decrease as i increases. This is the behavior expected from the exact solution (9.15). However, if $|1 - h\beta| > 1$, then w_i will increase as i increases. This behavior is opposite to the behavior of the exact solution. In this case, w_i are certainly wrong approximations, as even qualitatively, they do not reproduce the exact solution.

The surprising element is that, depending on the value of the step size we choose, the quantity $|1 - h\beta|$ can be lower or higher than 1. Consequently, depending on our choice of h, the approximations w_i will either reproduce or not reproduce qualitatively the solution. We have:

1. For $h < \dfrac{2}{\beta}$ then $|1 - h\beta| < 1$ and w_i decrease with increasing i as does the true solution y_i.
2. For $h > \dfrac{2}{\beta}$ then $|1 - h\beta| > 1$ and w_i increase with increasing i contrary to the true solution y_i.

In the first case, one says that Euler's method is stable, whereas in the second case, one says Euler's method is unstable. We observe that for only small enough values of the step size h, in our example for $h < \dfrac{2}{\beta}$, Euler's method is stable.

Example 9.2. Stability of Euler's Method

Let us discuss, with some examples, the stability of Euler's method for the initial value problem:

$$\begin{cases} \dfrac{dy}{dt} = -2y \\ y(0) = 1 \end{cases} \tag{9.19}$$

From our previous discussion, we know that Euler's method is instable for $h > \dfrac{2}{2} = 1$. Let us first consider such an example and let us choose $h = 2$. Using our Octave function `Euler` defined in Example 9.1, we obtain:

```
>> f = @(t,y) -2*y;
>> [t w] = Euler(f, to=0, yo=1, h=2, n=6);
>> [t, w]
```

```
ans =
     0     1
     2    -3
     4     9
     6   -27
     8    81
    10  -243
    12   729
```

Figure 9.2a confirms that the solution is unstable and doesn't reproduce the qualitative behavior of the true solution of the initial value Problem (9.19).

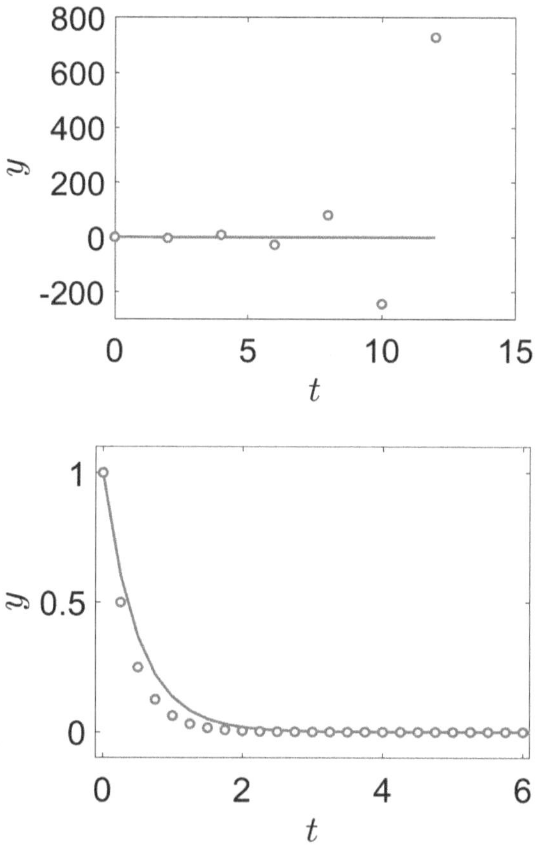

FIGURE 9.2 Numerical approximations (o) of the initial value Problem (9.19) using Euler's method with the step size (a) $h = 2$ and (b) $h = 0.25$ together with the true solution (–).

We note as well that the error is increasing dramatically with the number of steps i:

```
>> err = abs(w-exp(-2*t))
err =
         0
    3.0183
    8.9997
   27.0000
   81.0000
  243.0000
  729.0000
```

Let us now use a step size of $h = 0.25$, for which Euler's method will be stable. As shown in b, this time, the numerical approximation is stable (reproduces qualitatively the true solution).

9.2.3 Convergence of Euler's Method

From the discussion in the previous section, 9.2.2, we know that choosing a too-large step size h can result in unstable numerical approximations. Let us now, with an example, observe how Euler's method behaves as we reduce the step size. We consider the following initial value problem:

$$\begin{cases} \dfrac{dy}{dt} = ty + t^3 \\ y(0) = 1 \\ t \in [0, 1] \end{cases} \tag{9.20}$$

The reader is invited to verify that the true solution of (9.20) is:

$$y(t) = 3e^{t^2/2} - t^2 - 2 \tag{9.21}$$

Using our Octave function `Euler` defined in Example 9.1, we can produce numerical approximations for various step sizes $\sigma = 0.5$. Figure 9.3a illustrates some examples. We note that as the step size decreases, the numerical approximations become closer to the true solution. We can quantify this effect by computing the true error in $t = 1$ (i.e., the difference between the true solution $y(t = 1)$ and the corresponding approximation produced by Euler's algorithm). Using Octave:

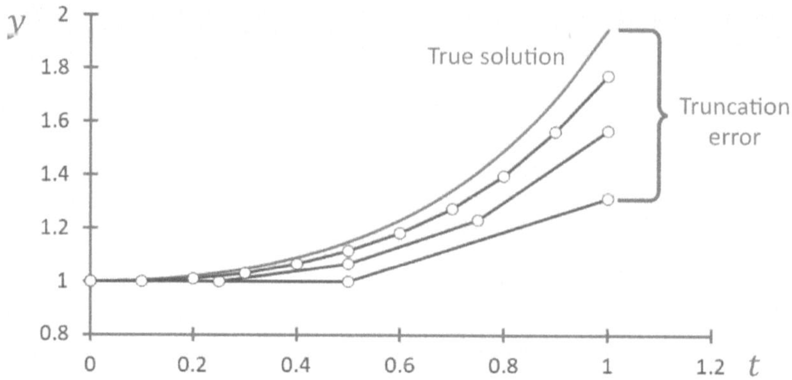

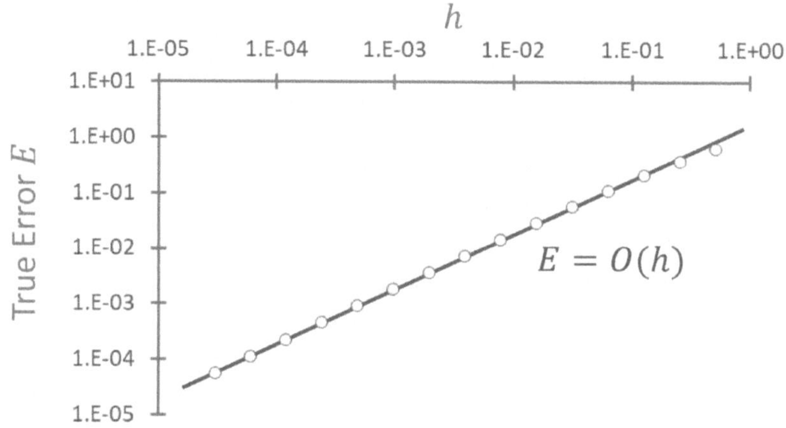

FIGURE 9.3 (a) Numerical approximations (o) of the initial value Problem (9.20) using Euler's method with the step sizes $h = 0.5, 0.25$ and 0.2 together with the true solution (9.21) (–) and (b) true error of the numerical approximations in $t = 1$ in function of the step size h.

```
>> f = @(t,y) t*y+t^3;
>> y = @(t) 3*exp(t^2/2)-t^2-2;
>> err=[]; h=[];
>> n=2;
>> for i = 1:15
>> [t w] = Euler(f, to=0, yo=1, 1/n, n);
>> err = [err; abs(w(end)-y(1))];
>> h = [h; 1/n];
>> n = 2*n;
>> endfor
>> loglog(h, err, 'o')
```

Figure 9.3b shows how the error is indeed decreasing with the step size h. It is interesting to determine the order of convergence of this error with h by fitting a power law:

```
>> A = [h.^0 log(h)];
>> a = A\log(err)
a =
  0.4340
  0.9740
```

which shows that the order of convergence is roughly linear, i.e., $O(h)$. This is rather surprising if we recall Equation (9.5) from Section 9.2.1. According to (9.5), we would rather expect a second order decrease. Section 9.3 aims to help us understand this apparent discrepancy.

9.3 ANALYSIS OF INITIAL VALUE PROBLEM SOLVERS

The example we studied in Section 9.2.3 showed an apparent discrepancy between the convergence order of Euler's method suggested to be $O(h^2)$ according to Equation (9.5), and the observed convergence $O(h)$ in our numerical example. This apparent disagreement comes from our misinterpretation of Equation (9.5). To understand this, we need to introduce the concepts of local and global truncation errors.

9.3.1 Local and Global Truncation Errors of Numerical IvP Solvers

To understand the difference between local and global truncation errors of numerical IvP solvers, let us discuss Figure 9.4. Figure 9.4 displays the solution

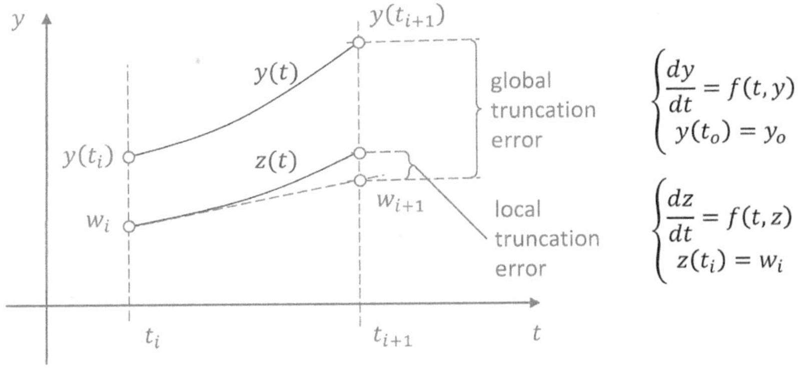

FIGURE 9.4 Schematic representation of the solution $y(t)$ of a first order initial value Problem (9.4) together with the numerical approximations w_i produced by a numerical IvP solver and the local solution $z(t)$ in (t_i, w_i).

$y(t)$ to a first order initial value problem (9.4). In $t = t_i$, a numerical IvP solver produced the approximation w_i. Let us now consider the following initial value problem.

$$\begin{cases} \dfrac{dz}{dt} = f(t, z) \\ z(t_i) = w_i \end{cases} \tag{9.22}$$

Problem (9.22) is based on the same differential equation than Problem (9.4). However, the initial value is different. The solution $z(t)$ of Problem (9.22) is the solution of the differential equation from Problem (9.4) passing by the point (t_i, w_i). This solution is called the local solution.

Definition 9.4. Local Solution in (t_i, w_i)

The local solution $z(t)$ in (t_i, w_i) of the initial value Problem (9.4), is the exact solution to Problem (9.22). In other words, it is the solution of the differential equation $y' = f(t, y)$ that passes by the point (t_i, w_i).

Figure 9.4 displays further the numerical approximation w_{i+1} of $y(t)$ in $t = t_{i+1}$. From the figure, we understand that one has to distinguish between the error between the true solution and the numerical approximation and the error between the local solution and the numerical approximation. This leads to the following definition.

Definition 9.5. Local and Global Truncation Error

For a numerical initial value problem solver producing approximations w_{i+1}, in t_{i+1}, one defines the global truncation error as the difference between the true solution $y(t)$ of the IvP and its numerical approximation:

$$\text{Global truncation error} = \left| y(t_{i+1}) - w_{i+1} \right| \tag{9.23}$$

and the local truncation error as the difference between the local solution $z(t)$ in (t_i, w_i) and its numerical approximation:

$$\text{Local truncation error} = \left| z(t_{i+1}) - w_{i+1} \right| \tag{9.24}$$

Based on Figure 9.4, we can now understand better the meaning of Equation (9.5). In particular, it is wrong to conclude that $w_{i+1} = w_i + hf(t_i, w_i) + \mathcal{O}(h^2)$. What, however, is correct is to write

$$z(t_i+h)=z(t_i)+hf(t_i,z(t_i))+\mathcal{O}(h^2) \tag{9.25}$$

And, as $z(t_i)=w_i$:

$$z(t_i+h)=w_i+hf(t_i,w_i)+\mathcal{O}(h^2) \tag{9.26}$$

Equation (9.26) shows that the local truncation error is in second order $\mathcal{O}(h^2)$. Using the explicit expression from the Taylor series expansion (9.5) of the solution $y(t)$, results in the following theorem:

Theorem 9.1. Local Truncation Error of Euler's Method

The local truncation error of Euler's method is in second order in the step size h:

$$\left|z(t_{i+1})-w_{i+1}\right|=\frac{h^2}{2}\frac{d^2}{dt^2}y(\xi) \tag{9.27}$$

with ξ some number in $[t_i,t_i+h]$. If one defines

$$C=\max_{t_o\leq\xi\leq t_f}\left|\frac{1}{2}\frac{d^2}{dt^2}y(\xi)\right| \tag{9.28}$$

it follows that

$$\left|z(t_{i+1})-w_{i+1}\right|\leq Ch^2 \tag{9.29}$$

for any t_{i+1} in $[t_i,t_i+h]$.

As we will discuss in the next section, the global truncation error of Euler's method, which is the error we try to estimate when solving numerically an IvP, is one order lower, i.e., of first order. We mention here that even though Theorem 9.1 is specific to Euler's method, similar results are valid for other numerical IvP solvers. In particular, if a numerical IvP solver has a local truncation error in order $\mathcal{O}(h^k)$, then there exists some number C such that for any t_{i+1} in in $[t_i,t_i+h]$

$$\left|z(t_{i+1})-w_{i+1}\right|\leq Ch^k \tag{9.30}$$

9.3.2 Lipschitz Theorems

There exists a general relation between the global and local truncation errors of a numerical initial value problem solver based on Lipschitz's theorem.[2] Before formulating this result, we must introduce a definition.

Definition 9.6. Lipschitz Function

A function $f(t, y)$ satisfies the Lipschitz condition in the variable t on the rectangle $S = [t_o, t_f] \times [y_1, y_2]$ if there exists a constant L (called Lipschitz constant) satisfying:

$$|f(t, y_1) - f(t, y_2)| \leq L|y_1 - y_2| \tag{9.31}$$

for each t in $[t_o, t_f]$.

There is a broad set of functions that are Lipschitz. For example, any function that is continuously differentiable over S will be Lipschitz. Indeed, from the mean value theorem, it follows for each interval $[y_1, y_2]$ that there exists a number c in that interval $[y_1, y_2]$ such as

$$f(t, y_1) - f(t, y_2) = \frac{\partial f}{\partial y}(t, c) \cdot (y_1 - y_2) \tag{9.32}$$

As f is continuously differentiable over S, the first derivative of f is bounded. We can choose a Lipschitz constant L as follows:

$$L = \max_{(t,y) \in S} \left| \frac{\partial f}{\partial y}(t, y) \right| \tag{9.33}$$

and the Lipschitz condition (9.31) is fulfilled for each t in $[t_o, t_f]$.

Lipschitz functions appear in two import theorems of the theory of ordinary differential equations that are relevant to our purpose. The first one is the existence and uniqueness theorem of initial value problems, also known as the Picard-Lipschitz-Cauchy theorem.[3]

Theorem 9.2. Existence and Uniqueness Theorem of Initial Value Problems

If the function $f(t, y)$ is Lipschitz, then the first order initial value Problem (9.4) has a solution, and this solution is unique.

To know that a given problem has a solution and further that the solution is unique, simplifies the analysis of numerical IvP solvers. The second theorem, the Lipschitz theorem, allows us to evaluate how two solutions of the same first order ordinary differential equation differ from each other's for increasing t.

Theorem 9.3. Lipschitz Theorem

Consider a function $f(t, y)$ satisfying the Lipschitz condition in the variable t on the rectangle $S = [t_o, t_f] \times [y_1, y_2]$. If $y(t)$ and $z(t)$ are two solutions of the differential equation

$$\frac{dy}{dt} = f(t, y) \tag{9.34}$$

with initial conditions $y(t_o) = y_o$ and $z(t_o) = z_o$ then

$$|y(t) - z(t)| \le e^{L(t-t_o)} |y_o - z_o| \tag{9.35}$$

The Lipschitz theorem has a very practical application. If two solutions $y(t)$ and $z(t)$ of the same first order ordinary differential equation differ in t_o by $|y_o - z_o|$, then they will differ in t, by a quantity bounded by $e^{L(t-t_o)}|y_o - z_o|$ (Figure 9.5). Even though Theorem 9.3 gives only an upper bound of the difference between the two solutions, it still gives us an intuitive feeling on how this difference will typically grow as t increases. Inspecting this upper bound, we conclude that first, two solutions of the same first order ordinary differential equation will differ increasingly as t grows. The growth is exponential in $t - t_o$. Furthermore, the larger the Lipschitz constant L of f is, the faster the difference grows. In fact, depending on the value of L, this growth may even be very fast, meaning that two solutions, initially close to each other, will rapidly become very different. This effect is what one could qualify as ill-conditioning of IvPs. The next example illustrates this remark.

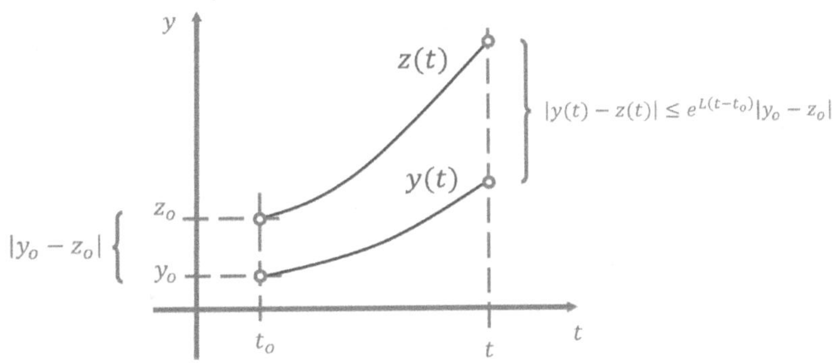

FIGURE 9.5 Two solutions $y(t)$ and $z(t)$ of the same first order ordinary differential equation. If they differ in t_o by $|y_o - z_o|$, then they will differ in t, according to Theorem 9.3, by a quantity bounded by $e^{L(t-t_o)}|y_o - z_o|$.

Example 9.3. Lipschitz Theorem

Consider the following initial value problem:

$$\begin{cases} \dfrac{dy}{dt} = ty + t^3 \\ y(0) = y_o \end{cases} \tag{9.36}$$

The reader can verify that the exact solution of (9.36) is

$$y(t) = (2 + y_o)e^{t^2/2} - t^2 - 2 \tag{9.37}$$

Let us now consider two solutions: $y(t)$ and $z(t)$. The first is the solution of (9.36) for the initial condition $y(0) = -2$, and the second is the solution of (9.36) for the initial condition $y(0) = -2.001$. In other words, both solutions are very close to each other at $t = 0$, with a difference of

$$|y_o - z_o| = 0.001 \tag{9.38}$$

But, at $t = 5$, this difference has grown significantly:

$$|y(t=5) - z(t=5)| \cong 268 \tag{9.39}$$

which is about $2.6 \cdot 10^5$ larger than the initial difference. Figure 9.6 illustrates this behavior on a graph. Even though $y(t)$ and $z(t)$ are nearly identical in t_o, they become rapidly very different from each other.

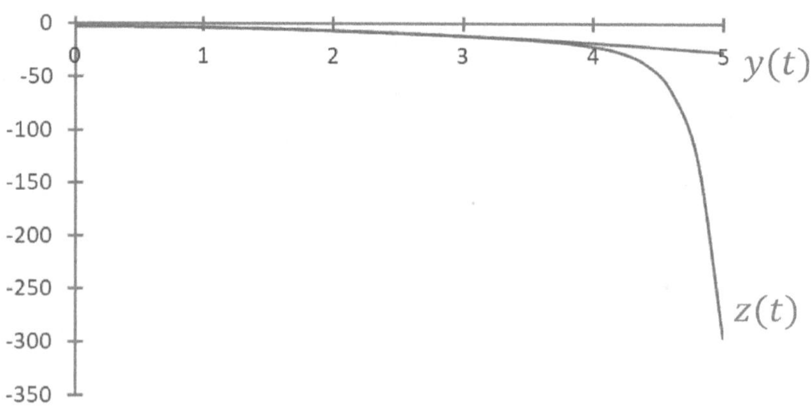

FIGURE 9.6 Two solutions $y(t)$ and $z(t)$ of the same first order ordinary differential Equation (9.36). The solutions are nearly identical in t_0, but become rapidly significantly different from each other.

Problem (9.36) has, in fact, a large Lipschitz constant. A possible value for L is given by

$$L = \max_{(t,y)\in S}\left|\frac{\partial f}{\partial y}(t,y)\right| = \max_{(t,y)\in S}|t| = |t_f| \tag{9.40}$$

In our case $t_f = 5$. From Theorem 9.3 it follows

$$|y(t=5) - z(t=5)| \le e^{25}|y_o - z_o| \approx 7\cdot 10^7 |y_o - z_o| \tag{9.41}$$

Of course, our estimation from Theorem 9.3 is an upper bound, and there may be smaller values for L than 5. However, it still explains why, in this case, the two initially very close solutions diverge dramatically in a very short time.

Problems similar to the one discussed in Example 9.3 are very challenging problems for numerical IvP solvers. The slightest error in their solution process will result in a very large error as t grows. In that sense, first order ordinary differential equations with high Lipschitz constants, or problems necessitating approximate solutions that are far from the initial condition, are termed ill-conditioned problems.

Theorem 9.3 can be used to understand how the local truncation errors will accumulate to build the global truncation error. The coming developments are rather technical. The reader may skip them in a first reading and jump directly to the result presented in Theorem 9.4.

Let us consider the initial value Problem (9.4) and a numerical IvP solver producing approximations w_i. We assume our numerical IvP solver has a local truncation error in order $\mathcal{O}(h^{k+1})$. For example, for Euler's method, this would be $\mathcal{O}(h^2)$.

At $t = t_o$, we have $y(t_o) = y_o$, and the global truncation error g_o will be:

$$g_0 = |w_0 - y_0| = |y_0 - y_0| = 0 \tag{9.42}$$

After computing a first step $i = 1$ with our numerical IvP solver, there is no accumulated error, and the global truncation error is equal to the local truncation error e_1:

$$g_1 = |w_1 - y_1| = e_1 \tag{9.43}$$

In the second step $i = 2$, the global truncation error g_2 will result from the local truncation error e_2 and the truncation error accumulated in the previous step, as we discussed in Figure 9.4.

To estimate the global truncation error g_2, we consider $z(t)$, the local solution in (t_2, w_2), and write (we use the notation $z_2 = z(t_2)$):

$$g_2 = |w_2 - y_2| = |w_2 - z_2 + z_2 - y_2| \le |w_2 - z_2| + |z_2 - y_2| \tag{9.44}$$

Using Theorem 9.3, we have

$$|z_2 - y_2| \le e^{Lh}|z_1 - y_1| = e^{Lh}|w_1 - y_1| = e^{Lh} g_1 \tag{9.45}$$

where L is the Lipschitz constant of the Problem (9.4) and h the step size $h = t_2 - t_1$. Combining this with Equations (9.43) and (9.44) results in:

$$g_2 \le |w_2 - z_2| + |z_2 - y_2| \le e_2 + e^{Lh} g_1 = e_2 + e^{Lh} e_1 \tag{9.46}$$

The same argument can be repeated for the next step $i = 3$:

$$g_3 = |w_3 - y_3| \le e_3 + e^{Lh} g_2 \le e_3 + e^{Lh} e_2 + e^{2Lh} e_1 \tag{9.47}$$

Or in general:

$$g_i = |w_i - y_i| \le e_i + e^{Lh} e_{i-1} + e^{2Lh} e_{i-2} + \ldots + e^{(i-1)Lh} e_1 \tag{9.48}$$

As our numerical IvP solver has a local truncation error in order $\mathcal{O}(h^{k+1})$ there exists a number C such that for all $e_1, e_2, \ldots, e_i$ and for small enough step sizes h (see discussion following Theorem 9.1):

$$e_j \le Ch^{k+1} \text{ for } j = 1, \ldots, i \tag{9.49}$$

This allows us to reformulate Equation (9.48) as

$$g_i = |w_i - y_i| \le Ch^{k+1}\left(1 + e^{Lh} + e^{2Lh} + \ldots + e^{(i-1)Lh}\right) = Ch^{k+1}\frac{e^{iLh} - 1}{e^{Lh} - 1} \tag{9.50}$$

$$\le Ch^{k+1}\frac{e^{L(t_i - t_o)} - 1}{Lh} = \frac{Ch^k}{L}\left(e^{L(t_i - t_o)} - 1\right)$$

This result shows that if the local truncation error of a numerical IvP solver is in order $\mathcal{O}(h^{k+1})$, then the global truncation error will be in order $\mathcal{O}(h^k)$:

Theorem 9.4. Global Truncation Error of a Numerical IVP Solver

If a numerical IvP solver has a local truncation error in order $\mathcal{O}(h^{k+1})$, then the global truncation error will be in order $\mathcal{O}(h^k)$. Specifically,

$$|w_i - y_i| \le \frac{Ch^k}{L}\left(e^{L(t_i - t_o)} - 1\right) \tag{9.51}$$

where C is a constant such that the local truncation error $e_j \le Ch^{k+1}$ for all $j = 1, \ldots, i$ and L the Lipschitz constant of the IvP for which the numerical solver is producing approximations w_i of the exact solution y_i.

From Theorem 9.4, we can conclude that the global truncation error of Euler's method is in order $\mathcal{O}(h)$, as we observed in Section 9.2.3. Theorem 9.4 is a key theorem in the theory of numerical IvP solvers. The order of the local truncation error of numerical IvP solvers can, in general, be found readily using Taylor Series expansion. With Theorem 9.4, it becomes possible as well to find the global truncation error (simply reduce by one the order of the local truncation error). As we will discuss in the upcoming section, this theorem also serves as the basis for a method to estimate the global truncation error, similar to how we estimated the truncation error of quadrature methods in Chapter 8.

9.4 RUNGE-KUTTA METHODS

In this section, we discuss a popular family of numerical initial value problem solvers: the Runge-Kutta methods.[4] In fact, Euler's method is an example of a Runge-Kutta IvP solver. Figure 9.7 provides an intuitive understanding of how Runge-Kutta IvP solvers operate. Let us first discuss the familiar Euler IvP solver. The algorithm operates with a step size h. In t_i it produced an approximation w_i of the exact solution $y(t_i)$. Passing through point (t_i, w_i) is the local solution $z(t)$ of our initial value problem, which has a slope represented by an arrow in Figure 9.7. To compute the approximation w_{i+1} in t_{i+1}, Euler's method extrapolates linearly w_i using the slope $f(t_i, w_i)$, which is, according to the differential equation of our IvP, the slope of the local solution in t_i.

Let us now turn our attention to a generic Runge-Kutta IvP solver as sketched in Figure 9.7 b. The principle is identical to Euler's method. The only difference is the slope used for the extrapolation of the approximation w_i. Runge-Kutta methods use an expression $\varphi(t_i, w_i)$. In the case of Euler's method, $\varphi = f$. The

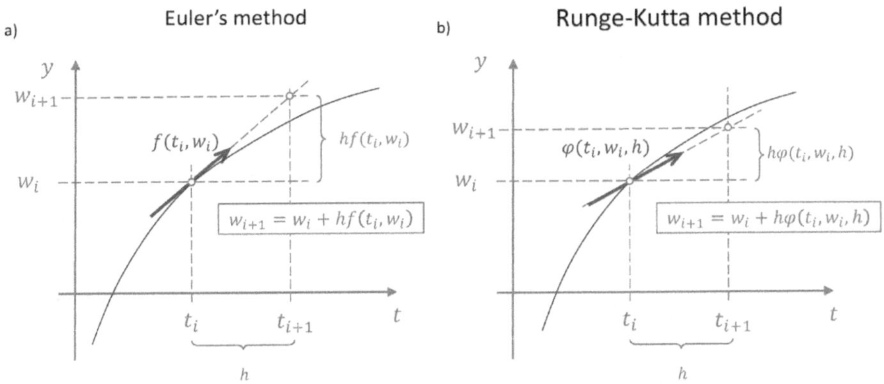

FIGURE 9.7 Schematic of (a) Euler's method and (b) a generic Runge-Kutta IvP solver. Here h is the step size of the method. The arrow represents the slope of the local solution $z(t)$ in t_i of the IvP problem.

following definition summarizes this aspect (note: many textbooks will have a more formal definition of Runge-Kutta methods than the one we present here):

Definition 9.7. Runge-Kutta IVP Solver

A Runge-Kutta initial value problem solver is a numerical IvP solver that writes as

$$w_{i+1} = w_i + h\varphi(t_i, w_i, h) \tag{9.52}$$

where $\varphi(t_i, w_i, h)$ is an estimation of the slope to be used to extrapolate the approximation w_i to w_{i+1}.

In the following section, we will provide some examples of Runge-Kutta methods.

9.4.1 Second Order Runge-Kutta Methods

As a first example of a Runge-Kutta method, we discuss the so-called trapezoid method. The trapezoid method is sketched in Figure 9.8. The method extrapolates the approximation w_i to w_{i+1} by using the average between the slopes of the local solutions in t_i and t_{i+1}. This means φ is computed as

$$\varphi(t_i, w_i, h) = \frac{f(t_i, w_i) + f(t_i, w_{i+1})}{2} \tag{9.53}$$

When w_i is the approximation in t_i and w_{i+1} would be the approximation in t_{i+1}. The problem is that in t_i, we do not yet know the approximation w_{i+1}, as this is

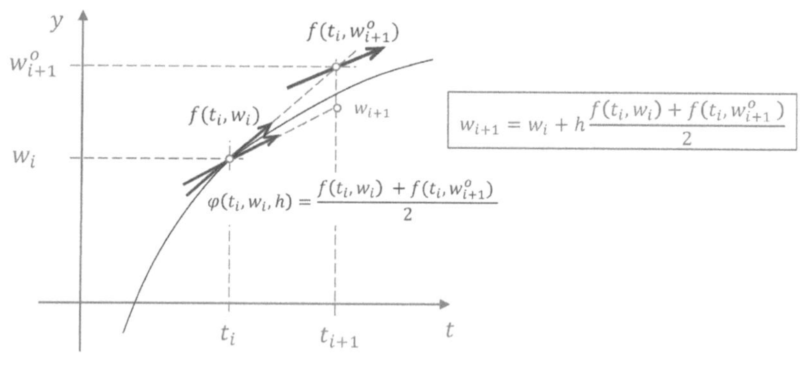

FIGURE 9.8 Schematic of the Trapezoid method. The method extrapolates the approximation w_i to w_{i+1} by using the average between the slopes of the local solutions in t_i and t_{i+1}.

exactly the quantity we want to compute. Consequently, one has to use a further estimation w_{i+1}^{o} of w_{i+1}. A possible way to estimate w_{i+1}^{o} is to use Euler's method, which is used by the trapezoid IvP solver. Without entering into technical details, the theory of IvP solves Expression (9.53), which is called a *corrector,* and the expression is used to estimate w_{i+1}^{o} a *predictor*. Combining predictors and correctors results then in even more general IvP solvers than Runge-Kutta methods (the so-called predictor-corrector solvers). In the case of the trapezoid solver, this results in Heun's method.[5]

Algorithm 9.2. Heun's IVP Solver

To produce approximations w_i of a first order initial value Problem (9.4), we proceed as follows:

1. Predict w_{i+1}^{o} using Euler's method:

$$w_{i+1}^{o} = w_i + hf(t_i, w_i) \tag{9.54}$$

2. Correct w_{i+1}^{o} to obtain w_{i+1} using

$$w_{i+1} = w_i + h\frac{f(t_i, w_i) + f(t_i, w_{i+1}^{o})}{2} \tag{9.55}$$

To rewrite this method in the Runge-Kutta way, we have to combine equations (9.54) and (9.55). For convenience, we introduce two variables k_1 and k_2, as seen in Algorithm 9.3:

Algorithm 9.3. Trapezoid IVP Solver

To produce approximations w_i of a first order initial value Problem (9.4), we proceed as follows:

$$w_{i+1} = w_i + h\left(\frac{1}{2}k_1 + \frac{1}{2}k_2\right) \tag{9.56}$$

with $k_1 = f(t_i, w_i)$ and $k_2 = f(t_i + h, w_i + k_1 h)$.

The reader may have noted the similarity between Equation (9.53) and the trapezoidal rule to approximate definite integrals discussed in Section 8.1. This is not an accident (and, in fact, it gives the name to Algorithm 9.3). Indeed, we can write, by reformulating the differential equation of the first order initial value Problem (9.4) (the precise formulation would be the Picard iteration algorithm):

$$w_{i+1} \cong w_i + \int_{t_i}^{t_{i+1}} f(t, w_i)dt \tag{9.57}$$

Using the trapezoidal rule to approximate the definite integral, we obtain Expression (9.53).

Example 9.4. Trapezoid IVP Solver

Let us approximate the initial value Problem (9.20) with the trapezoid IvP solver.

Algorithm 9.3 can be implemented in Octave with a simple function:

```
function [t, w] = RK_trap(f, to, yo, h, n)
 w=[yo];
 t=[to];
 for i = 1:n
 k1=f(t(i), w(i));
 k2=f(t(i)+h, w(i)+k1*h);
 w=[w ; w(i)+h/2*(k1+k2)];
 t=[t ; t(i)+h];
 endfor
endfunction
```

If we choose the step size $h = 0.2$, the algorithm computes the following approximations:

```
>> f = @(t,y) t.*y+t.^3;
>> [t w] = RK_trap(f, to=0, yo=1, h=0.2, n=5);
>> [t, w]
ans =
          0    1.0000
     0.2000    1.0208
     0.4000    1.0909
     0.6000    1.2340
     0.8000    1.4949
     1.0000    1.9494
```

Note the usage of the. * and. ^ operators in the definition of the function f. Since the exact solution (9.21) of the Problem (9.20) is known, we can compute the true error. We compute this error for the approximations in $t = 1$ for various step sizes h following the same method as in Section 9.2.3:

```
>> y = @(t) 3*exp(t^2/2)-t^2-2;
>> err=[]; h=[]; n=2;
```

```
>> for i = 1:15
>> [t w] = RK_trap(f, to=0, yo=1, 1/n, n);
>> err = [err; abs(w(end)-y(1))];
>> h = [h; 1/n];
>> n = n*2;
>> endfor
```

We can further determine the order of the global truncation error of the algorithm in h:

```
>> A = [h.^0 log(h)];
>> a = A\log(err)
a =
   -2.5643
    1.9445
```

The trapezoid method is in order $\mathcal{O}(h^2)$ for the global truncation error, as confirmed by Figure 9.9.

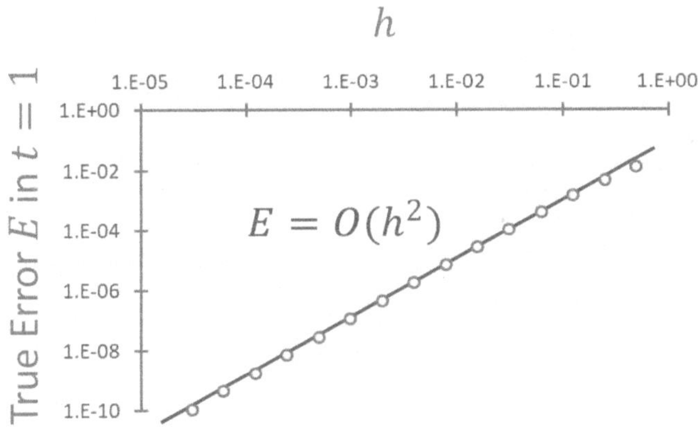

FIGURE 9.9 True error of the approximations produced in $t = 1$ of the initial value Problem (9.20) using the trapezoid IvP solver.

Let us now turn to a second example of a Runge-Kutta IvP solver, the midpoint IvP solver. The principle is sketched in Figure 9.10. The method extrapolates the approximation w_i to w_{i+1} by using the slope of the local solutions in $t_{i+1/2} = t_i + h/2$. As with the trapezoid IvP solver, this slope cannot be computed directly, as we would have to know the approximation $w_{i+1/2}$ in $t_{i+1/2}$. We need a predictor to estimate $w^o_{i+1/2}$. Euler's method can be used for this:

$$w^o_{i+1/2} = w_i + \frac{h}{2} f(t_i, w_i) \tag{9.58}$$

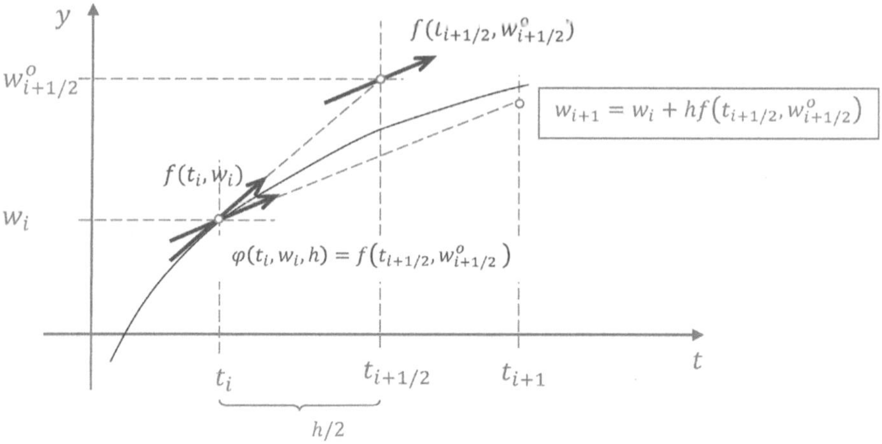

FIGURE 9.10 Schematic of the Midpoint method. The method extrapolates the approximation w_i to w_{i+1} by using the slope of the local solutions in $t_{i+1/2} = t_i + h/2$.

The corrector, to compute w_{i+1}, writes:

$$w_{i+1} = w_i + hf\left(t_{i+1/2}, w^o_{i+1/2}\right) \tag{9.59}$$

As with the Trapezoid IvP solver, Equations (9.58) and (9.59) can be combined and written in the Runge-Kutta form.

Algorithm 9.4. Midpoint IVP Solver

To produce approximations w_i of a first order initial value Problem (9.4), we proceed as follows:

$$w_{i+1} = w_i + hk_2 \tag{9.60}$$

with $k_1 = f(t_i, w_i)$ and $k_2 = f\left(t_i + \frac{h}{2}, w_i + \frac{1}{2}k_1 h\right)$.

The following example illustrates the application of the midpoint IvP solver.

Example 9.5. Midpoint IVP Solver

Let us approximate the initial value Problem (9.20) with the midpoint IvP solver.

Algorithm 9.4 can be implemented in Octave with a simple function:

```
function [t, w] = RK_mid(f, to, yo, h, n)
 w=[yo];
 t=[to];
 for i = 1:n
 k1=f(t(i), w(i));
 k2=f(t(i)+h/2, w(i)+h/2*k1);
 w=[w ; w(i)+h*k2];
 t=[t ; t(i)+h];
 endfor
endfunction
```

If we choose the step size $h = 0.2$, the algorithm computes the following approximations:

```
>> f = @(t,y) t.*y+t.^3;
>> [t w] = RK_mid(f, to=0, yo=1, h=0.2, n=5);
>> [t, w]
ans =
         0    1.0000
    0.2000    1.0202
    0.4000    1.0881
    0.6000    1.2269
    0.8000    1.4806
    1.0000    1.9234
```

Note the usage of the. * and. ^ operators in the definition of the function f. As for Example 9.4, we can compute the true error based on the knowledge of the true solution (9.21) of Problem (9.20):

```
>> y = @(t) 3*exp(t^2/2)-t^2-2;
>> err=[]; h=[]; n=2;
>> for i = 1:15
>> [t w] = RK_mid(f, to=0, yo=1, 1/n, n);
>> err = [err; abs(w(end)-y(1))];
>> h = [h; 1/n];
>> n = n*2;
>> endfor
```

We can further determine the order of the global truncation error of the algorithm in h:

```
>> A = [h.^0 log(h)];
>> a = A\log(err)
```

```
a =
   -0.5939
    1.9764
```

The midpoint method is in order $\mathcal{O}(h^2)$ for the global truncation error, as confirmed in Figure 9.11.

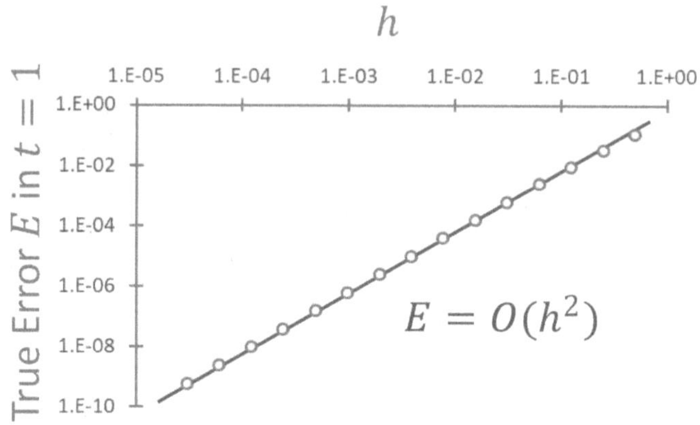

FIGURE 9.11 True error of the approximations produced in $t = 1$ of the initial value Problem (9.20) using the trapezoid IvP solver.

Both Runge-Kutta solvers, the trapezoid and midpoint IvP solver, are of order two in the step size h. They are referred to as RK2 methods. All RK2 methods have expressions similar to the ones of the trapezoid and midpoint methods and involve calculations of two auxiliary variables k_1 and k_2 in each step. Note that each variable k_i requires one function evolution of f. This is the case for all RK2 methods, and all RK2 methods are of second order in the step size h for the global truncation error.

9.4.2 Fourth Order Runge-Kutta Methods

In Section 9.4.1, we discussed two examples of the family of the RK2 IvP solvers. A general RKN IvP solver is expressed as follows:

$$w_{i+1} = w_i + h\left(a_1 k_1 + a_2 k_2 + \cdots a_N k_N\right) \tag{9.61}$$

with a_i some fixed constants. The N auxiliary variables k_i will have to be computed in each step using the function f defining the first order differential equation of our initial value problem. For a given N, there exists, in principle, an infinite number of RKN solvers. Among the most popular RKN methods is the classical RK4 method.

Algorithm 9.5. Classical RK4 IVP Solver

To produce approximations w_i of a first order initial value Problem (9.4), we proceed as follows:

$$w_{i+1} = w_i + \frac{h}{6}(k_1 + 2k_2 + 2k_3 + k_4) \tag{9.62}$$

$$\text{with: } k_1 = f(t_i, w_i)$$

$$k_2 = f\left(t_i + \frac{h}{2}, w_i + \frac{h}{2}k_1\right)$$

$$k_3 = f\left(t_i + \frac{h}{2}, w_i + \frac{h}{2}k_2\right)$$

$$k_4 = f(t_i + h, w_i + hk_3)$$

The approximations of the global truncation errors are in order $\mathcal{O}(h^4)$.

The popularity of RK4 comes from the trade-off between the precision of the algorithm, which produces approximations with a global truncation error in order $\mathcal{O}(h^4)$, and the cost of the algorithm, which requires four function evaluations in each step.

Example 9.6. Classical RK4 IVP Solver

Let us approximate the initial value Problem (9.20) with the classical RK4 IvP solver.

Algorithm 9.5 can be implemented in Octave with a simple function:

```
function [t, w] = RK4(f, to, yo, h, n)
 w=[yo];
 t=[to];
 for i = 1:n
 k1=f(t(i),w(i));
 k2=f(t(i)+h/2,w(i)+k1*h/2);
 k3=f(t(i)+h/2,w(i)+k2*h/2);
 k4=f(t(i)+h,w(i)+k3*h);
 w=[w ; w(i)+h/6*(k1+2*k2+2*k3+k4)];
 t=[t ; t(i)+h];
 endfor
endfunction
```

If we choose the step size $h=0.2$, the algorithm computes the following approximations:

```
>> f = @(t,y) t.*y+t.^3;
>> [t w] = RK4(f, to=0, yo=1, h=0.2, n=5);
>> [t, w]
ans =
         0    1.0000
    0.2000    1.0206
    0.4000    1.0899
    0.6000    1.2316
    0.8000    1.4914
    1.0000    1.9461
```

Note the usage of the. * and. ^ operators in the definition of the function f. As for Example 9.4, we can compute the true error based on the knowledge of the true solution (9.21) of Problem (9.20):

```
>> y = @(t) 3*exp(t^2/2)-t^2-2;
>> err=[]; h=[]; n=2;
>> for i = 1:15
>> [t w] = RK4(f, to=0, yo=1, 1/n, n);
>> err = [err; abs(w(end)-y(1))];
>> h = [h; 1/n];
>> n = n*2;
>> endfor
```

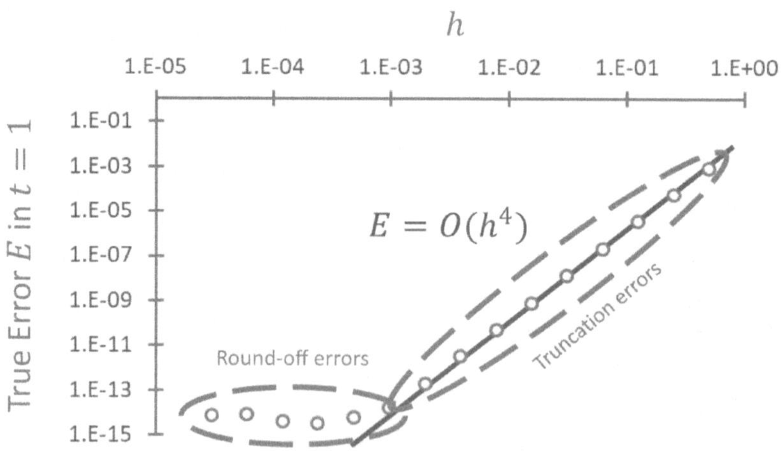

FIGURE 9.12 True error of the approximations produced in $t=1$ of the initial value Problem (9.20) using the trapezoid IvP solver.

Figure 9.12 shows in a log-log plot the true error in the function of the step size h. Two regions can be identified. For large values of h, larger than 10^{-3}, the true error decreases, according a power law. The order of this power law can be determined. For this, we first need to find the entries in vector `h` that are larger than 10^{-3}:

```
>> length(find(h>1E-3))
ans = 9
```

This means the first nine entries in `h` are larger than 10^{-3}. The reader can verify this by computing `h([1:9])`, which will show these entries. We can now fit a power law to the first nine entries of h and the true error:

```
>> A = [h([1:9]).^0 log(h([1:9]))];
>> a = A\log(err([1:9]))
a =
   -4.2106
    4.0078
```

The RK4 method is in order $\mathcal{O}(h^4)$ for the global truncation error.

For values h smaller than 10^{-3}, the global truncation error no longer follows the power law $\mathcal{O}(h^4)$. This is due to the accumulation of round-off errors, similar to what we observed with numerical quadrature methods in Chapter 8.

9.4.3 Estimating the Global Truncation Error of Numerical IvP Solvers

In a similar way to the discussion in Section 8.3.5, covering quadrature methods, it is possible to derive a method to estimate the global truncation error of a numerical IvP solver. We can, for a given time step t_i, produce two approximations with two different step sizes h_1 and h_2:

$$y(t_i) = w_i(h_1) + E_i(h_1) = w_i(h_2) + E_i(h_2) \tag{9.63}$$

where $E_i(h_1)$ is the global truncation error in t_i that we obtain when using a step size h_1. $E_i(h_2)$ is the global truncation error in t_i that we obtain when using the step size h_2. If the global truncation error of the IvP solver is in order $\mathcal{O}(h^q)$, and both step sizes h_1 and h_2 are small enough that they follow a power law (but large enough that round-off errors are not yet significant compared to the truncation errors), then we can write:

$$E_i(h) \cong Ch^q \tag{9.64}$$

And, consequently,

$$\frac{E_i(h_1)}{E_i(h_2)} \cong \frac{h_1^q}{h_2^q} \tag{9.65}$$

Combining Equations (9.63) and (9.65) yields

$$w_i(h_1) + \frac{h_1^q}{h_2^q} E_i(h_2) \cong w_i(h_2) + E_i(h_2) \tag{9.66}$$

Solving Equation (9.66) for $E_i(h_2)$ results in Richardson's formula for IvP solvers:

Theorem 9.5. Richardson's Truncation Error Estimation for Numerical IVP Solvers

If $w_i(h_1)$ and $w_i(h_2)$ are two approximations produced with two step sizes h_1 and h_2 by a numerical IvP solver producing approximations with a global truncation error in order $\mathcal{O}(h^q)$, then the global truncation error $E_i(h_2)$ of $w_i(h_2)$ is given by

$$E_i(h_2) \cong \left| \frac{w_i(h_2) - w_i(h_1)}{\left(\frac{h_1}{h_2}\right)^q - 1} \right| \tag{9.67}$$

as long as h_1 and h_2 are such that the global truncation errors can be approximated reasonably as $E_i(h) = Ch^q$.

Similar to quadrature methods, Richardson's error formula is of priceless practical value to estimate the global truncation errors of approximations produced by numerical IvP solvers.

Example 9.7. Richardson's Error Formula

Solve the following initial value problem with the classical RK4 method

$$\begin{cases} \dfrac{dy}{dt} = -0.6y + 10e^{-100(t-2)^2} \\ y(0) = 0.5 \end{cases} \tag{9.68}$$

and provide an estimation of the global truncation error.

Using the Octave function RK4 from Example 9.6, we can produce approximations of Problem (9.68). Our main concern is to identify the step size h to be used. When using Richardson's error formula, we need to choose a step size satisfying the hypothesis of Theorem 9.5. This means a step size for which $E_i(h) = Ch^4$ is valid (recall that the RK4 method is in order $q = 4$). To find out the range of h for which this condition is satisfied, we compute approximations of (9.68) and plot the estimated truncation errors with Richardson's formula versus h.

Approximations w4 in $t = 4$ of Problem (9.68) can be computed with the function RK4:

```
>> f = @(t,y) -0.6*y+10*exp(-100*(t-2).^2);
>> n = 1; w4 = []; h = [];
>> for i = 1:16
>> [t, w] = RK4(f, to=0, yo=0.5, 4/n, n);
>> w4 = [w4; w(end)];
>> h = [h; 4/n];
>> n = n*2;
>> endfor
```

We then compute the Richardson error estimations err in $t = 4$:

```
>> err = [];
>> for i = 2:length(w4)
>> err = [err; abs(w4(i)-w4(i-1))/(2^4-1)];
>> endfor
```

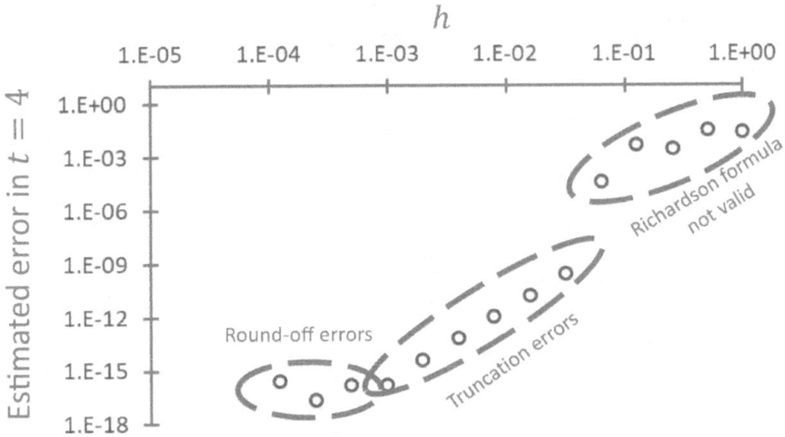

FIGURE 9.13 Richardson error estimations of the approximations produced in $t = 4$ of the initial value Problem (9.68) using the classical RK4 IvP solver.

Figure 9.13 shows, in a log-log plot, Richardson's error estimations of the approximations in $t = 4$ in function of the step size h. Three regions can be identified. For large values of h, up to about $h = 10^{-1}$, the hypothesis $E_i(h) = Ch^4$ is not valid as h is too large.

For very small values for $h < 10^{-3}$, the round-off errors are not negligible, and Richardson's error formula, which estimates only the truncation error, will not estimate correctly the error.

For intermediate values (about $10^{-3} < h < 10^{-1}$), the estimated error follows a power law. The order of convergence can be estimated. First, we find the indexes of the values h that are in the correct range:

```
>> range = find((h>1E-3) & (h<0.5*1E-1));
```

Then, we fit a power law to the error estimations:

```
>> A = [h(range).^0, log(h(range))];
>> a = A\log(err(range))
a =
   -10.3707
     4.1413
```

which confirms that the order of convergence of the error is $\mathcal{O}(h^4)$, as expected from the RK4 method. This means that for $10^{-3} < h < 10^{-1}$, the classical RK4 method operates as expected, and Richardson's error formula can be used to estimate the truncation error. For example, if we choose $h = 0.005$ (which corresponds to $n = 800$ in our function RK4 if we want to reach $t = 4$), the error E_{800} in $t = 4$ will be less than 10^{-12}, according to Figure 9.13. Figure 9.14 depicts the numerical approximation of the initial value Problem (9.68) using the classical RK4 IvP solver with the step size of $h = 0.005$.

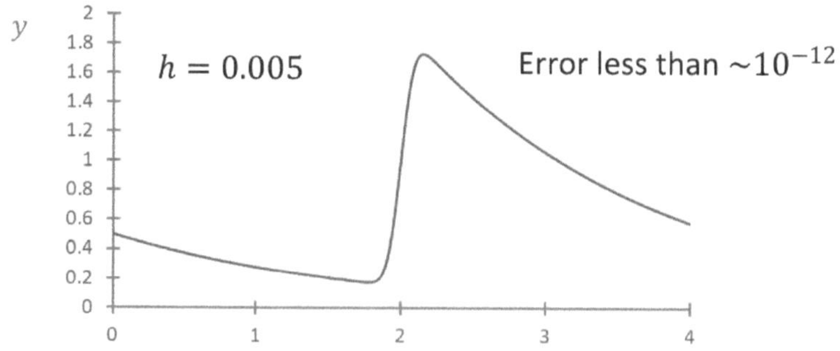

FIGURE 9.14 Numerical approximation of the initial value Problem (9.68) using the classical RK4 IvP solver with a step size of $h = 0.005$. The estimated error in $t = 4$ is lower than 10^{-12}.

9.5 SYSTEMS AND HIGHER ORDER INITIAL VALUE PROBLEMS

9.5.1 Systems of First Order Ordinary Differential Equations

The algorithms discussed so far in this chapter provide numerical approximations to the first order initial value Problem (9.4). In Definition 9.2, the problem was introduced for a first odder ordinary differential equation involving a single unknown function $y(t)$. This definition applies, however, equally to first order ordinary differential equations involving several unknown functions. Indeed, a system of to first order ordinary differential equations writes:

$$\begin{cases} \frac{dy_1}{dt} = f_1(t, y_1, y_2, \dots, y_n) \\ \frac{dy_2}{dt} = f_2(t, y_1, y_2, \dots, y_n) \\ \vdots \\ \frac{dy_n}{dt} = f_n(t, y_1, y_2, \dots, y_n) \end{cases} \tag{9.69}$$

where $y_1(t), y_2(t), \dots, y_n(t)$ are the n unknown functions in the single dependent variable t. If we introduce, as we did for algebraic systems of equations in Chapter 4, vector notations, system (9.69) becomes formally equivalent to Equation (9.4). For this, we define the column vector $\underline{y}(t)$ of the unknown functions as

$$\underline{y}(t) = \begin{bmatrix} y_1(t) \\ \vdots \\ y_n(t) \end{bmatrix} \tag{9.70}$$

and the right-hand side vector $\underline{f}(t, \underline{y})$ as

$$\underline{f}(t, \underline{y}) = \begin{bmatrix} f_1(t, y_1, \dots, y_n) \\ \vdots \\ f_n(t, y_1, \dots, y_n) \end{bmatrix} \tag{9.71}$$

With these notations, an initial value problem becomes then

$$\begin{cases} \frac{d}{dt}\underline{y} = \underline{f}(t, \underline{y}) \\ \underline{y}(t_o) = \underline{y}_o = \begin{bmatrix} y_{1,o} \\ \vdots \\ y_{n,o} \end{bmatrix} \end{cases} \tag{9.72}$$

Definition 9.2 appears now to be a particular case of (9.72) when the number of equations is equal to one.

Not only can Definition 9.2 be generalized using this vector notation, but, in fact, all algorithms we discussed so far. Indeed, all numerical IvP solvers we presented in the chapter involve only operations that can be rewritten in vector notations. For example, Euler's method rewrites as

$$\underline{w}_{i+1} = \underline{w}_i + h\underline{f}\left(t_i, \underline{w}_i\right) \tag{9.73}$$

with the vector $\underline{w}_i$ of the approximations of $\underline{y}\left(t_i\right)$:

$$\underline{w}_i = \begin{bmatrix} w_{1,i} \\ \vdots \\ w_{n,i} \end{bmatrix} \tag{9.74}$$

Practically, this means one applies Euler's algorithm to each individual equation of the system (9.69).

Example 9.8. Euler's Method for IVP of Systems of Equations

Let us approximate numerically, using Euler's algorithm with a step size $h = 0.1$ on the interval $\left[0, 0.6\right]$, the following initial value problem

$$\begin{cases} \dfrac{dy_1}{dt} = y_2^2 - 2y_1 \\ \dfrac{dy_2}{dt} = y_1 - y_2 - ty_2^2 \\ y_1\left(0\right) = 0 \\ y_2\left(0\right) = 1 \end{cases} \tag{9.75}$$

In Example 9.1, we developed an Octave function to apply Euler's algorithm in the case of an IvP involving a single unknown function. It is simple to generalize that function to the case of IvPs involving several unknown functions by taking advantage of Octave's vector notations.

```
function [t, w] = Euler(f, to, yo, h, n)
 w = [yo];
 t = [to];
 for i = 1:n
 w = [w ; w(i,:) + h*f(t(i), w(i,:))];
 t = [t ; t(i) + h];
 endfor
endfunction
```

Comparing this code with the one from Example 9.1 reveals only minor differences. In fact, the only difference is the way the approximations wi are handled. For any code implementing numerical IvP solvers for systems of equations, the way the data is handled needs to be chosen. In the present code, we chose to represent wi as a line vector. The quantity w, which holds all the computed approximations, becomes a matrix. Each line of w contains the approximations of the functions y_1 to y_n in t_i. The fist line will contain the initial conditions, the second the approximations in t_1, and so on.

Applied to our Problem (9.75):

```
>> function z = f(t,y)
>> z(1) = y(2)^2 - 2*y(1);
>> z(2) = y(1) - y(2) - t*y(2)^2;
>> endfunction
>> [t w] = Euler(@f, to=0, to=[0 1], h=0.1, n=3)
t =
     0.0000
     0.1000
     0.2000
     0.3000
w =
     0.0000    1.0000
     0.1000    0.9000
     0.1610    0.8119
     0.1947    0.7336
```

To save space, we show the numerical approximations only up to $t = 0.3$. To obtain the approximations up to $t = 1$, the function Euler should be called with n=10.

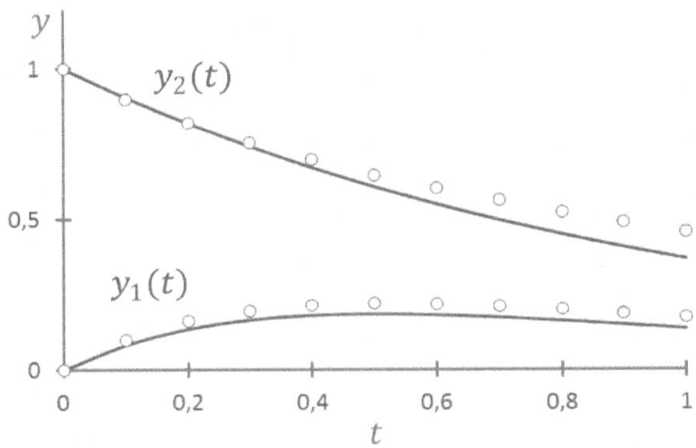

FIGURE 9.15 Numerical approximations (o) of the initial value Problem (9.75) using Euler's IvP solver with a step size of $h = 0.1$ and the true solutions (–).

The numerical approximations can be plotted in a graph (Figure 9.15):

```
>> plot(t, w(:,1), 'o', t, w(:,2), 'o')
```

where `w(:,1)` are the numerical approximations of $y_1(t)$ and `w(:,2)` are the numerical approximations of $y_2(t)$.

In Figure 9.15, besides the numerical approximations produced by Euler's IvP solver, the true solutions of (9.75) are displayed:

$$\begin{cases} y_1(t) = te^{-2t} \\ y_2(t) = e^{-t} \end{cases} \tag{9.76}$$

9.5.2 Higher Order Ordinary Differential Equations

Initial value problems involving higher order differential equations can be approximated by the methods we discussed so far. Consider a general n-order ordinary differential equation:

$$\frac{d^n}{dt^n} y(t) = f\left(t, y, y', y^{(2)}, \dots, y^{(n-1)}\right) \tag{9.77}$$

where $y^{(i)}$ stands for the i-th order derivative of the function $y(t)$. Equation (9.77) can be transformed into a system of first order ordinary differential equations by introducing the new unknowns y_i as follows:

$$\begin{aligned} y_1 &= y \\ y_2 &= y' \\ &\vdots \\ y_n &= y^{(n-1)} \end{aligned} \tag{9.78}$$

With these new variables, Equation (9.77) writes

$$\begin{aligned} y_1 &= y \\ y_2' &= y_1 \\ &\vdots \\ y_{n-1}' &= y_n \\ y_n' &= f(t, y_1, y_2, \dots, y_n) \end{aligned} \tag{9.79}$$

which is a first order system of ordinary differential equations for the unknowns $y_1, y_2, \dots, y_n$, which can be solved as discussed in Section 9.5.1.

Example 9.9. Mathematical Pendulum

From elementary mechanics, we know that the differential equation describing the evolution of a pendulum is

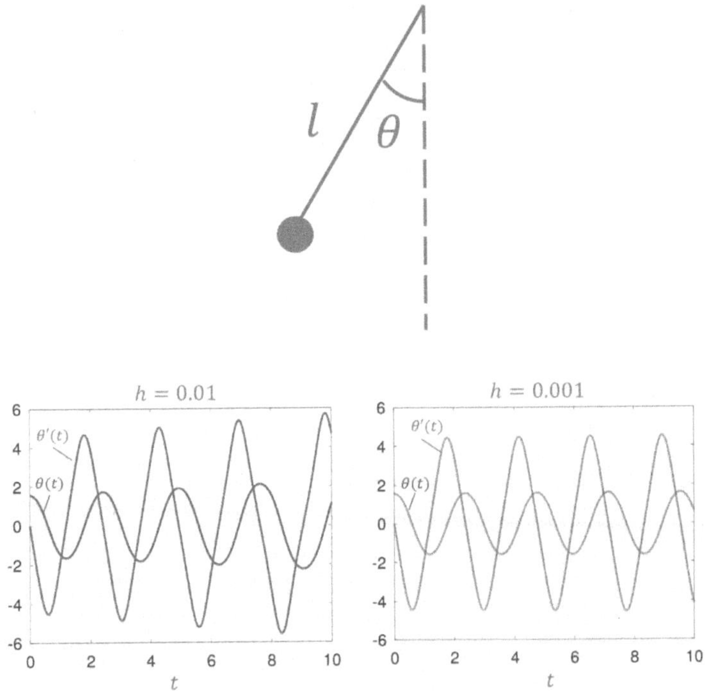

FIGURE 9.16 (a) Mathematical pendulum and (b) numerical approximations of the time evolution of the angle θ computed with Euler's IvP solver for two-step sizes ($h = 0.01$ and $h = 0.001$),

$$\frac{d^2}{dt^2}\theta(t) = -\frac{g}{l}\sin\theta \tag{9.80}$$

where θ is the angle between the pendulum and the vertical (Figure 9.16a). We will approximate the solution of (9.80) for the particular case where the length of the pendulum $l = 1$ meter and initial conditions:

$$\theta(0) = \frac{\pi}{2} \text{ and } \frac{d}{dt}\theta(0) = 0 \tag{9.81}$$

First, we transform the original Equation (9.80) into a system of first order differential equations by introducing

$$\begin{aligned} y_1 - \theta \\ y_2 = \theta' \end{aligned} \tag{9.82}$$

Equation (9.80) rewrites as

$$\begin{aligned} y_1' &= y_2 \\ y_2' &= -\frac{g}{l}\sin y_1 \end{aligned} \tag{9.83}$$

The numerical approximations of Equation (9.83) can, using our function `Euler` from Example 9.8, be obtained (for example for $h = 0.01$) as follows:

```
>> function z=f(t,y)
>> z(1) = y(2);
>> z(2) = -9.81*sin(y(1));
>> endfunction
>> [t w]=Euler(@f, to=0, yo=[pi/2 0], h=0.01,
n=1000);
>> plot(t, w(:,1), t, w(:,2))
```

Similarly, one can obtain approximations using $h = 0.001$. Figure 9.16b) plots these approximations. For the case where $h = 0.01$, the amplitude of the angle θ and its time derivative increase with time. The mathematical pendulum is a system that conserves mechanical energy and, consequently, this behavior can not be correct. The situation for $h = 0.001$ seems better.

This method, which consists in transforming a higher ordinary differential equation into a system of first order differential equations with the help of some additional variables, can as well be used for systems of higher order differential equations as illustrated by the next example.

Example 9.10. Kepler Motion

Newton's law of motion applied to a satellite moving around a planet of a much larger mass M placed at the origin of a coordinate system writes (Figure 9.17a):

$$\begin{cases} \dfrac{d^2}{dt^2}x(t) = -G\dfrac{M \cdot x}{\left(x^2+y^2\right)^{3/2}} \\ \dfrac{d^2}{dt^2}y(t) = -G\dfrac{M \cdot y}{\left(x^2+y^2\right)^{3/2}} \end{cases} \tag{9.84}$$

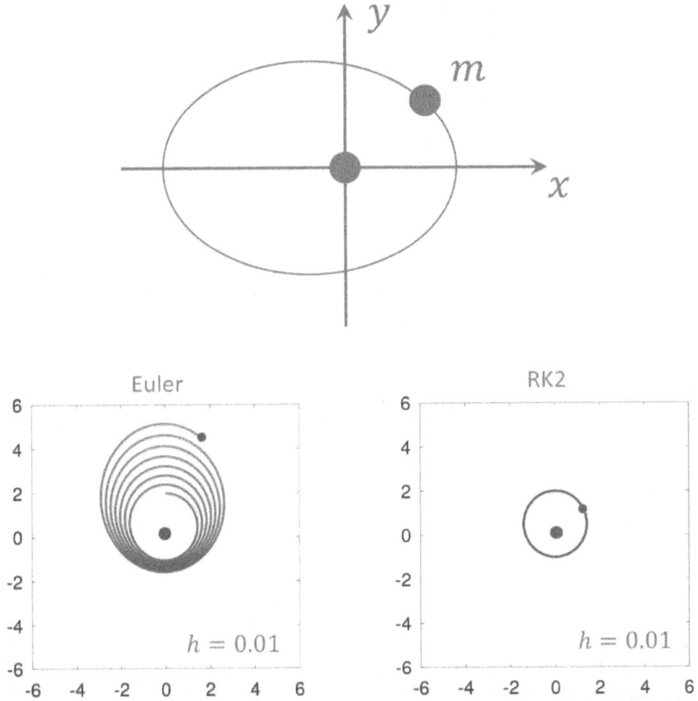

FIGURE 9.17 (a) A satellite of mass m orbiting around a significantly larger mass placed at the center of the coordinate system and (b) numerical approximations of the time evolution of the satellite computed with Eulers' and the trapezoid IvP solvers for a step size of $h = 0.01$.

with G the universal gravitational constant. As the system of Equations (9.84) consists of two second order ordinary differential equations, four auxiliary variables are required to transform this system into a system of first order ordinary differential equations:

$$\begin{aligned} y_1 &= x & y_3 &= y \\ y_2 &= \frac{dx}{dt} & y_4 &= \frac{dy}{dt} \end{aligned} \tag{9.85}$$

With the auxiliary variables (9.85), system (9.84) can be rewritten as a system of first order ordinary differential equations:

$$\begin{aligned} y_1' &= y_2 & y_3' &= y_4 \\ y_2' &= -G\frac{M \cdot y_1}{\left(y_1^2 + y_3^2\right)^{3/2}} & y_4' &= -G\frac{M \cdot y_3}{\left(y_1^2 + y_3^2\right)^{3/2}} \end{aligned} \tag{9.86}$$

As an example, we compute the approximations of (9.86) for the case where $G \cdot M = 3$ and the following initial conditions:

$$\begin{aligned} y_1(0) &= x(0) = 0 & y_3(0) &= y(0) = 2 \\ y_2(0) &= \frac{dx}{dt}(0) = 1 & y_4(0) &= \frac{dy}{dt}(0) = 0 \end{aligned} \tag{9.87}$$

We compare the approximations produced by Euler's IvP solver and the trapezoid IvP solver. For Euler's method, we can use our function defined in Example 9.8. For The trapezoid IvP solver, we define the following function, which is adapted from Example 9.4 for systems of ordinary differential equations:

```
function [t, w] = RK_trap(f, to, yo, h, n)
 w = [yo];
 t = [to];
 for i = 1:n
 k1 = f(t(i), w(i,:));
 k2 = f(t(i) + h, w(i,:) + k1*h);
 w = [w ; w(i,:) + h/2 * (k1 + k2)];
 t = [t ; t(i) + h];
 endfor
endfunction
```

System (9.86) is implemented as follows:

```
function z = orb(t, y)
z(1) = y(2);
z(2) = -3*y(1) / (y(1)^2 + y(3)^2)^1.5;
z(3) = y(4);
z(4) = -3*y(3) / (y(1)^2 + y(3)^2)^1.5;
endfunction
```

To compute and plot the numerical approximations, for example, using Euler's IvP solver:

```
>> [t w] = Euler(@orb, 0, [0 1 2 0], 0.01, 10000);
>> plot(w(:,1), w(:,3));
>> axis([-6, 6, -6, 6], "square");
```

Figure 9.17 b shows the approximated orbits of the satellite using Euler's and the trapezoid IvP solver. According to Kepler's law, we know that the orbit should be an ellipse. Euler's method does not produce a correct approximation, whereas the trapezoid IvP solver seems indeed to produce an ellipse as an approximation of the orbit.

9.5.3 Estimating Errors of Numerical IvP Solvers Involving Systems of Equations

Estimating errors of approximations computed by a numerical IvP solver for a system of equations can be achieved using Richardson's error formula (9.67). However, we have to apply the formula for each approximation individually. To illustrate this approach, let us consider the following initial value problem:

$$\begin{cases} y_1' = y_1 + y_2 \\ y_2' = -y_1 + y_2 \\ y_1(0) = 1 \\ y_2(0) = 0 \end{cases} \tag{9.88}$$

The exact solutions are

$$\begin{cases} y_1(t) = e^t \cos t \\ y_2(t) = -e^t \sin t \end{cases} \tag{9.89}$$

We will use the classical RK4 method to approximate the solutions of Problem (9.88). A possible Octave implementation of the classical RK4 method for a system of equations is as follows:

```
function [t, w] = RK4(f, to, yo, h, n)
 w = [yo];
 t = [to];
 for i = 1:n
 k1 = f(t(i), w(i,:));
 k2 = f(t(i) + h/2, w(i,:) + k1*h/2);
 k3 = f(t(i) + h/2, w(i,:) + k2*h/2);
 k4 = f(t(i) + h,w(i,:) + k3*h);
 w = [w ; w(i,:) + h/6*(k1 + 2*k2 + 2*k3 + k4)];
 t = [t ; t(i)+h];
 endfor
endfunction
```

Figure 9.18 shows the true errors in $t = 1$ of the numerical approximations of the Problem (9.88) obtained by the classical RK4 method for different values of the step size h. Two regions can be identified. For step sizes small enough (lower than about 10^{-3}), the errors decrease in order $\mathcal{O}(h^4)$ as expected from the RK4 IvP solver. For smaller values of the step size, round-off errors become significantly higher than the truncation errors, and the total error no longer decreases according $\mathcal{O}(h^4)$.

The interesting element to read from Figure 9.18 is that the errors of the approximations of both solutions $y_1(t)$ and $y_2(t)$ of Problem (9.88) decrease in $\mathcal{O}(h^4)$, but their absolute values are not identical. In our case, the approximations of $y_1(t)$

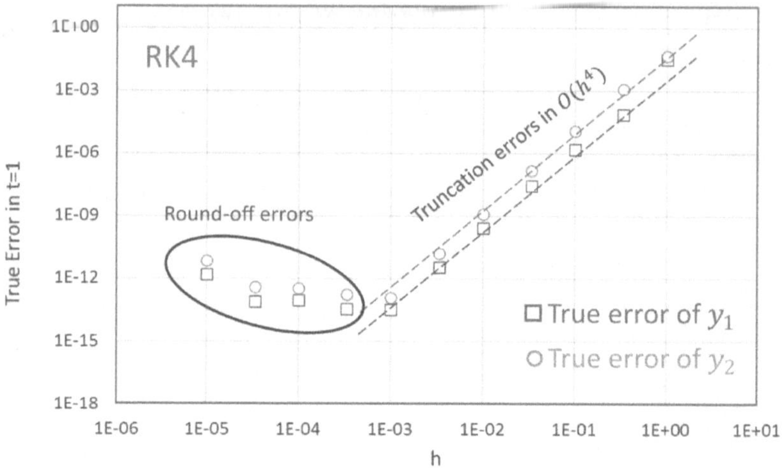

FIGURE 9.18 True errors in $t = 1$ of the numerical approximations of the Problem (9.88) obtained by the classical RK4 method.

have an error of roughly 5 times less than the approximations of $y_2(t)$ (note the logarithmic scale). These errors can be estimated with Richardson's error formula (9.68).

We start by computing two approximations with two different step sizes. First, we define the IvP (9.88):

```
function z = f(t, x)
 z(1) = x(1) + x(2);
 z(2) = -x(1) + x(2);
endfunction
```

The approximations in $t = 1$ for a step size of $h = 1/3$ are:

```
>> [t w] = RK4(@f, 0, [1 0], 1/3, 3);
>> w(end, :)
ans =
   1.4688   -2.2886
```

whereas for $h = 1/6$:

```
>> [t w] = RK4(@f, 0, [1 0], 1/6, 6);
>> w(end, :)
ans =
   1.4687   -2.2874
```

The estimated error of the approximation of $y_1(t = 1)$, according to Richardson, will be:

$$E(t=1) \cong \left| \frac{1.4687 - 1.4688}{\left(\frac{1/3}{1/6}\right)^4 - 1} \right| \cong 6 \cdot 10^{-6} \tag{9.90}$$

and the estimated error of the approximation of $y_2(t=1)$:

$$E(t=1) \cong \left| \frac{-2.2874 + 2.2886}{\left(\frac{1/3}{1/6}\right)^4 - 1} \right| \cong 8 \cdot 10^{-5} \tag{9.91}$$

The reader may want to compare the estimated errors with the true errors, using the known true solutions (9.89). An excellent agreement exists between the estimated and true errors.

Example 9.11. Orbital Motion in a Perturbated Gravitational Field

Example 9.10 approximated numerically the motion of a satellite around a massive object. In this example, we will approximate a more realistic case and control the error of the numerical approximations.

One way to study the motion of a satellite around a massive object while considering the presence of other objects is to use a perturbed gravitational force. In such case, the force is no longer inversely proportional to the square of the relative distance r of the objects but inversely proportional to the distance to a power α, where α is different than two:

$$F = G\frac{m_1 m_2}{r^\alpha} \tag{9.92}$$

Equations (9.84) become:

$$\begin{cases} \dfrac{d^2}{dt^2} x(t) = -G\dfrac{M \cdot x}{\left(x^2 + y^2\right)^{(\alpha+1)/2}} \\ \dfrac{d^2}{dt^2} y(t) = -G\dfrac{M \cdot y}{\left(x^2 + y^2\right)^{(\alpha+1)/2}} \end{cases} \tag{9.93}$$

In the first step, we transform the system (9.93) into a system of first order ordinary differential equations using the auxiliary variables (9.85) from Example 9.10. The system becomes:

$$y_1' = y_2 \qquad\qquad y_3' = y_4$$

$$y_2' = -G\frac{M \cdot y_1}{\left(y_1^2 + y_3^2\right)^{(\alpha+1)/2}} \qquad y_4' = -G\frac{M \cdot y_3}{\left(y_1^2 + y_3^2\right)^{(\alpha+1)/2}} \tag{9.94}$$

As an example, we will approximate the orbit of the satellite for the case where $G \cdot M = 3$ and α=1.5. We use the initial conditions (9.87) as in Example 9.10.

We define our problem in Octave:

```
function z = f(t, y)
 z(1) = y(2);
 z(2) = -3*y(1) / (y(1)^2 + y(3)^2)^1.25;
 z(3) = y(4);
 z(4) = -3*y(3) / (y(1)^2 + y(3)^2)^1.25;
endfunction
```

To estimate the error, we will use the errors on the x and y position of the satellite in $t = 20$. Using the classical RK4 method and Richardson's error formula, we can compute these error estimations for some step sizes h:

```
>> err = []; h = [];
>> for i = 0:5
>> n = 10^i;
>> [t w] = RK4(@f, 0, [0 1 2 0], 20/n, n);
>> n = 2*10^i;
>> h = [h ; 1/n];
>> [t ww] = RK4(@f, 0, [0 1 2 0], 20/n, n);
>> err = [err ; abs(w(end,1) - ww(end,1)) / (2^4-
1), abs(w(end,3) - ww(end,3)) / (2^4-1)];
>> endfor
```

And plot them in a log-log plot (Figure 9.19a)

```
>> loglog(h, err)
```

In this example, the errors in x and y are very similar, which was to be expected considering the symmetry between these two variables. From Figure 9.19a, we can conclude that for step sizes roughly from 10^{-4} to $5 \cdot 10^{-2}$, the error follows the $\mathcal{O}\left(h^4\right)$ trend of the classical RK4 IvP solver. In this region, the application of the Richardson error formula is valid. If we chose $h = 3 \cdot 10^{-4}$ (i.e., $n = 67000$ steps), we can expect an error of about 10^{-8}. The corresponding orbit is shown in Figure 9.19b. The trajectory is essentially an ellipse but an ellipse that is rotating over time. The angle β defined in

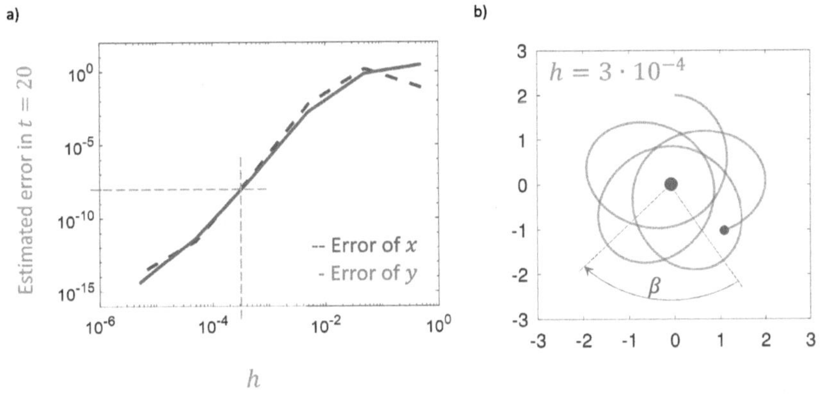

FIGURE 9.19 (a) Richardson error estimates of the approximated x and y position of a satellite subject to Equations (9.94) computed using the classical RK4 IvP solver. (b) Orbit computed using the RK4 solver with $h = 3 \cdot 10^{-4}$, for which the resulting error will be lower than 10^{-8}.

Figure 9.19b is called the advance of the perihelion. The perihelion is defined as the position where the satellite is the furthest away from the massive object around which it is orbiting.

The advance of the perihelion is indeed observed for all planets in our solar system and can be explained by the perturbations created by the presence of all these masses moving around the sun. Only for Mercury is the observed advance of the perihelion larger than the one predicted by Newton's law of motion and gravity. For a long time, astronomers had no convincing explanation for this anomaly. It wasn't until 1918, when Einstein[6] reformulated the law of gravity within the framework of his general theory of relativity, that this puzzle was solved. One of the consequences of Einstein's theory of gravity is that the gravitational force does not follow an inverse square law in relation to the relative distance. Instead, it adheres to an equation resembling (9.92) when close to massive objects.

Recomputing the trajectory of Mercury with Einstein's equations allows for predicting the observed advance of the perihelion very precisely. It was one of the first experimental poofs of Einstein's theory.

9.6 ADAPTIVE METHODS

So far, we have considered numerical IvP solvers that belong to the family of fixed step size solvers, as they all keep the step size h constant during the approximation process of the solution of the IvP. There are, however, some drawbacks by forcing the IvP solvers to keep the step size h constant. To get a better feeling about these drawbacks, let us examine an example.

9.6.1 Introductory Example

Consider the following second order ordinary differential equation

$$\frac{d^2x}{dt^2} - \mu\left(1 - x^2\right)\frac{dx}{dt} + x = 0 \tag{9.95}$$

where μ is a positive parameter. Equation (9.95) is known as the Van der Pol oscillator.[7] It was originally studied by the Dutch physicist Balthasar van der Pol while working at Philips in the early past century. It describes some nolinear electrical circuits involving vacuum tubes. Since then, several additional applications, ranging from biology to seismology, are known.

The Van der Pol oscillator (9.95) is a second order differential equation very similar to the linear damped oscillator, such as a mass suspended by a spring and a damping element. The difference is the nonlinear coefficient of friction $\mu\left(1 - x^2\right)$ in front of the first derivative of x. To compute some numerical approximations of the solutions of (9.95), we start by transforming the second order equation into a system of first order equations by defining

$$\begin{aligned} y_1 &= x \\ y_2 &= \frac{dx}{dt} \end{aligned} \tag{9.96}$$

With these auxiliary variables, Equation (9.95) becomes a first order system of ordinary differential equations:

$$\begin{aligned} y_1' &= y_2 \\ y_2' &= \mu\left(1 - y_1^{\,2}\right)y_2 + y_1 \end{aligned} \tag{9.97}$$

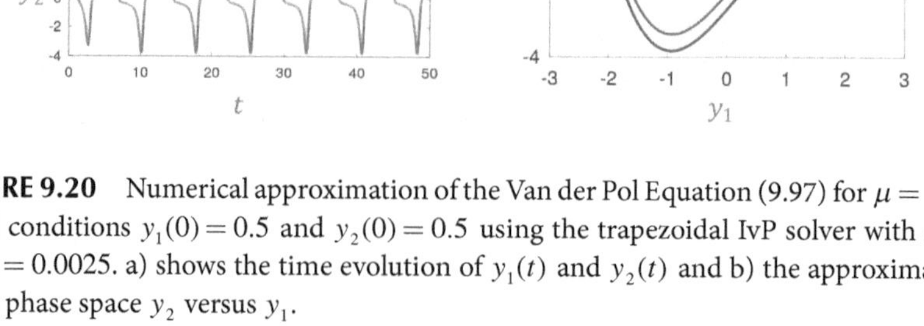

FIGURE 9.20 Numerical approximation of the Van der Pol Equation (9.97) for $\mu = 2$ and initial conditions $y_1(0) = 0.5$ and $y_2(0) = 0.5$ using the trapezoidal IvP solver with a step size $h = 0.0025$. a) shows the time evolution of $y_1(t)$ and $y_2(t)$ and b) the approximations in the phase space y_2 versus y_1.

Figure 9.20 depicts the numerical approximation of the Van der Pol Equation (9.97) for $\mu = 2$ and initial conditions $y_1(0) = 0.5$ and $y_2(0) = 0.5$ using the trapezoidal IvP solver with a step size $h = 0.0025$. Similarly to the linear oscillator case, we obtain periodic solutions. However, surprisingly, despite the fact that the Van der Pol Equation (9.97) contains a friction term, these oscillations do not seem to lose energy (the amplitudes of $y_1(t)$ and $y_2(t)$ remain constant over time). In fact, by plotting the numerical approximations in the so-called phase space, we observe this more clearly. The phase space representation is a plot of y_2 versus y_1 obtained by plotting on the figure for each time instant the points $(y_1(t), y_2(t))$. As depicted in Figure 9.20b, after some initial trajectory, the approximations rapidly follow a similar trajectory, called a limit cycle in the frame of nonlinear systems. Our aim is not to understand how a system involving friction can end up with such behavior. Rather, we want to understand if this observation is correct or an artifact due to numerical errors.

To estimate the errors of our approximation, we proceed as discussed in Section 9.5.3. We compute two approximations using different step sizes. We first use a step size $h_1 = 0.0050$ then a step size $h_2 = 0.0025$. Using Richardson's error formula (9.67), the error in $t = 50$ is estimated to be about $1 \cdot 10^{-4}$ for y_1 andabout $4 \cdot 10^{-5}$ for y_2. These are the error estimations in only $t = 50$. One may wonder how are the errors in the other values of t in the interval $[0, 50]$.

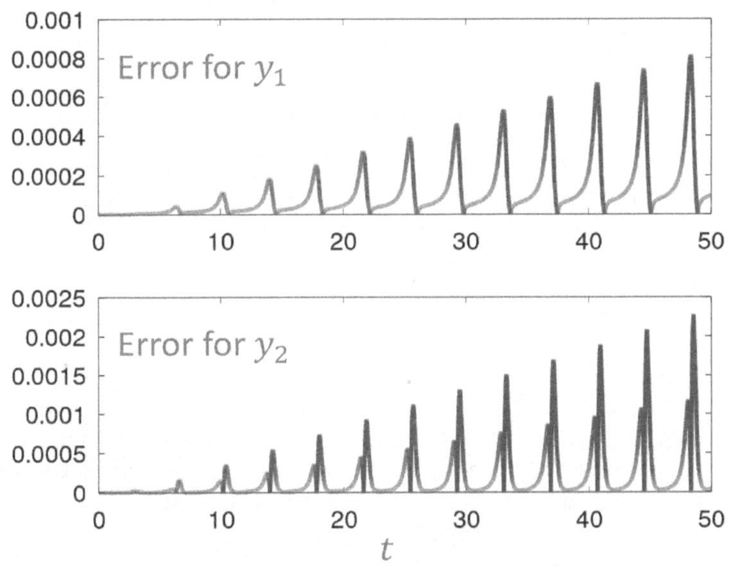

FIGURE 9.21 Estimated errors, according to Richardson, of the numerical approximation of the Van der Pol Equation (9.97) for $\mu = 2$ and initial conditions $y_1(0) = 0.5$ and $y_2(0) = 0.5$ obtained with the trapezoidal IvP solver with a step size $h = 0.0025$ (Figure 9.20).

Figure 9.21 displays the estimated Richardson errors in function of t in the interval $[0, 50]$. From the figure, we conclude that, for this example, the error is not the largest in $t = 50$. In fact, the error varies and reaches values as high as 0.002 in the case of y_2. Two conclusions can be drawn.

First, the numerical approximations of the Van Der Pol oscillator are with an error less than 0.001 for both y_1 and y_2 over the complete time interval of 0 to 50. That means the behavior of the limit cycle observed in Figure 9.20b is not an artifact from the numerical errors.

Second, and this is of greater interest for the study of numerical IvP solvers, we observe that the errors of numerical approximations produced by numerical IvP solvers change over a time interval t_o to t_f. In the particular case of the Van der Pol oscillator, the errors grow on average. This is expected from Theorem 9.3. However, unfortunately, if one aims to control the error of numerical approximations of the IvP below a certain tolerance over the complete interval $[t_o, t_f]$, it is not enough to consider the error at the endpoint t_f, as this error is not necessarily the largest. In fact, we have to consider the error over a complete solution interval $[t_o, t_f]$. Obviously, this becomes a lengthy process. To make things worse, in case the error is at some point larger than the desired tolerance, we have to restart all calculations with a smaller step size and estimate the errors again. A further drawback of fixed step sizes methods becomes obvious from Figure 9.21. As the error grows on average over time, a consequence of Theorem 9.3, we are forced to use a small step size imposed by the approximations towards the end of the interval. But in the first part of the time interval, a larger step size would probably do perfectly fine.

9.6.2 Adaptive Initial Value Problem Solvers

The observations made by solving the Van der Pol oscillator in Section 9.6.1 allow a more precise formulation of how an ideal IvP solver should operate. An ideal initial value problem solver should produce approximations w_i of the exact solution such that:

$$|w_i - y(t_i)| \leq TOL \tag{9.98}$$

for all t_i in the solution interval $[t_o, t_f]$ where TOL is the maximal error we can accept. To reach this goal, we have to estimate the error of each approximation w_i. So far, we have discussed only Richardson's method to estimate the global truncation error. In principle, one could estimate the global truncation error and check if condition (9.98) is always satisfied in the interval $[t_o, t_f]$. This is, however, a very costly process. Indeed, two approximations computed over the full interval $[t_o, t_f]$ with two different step sizes are required to apply Richardson's error formula. If it turns out that the errors are too large, we have to recalculate another approximation over the full interval $[t_o, t_f]$ with a smaller step size. Luckily, there exists an alternative, more efficient method.

As will be discussed later in this section, it is possible to estimate the global truncation error from the local truncation error. The benefit of this approach is that the local truncation error can be estimated when computing w_{i+1} from w_i. In other words, to estimate the local truncation error, it will not be required to compute the approximations over the full interval $[t_o, t_f]$. A decision to change the step size can, therefore, be taken if it turns out that the error becomes too large. The remainder of this section will outline this approach. Essentially, three steps need to be worked out:

1. Estimate the local truncation error when computing w_{i+1} from w_i.
2. Estimate the global truncation error from the local truncation error.
3. Adapt, if required, the step size h to guarantee the approximations satisfy conditions (9.98).

We start by explaining how the local truncation error can be estimated. There exist several techniques. A popular one is to use two IvP solvers, one of low order (producing the approximations w_i) and one of high order (producing the approximations $\tilde{w}_i$). The local truncation error e_i is then estimated as

$$e_i \cong |\tilde{w}_i - w_i| \tag{9.99}$$

This approach assumes that when IvP solvers proceed from t_i to t_{i+1}, the high order approximation $\tilde{w}_{i+1}$ will be almost equal to the local solution $z(t_{i+1})$. This approach, which consists of running two fixed-step size IvP solvers in parallel, is called the embedded approach to estimate the local truncation error.

Next, we explain how the global truncation error $g_i = |w_{i+1} - y(t_{i+1})|$ can be estimated from the local truncation error $e_i = |w_{i+1} - z(t_{i+1})|$. If the IvP solver is of order q in the global truncation error, and the step size is small enough (but not too small to avoid round-off errors), it follows:

$$|w_{i+1} - y(t_{i+1})| \cong \tilde{K}h^q \tag{9.100}$$

with some unknown constant $\tilde{K}$. According to Theorem 9.4, if the global truncation error is in order q, the local truncation error is in order $q+1$. Consequently, one can write

$$|w_{i+1} - z(t_{i+1})| \cong Kh^{q+1} \tag{9.101}$$

with some unknown constant K. Let us assume an estimation Δ of the local truncation error. From Equation (9.101) it follows

$$|w_{i+1} - z(t_{i+1})| \cong \Delta \cong Kh^{q+1} \tag{9.102}$$

Solving (9.102) for K it follows

$$K \cong \frac{\Delta}{h^{q+1}} \tag{9.103}$$

Let us now assume that K and $\tilde{K}$ are about equal. The reader should be aware that this is a very strong assumption that is not always correct. If we accept this assumption, it is possible to estimate the global truncation error using (9.102) as

$$\left|w_{i+1} - y\left(t_{i+1}\right)\right| \cong \tilde{K}h^q \cong Kh^q = \frac{\Delta}{h} \tag{9.104}$$

We can now turn our attention to the last step of adaptive IvP solvers: how to adapt the step size h in order to guarantee condition (9.98) remains valid. The idea is simple. If the global truncation error estimation (9.104) turns out to be larger than the desired tolerance TOL, the step size needs to be adjusted and replaced by a smaller one such that the global truncation error will be less than the set tolerance.

Figure 9.22 illustrates schematically the principle behind adaptive IvP solvers. An adaptive IvP solver uses a fixed step size IvP solver to first compute an approximation w_{i+1} in t_{i+1} of the true solution $y\left(t_{i+1}\right)$ using a step size h. Based on the estimation Δ of the local truncation error in t_{i+1}, the adaptive solver estimates the global truncation error in t_{i+1} using (9.104). If this global truncation error estimation is larger than the acceptable tolerance, a new step size h_{new} will be determined such that the new truncation error estimation in t_i+h_{new} will be less than the acceptable tolerance.

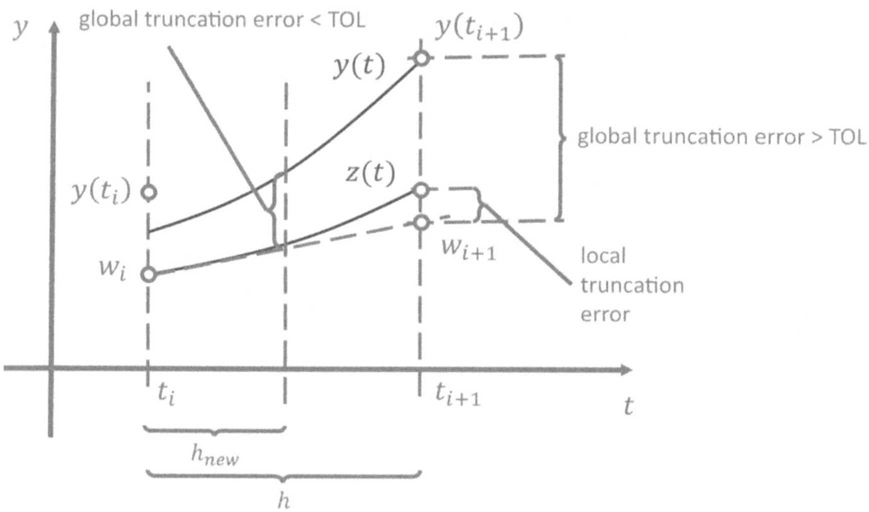

FIGURE 9.22 Schematic on how to operate adaptive IvP solvers. Adaptive IvP solvers adapt the current step size h to a new one h_{new}, such as the estimation of the global truncation error obtained by extrapolating the local truncation error according to (9.104) is below a desired tolerance.

Let us work out how to determine the new step size h_{new}. As the fixed step size IvP solver used by our adaptive solver is of order q one can write:

$$\left|w_{i+1}(\text{computed with } h_{new}) - y(t_{i+1})\right| \cong Kh_{new}^q \tag{9.105}$$

Further, we want that

$$\left|w_{i+1}(\text{computed with } h_{new}) - y(t_{i+1})\right| \le TOL \tag{9.106}$$

Combining Equations (9.105) and (9.106):

$$Kh_{new}^q \le TOL \tag{9.107}$$

Using further Equation (9.103) to estimate K and solve for h_{new}, one obtains

$$h_{new} \le h\left[\frac{TOL \cdot h}{\Delta}\right]^{1/q} \tag{9.108}$$

Equation (9.108) gives a condition h_{new} that has to be satisfied in order for condition (9.98) to be fulfilled. In practice, we often choose h_{new} with some safety factor. Algorithm 9.6 outlines how to operate an adaptive IvP solver. As the local truncation error is estimated using Equation (9.99) by running two fixed-step size IvP solvers in parallel, the algorithm is called an embedded adaptive IvP solver.

Algorithm 9.6. Embedded Adaptive IVP Solver

To produce an approximation of an IvP problem in $t_{i+1} = t_i + h_{new}$ such that the estimated global truncation error in t_{i+1} will be lower than an acceptable tolerance TOL, we proceed as follows:

1. Starting from a point t_i, we compute two approximations w_{i+1} and $\tilde{w}_{i+1}$ in t_{i+1} using two IVP solvers of order q and $q+1$ with step size h
2. Using these two approximations, we choose a new step size h_{new} such to satisfy the relation

$$h_{new} \le h\left[\frac{TOL \cdot h}{\Delta}\right]^{\frac{1}{q}}$$

   with $\Delta = \left|\tilde{w}_{i+1} - w_{i+1}\right|$
3. Using the new step size, we compute the new approximation w_{i+1} in $t_{i+1} = t_i + h_{new}$

In principle, any two fixed-step size IvP solvers can be used in Algorithm 9.6 to compute the approximations w_{i+1} and $\tilde{w}_{i+1}$. A common choice is to use two Runge-Kutta IvP solvers. These solvers can further be designed in order to minimize calculation efforts. A popular example is the Runge-Kutta-Fehlberg[8] RKF45 method which combines an RK4:

$$w_{i+1} = w_i + \frac{25}{216}k_1 + \frac{1408}{2665}k_3 + \frac{2197}{4104}k_4 - \frac{1}{5}k_5 \tag{9.109}$$

and an RK5 IvP solver

$$\tilde{w}_{i+1} = w_i + \frac{16}{135}k_1 + \frac{6656}{12825}k_3 + \frac{28561}{56430}k_4 - \frac{9}{50}k_5 + \frac{2}{55}k_6 \tag{9.110}$$

These two RK solvers are chosen so that the various coefficients k_i of the RK4 and RK5 methods are identical. They are given by:

$$k_1 = hf\left(t_i, w_i\right)$$

$$k_2 = hf\left(t_i + \frac{h}{4}, w_i + \frac{1}{4}k_1\right)$$

$$k_3 = hf\left(t_i + \frac{3h}{8}, w_i + \frac{3}{32}k_1 + \frac{9}{32}k_2\right)$$

$$k_4 = hf\left(t_i + \frac{12h}{13}, w_i + \frac{1932}{2197}k_1 - \frac{7200}{2197}k_2 + \frac{7296}{2197}k_3\right)$$

$$k_5 = hf\left(t_i + h, w_i + \frac{439}{216}k_1 - 8k_2 + \frac{3680}{513}k_3 - \frac{845}{4104}k_4\right)$$

$$k_6 = hf\left(t_i + \frac{h}{2}, w_i - \frac{8}{27}k_1 + 2k_2 - \frac{3544}{2565}k_3 + \frac{1859}{4104}k_4 - \frac{11}{40}k_5\right) \tag{9.111}$$

Due to this smart choice of the RK4 and RK5 method, in each step, six function evaluations are required and not, as one may expect, about nine function evaluations (four for the RK4 method and 5 for the RK5 method).

Example 9.12. RKF45 IVP Solver

To observe the RKF45 method in action, we approximate the solution of the Van der Pol Oscillator (9.97) with an error tolerance of 10^{-6}.

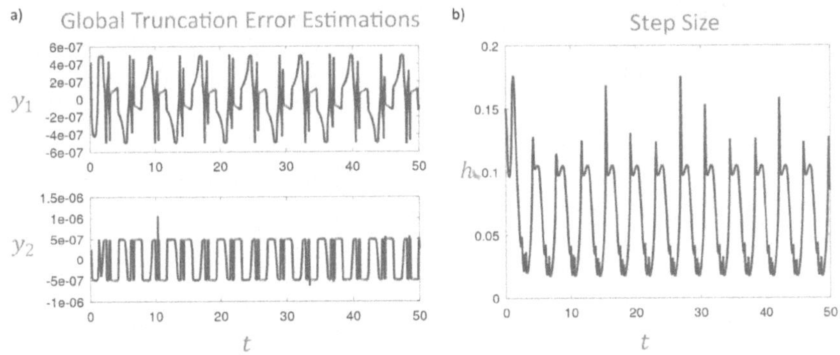

FIGURE 9.23 (a) Global truncation errors of the approximations of the solution of the Van der Pol Oscillator (9.97) produced by the RKF45 method with an error tolerance of 10^{-6}. (b) Step sizes used by the RK45 method during the calculations.

Figure 9.23a displays the global truncation error estimations as the RKF45 method is proceeding. As desired, these errors remain below 10^{-6}. The RKF45 had to adapt the step sizes to keep the global truncation error below this desired maximal value. Figure 9.23b illustrates how this step size was adapted. In all cases, the step size remained above 0.01, which is larger than the step size of 0.0025 used with the trapezoidal IvP solver in Section 9.6.1 (Figure 9.21). Despite using larger step sizes, the global truncation error of the approximations from the RKF45 remains significantly lower than the global truncation errors of the trapezoidal IvP solver approximations, where they reached values as large as 0.002.

This illustrates the benefit of the RKF45 adaptive method. Not only does it control the global truncation error to remain below our chosen maximal value, but it needs fewer calculation steps.

9.7 EXERCISES

9.7.1 EULER'S ALGORITHM

1. Solve the following initial value problems using Euler's algorithm and a step size of $h = 0.1$. For each of them, plot the numerical approximation together with the provided true solution over an interval of length 0.5.

   a. $\frac{d}{dt}y = ty^2$ with $y(1) = 1$. True solution: $y = \frac{2}{3-t^2}$

   b. $\frac{d}{dt}y = \frac{t-y}{t+y}$ with $y(1) = 0$. True solution for $t > 1$: $y = \sqrt{2t^2-1} - t$

2. Repeat exercise 1a, but over an interval $[1, 2]$ instead of $[1, 1.5]$. What do you observe?

3. Solve the following initial value problems using Euler's algorithm for various step sizes. For each of them, following the method presented in Section 9.2.3, plot the true error versus the step size and determine the convergence order.

   a. $\frac{d}{dt}y = te^{3t} - 2y$ with $y(0) = 0$. Interval to compute the approximations $[0, 1]$.

      True solution: $y = \frac{1}{5}te^{3t} - \frac{1}{25}te^{3t} + \frac{1}{25}te^{-2t}$

   b. $\frac{d}{dt}y = 1 + (t - y)^2$ with $y(2) = 1$. Interval to compute the approximations $[2, 3]$.

      True solution: $y = 1 + (1 - t)^{-1}$

   c. $\frac{d}{dt}y = 1 + \frac{y}{t}$ with $y(1) = 2$. Interval to compute the approximations $[1, 2]$.

      True solution: $y = t \cdot \ln(t) + 2t$

   d. $\frac{d}{dt}y = \cos(2t) + \sin(3t)$ with $y(0) = 1$. Interval to compute the approximations $[0, 1]$.

      True solution: $y = \frac{1}{2}\sin(2t) - \frac{1}{3}\cos(3t) + \frac{4}{3}$

9.7.2 Runge-Kutta Methods

4. Repeat Exercise 1 using Heun's, the Trapezoid, the Midpoint, and the classical RK4 algorithm

5. Repeat Exercise 3 using Heun's, the Trapezoid, the Midpoint, and the classical RK4 algorithm

6. Repeat Exercise 3 using Heun's, the Trapezoid, the Midpoint, and the classical RK4 algorithm, but instead of using the true error, use Richardson's error formula to estimate the error.

   Present your results in a log-log plot showing the estimated and the true error versus the step size.

7. For the following initial value problems, provide an answer with an absolute error below the specified accuracy. Use the classical RK4 method.

To answer these questions, follow the method outlined in Example 9.7. In short, you first need to determine the range of step sizes in which Richardson's error formula is valid. Afterward, select a step size that will allow you to reach the desired accuracy, and finally compute the requested answer.

a. $\frac{d}{dt}y=(1-t)y+\sin(y)$ with $y(-1)=0.5$. Compute an approximation of $y(2)$ with an absolute error below 10^{-8}.

b. $\frac{d}{dt}y=-\cos(ty)y$ with $y(0)=3.5$. Compute an approximation of $y(10)$ with an absolute error below 10^{-8}.

c. $\frac{d}{dt}y=-\cos(t^2y)\ln(y)$ with $y(1)=1.5$. Compute an approximation of $y(6)$ with an absolute error below 10^{-9}.

9.7.3 Systems and Higher Order Initial Value Problems

8. Solve the following initial value problems after having rewritten them as systems of first order differential equations using Euler's algorithm for $h=0.1$. Plot the approximations together with the provided true solution.

a. $\frac{d^2}{dt^2}y-2\frac{d}{dt}y+y=te^t-t$ with $y(0)=0$ and $\frac{d}{dt}y(0)=0$. Interval to compute the approximations $[0,1]$.

True solution: $y(t)=\frac{1}{6}t^3e^t-te^t+2e^t-t-2$

b. $t^2\frac{d^2}{dt^2}y+t\frac{d}{dt}y-t(t+1)y=2te^t$ with $y(1)=0$ and $\frac{d}{dt}y(1)=e$. Interval to compute the approximations $[1,2]$.

True solution: $y(t)=e^t\ln(t)$

c. $\frac{d^3}{dt^3}y=-y^4$ with $y(1)=-1$, $\frac{d}{dt}y(1)=-1$ and $\frac{d^2}{dt^2}y(1)=2$. Interval to compute the approximations $[1,1.5]$.

True solution: $y(t)=\frac{1}{t-2}$

d. $\frac{d^3}{dt^3}y+2\frac{d^2}{dt^2}y-\frac{d}{dt}y-2y=e^t$ with $y(0)=1$, $\frac{d}{dt}y(0)=2$ and $\frac{d^2}{dt^2}y(0)=0$. Interval to compute the approximations $[0,2]$.

True solution: $y(t)=\frac{43}{36}e^t+\frac{1}{4}e^{-t}-\frac{4}{9}e^{-2t}+\frac{1}{6}te^t$

9. Repeat Exercise 8 with Heun's, the Trapezoid, the Midpoint, and the classical RK4 algorithm.

10. Compute the approximations of the initial value problems of exercise 8 and estimations of the errors using Richardson's formula. For this, use two step sizes (h=0.1 and h=0.05, for example). Compare the estimated errors with the true errors. Repeat these calculations for Euler's, Heun's, the Trapezoid, the Midpoint, and the classical RK4 algorithm.

11. We want to approximate the solution of $\frac{d^2}{dt^2}y = \frac{d}{dt}\cos(t) - y\sin(t)$ with $y(0) = 1$ and $\frac{d}{dt}y(0) = 1$. Compute and with an absolute error lower than (for both of them). Use the classical RK4 method.

    To answer this question, follow the method outlined in Example 9.11. In short, you first need to determine the range of step sizes in which Richardson's error formula is valid. Afterward, select a step size that will allow you to reach the desired accuracy, and finally compute the requested answer.

9.7.4 Adaptive Methods

12. Solve the initial value problems from Exercise 3 using the RKF45 algorithm as outlined in Equation (9.111) with an error tolerance of .

    Compute the true errors and verify if they are below or not.

13. Solve the initial value problems from Exercise 8 using the RKF45 algorithm as outlined in Equation (9.111) with an error tolerance of .

    Compute the true errors and verify if they are below or not.

NOTES

1 Leonhard Euler (1707–1783), Swiss mathematician, physicist, astronomer, geographer, logician, and engineer.
2 Rudolf Otto Sigismund Lipschitz (1832–1903), German mathematician.
3 Charles Émile Picard (1856–1941), French mathematician.
4 Carl David Tolmé Runge (1856–1927), German mathematician, physicist, and spectroscopist.
Martin Wilhelm Kutta (1867–1944), German mathematician.
5 Karl Heun (1859–1929), German mathematician.
6 Albert Einstein (1879–1955), German-born theoretical physicist.
7 Balthasar van der Pol (1889–1959), Dutch physicist.
8 Erwin Fehlberg (1911–1990), German mathematician.

Part II

Numerical Methods for Data Analysis

10 Machine Learning

In this second part, we present an introduction to the field of machine learning. Machine learning is a very vast and fast-growing field. It is impossible to cover, even partially, this field in a single book. Our aim is to present some fundamental concepts and examples and explain how they relate to the numerical algorithms we have discussed so far. The term *machine learning* was first introduced in 1959 by Arthur Samuel.[1] At that time, the terminology *self-teaching computers* were equally used. As we will discuss in more detail, the aim of machine learning is to develop algorithms that can learn from data and generalize to unseen data. Predictions can include numerical values (for example, predicting the market value of a house based on some characteristics) or categorical data (for example, classifying an email into spam and legitimate emails). The process of using data to tune algorithms, with the aim of making predictions of acceptable quality, is referred to as *training*.

Machine learning has wide and constantly growing fields of applications, including large language models, speech recognition, and email filtering. In engineering, popular fields of applications include, among numerous others, computer vision and robotics. Increasingly, medicine benefits from this tool, as well as many other fields, including economy, humanity, art, and others. In general, wherever exist tasks that can be automated, machine learning may be employed. Recently, and rightfully so in the authors' opinion, concerns have emerged about the risks involved with the applications of machine learning. While we can't delve into them here, we encourage readers to follow this important discussion and be mindful that, like any technology, machine learning brings not only benefits but also risks. Today, every engineer and scientist can benefit from and will inevitably be confronted with machine learning. Besides attempting to predict unseen data, machine learning algorithms can be valuable tools in assisting with data analysis tasks. The methods we choose to discuss in this textbook belong to the so-called supervised learning approaches. This classification is related to the approach used to train the algorithms. In supervised learning, the data used in training includes the so-called labels (the values to be predicted). For example, a set of emails labeled as spam or legitimate will be used. Two other approaches not discussed here include unsupervised learning and reinforced learning.

In unsupervised learning, the algorithms will not only predict some outcomes but will also try to discover, to a certain extent, what needs to be predicted. A typical example is image classification. The algorithm will first create categories of similar images. These categories will be assigned to actual labels afterward (e.g., images of animals, landscapes, humans, cars, etc.).

In reinforced learning, in a very simplified way, the algorithm attempts to perform a certain task, such as grabbing an object with a robot. Using feedback from

DOI: 10.1201/9781003262121-12

its environment, the algorithm, based on some rules, will attempt to improve its way of solving this task.

In this first chapter on machine learning, we introduce some general vocabulary and concepts, and we spend some time discussing what it means to make predictions of acceptable quality. This is the heart of our discussions so far on numerical methods.

10.1 THE DATA, THE MODEL, AND THE PERFORMANCE

Deploying machine learning algorithms typically requires three main ingredients: data, a model, and a performance measure. This is exemplified by the famous formal definition of machine learning provided by Tom Mitchel.[2]

Definition 10.1. Machine Learning

Machine learning is the study of algorithms that improve performance on a given task based on some experience.

To illustrate Definition 10.1, let us revisit linear regression. In linear regression, we have a dataset (the experience) and a linear model with parameters to be adjusted. The task consists of predicting the outcomes of inputs. The quality of this prediction (performance) is measured by, for example, the mean squared error. The improvement of the task (the prediction of outcomes) is achieved by fine-tuning the model parameters of the linear model. Linear regression is, in fact, one among many machine learning algorithms.

Based on this, machine learning seems to be just another word for general curve fitting: the goal is to adjust model parameters to a given dataset by minimizing a chosen loss function (the performance), such as the mean squared error for example. Curve fitting is indeed part of machine learning. But machine learning goes a step further. On one side, machine learning includes approaches that are quite different from curve fitting (for example, in unsupervised or reinforcement learning). But, even when considering algorithms such as linear regression, machine learning has an additional key objective. Machine learning aims to predict outcomes for unseen data and attempts to give methods that allow the tuning of model parameters, not only with the aim of reproducing as well as possible a given dataset but adjust a model with the aim to have as good predictions as possible on unseen data. This task is extremely challenging and is referred to as the *generalization of new instances* or *generalizing data*.

In fact, the problem of generalization becomes obvious if we think about the problem of regression. Suppose we decide to use a linear model to be trained[3] on some dataset. In fact, by making that choice, we implicitly assume that the dataset can be represented by a linear model. In other words, we assume that the outcomes we desire to predict are actually modeled by a linear model and that any deviation from a linear model present in our dataset is attributed to statistical noise. But what if we do not make such an assumption (or we don't have any real reason that would allow us to state that the data we observe follows a linear model)? We could use other models, such as a cubic polynomial (or really any other model we could think of). How can we know which one of these models is better in generalizing the data? In 1996, David Wolpert[4] asked this question in his famous paper "The Lack of A Priori Distinctions Between Learning Algorithms". Wolpert showed that if you make absolutely no assumption about your data, then there is no objective reason to prefer one over another algorithm. This result is often referred to as the "No Free Lunch Theorem" (even though there is, in fact, not a single but a family of theorems behind that terminology). In fact, we would have to evaluate all possible and imaginable models, which, of course, in practice is not possible. This shows clearly the difficulty in assessing the generalization value of a model.

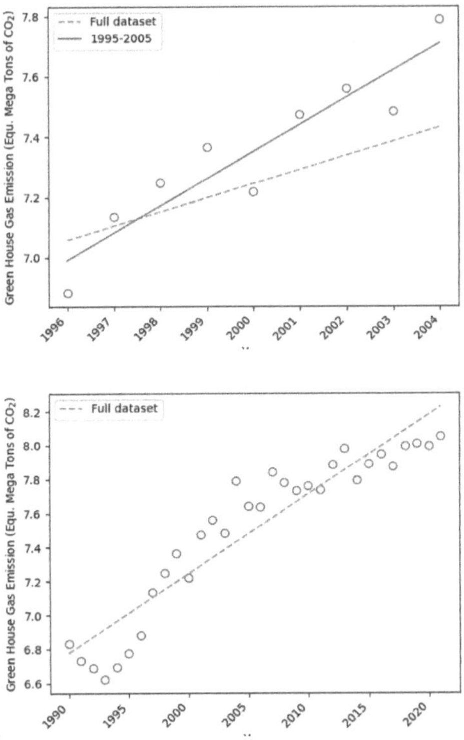

FIGURE 10.1 (a) Greenhouse gas emissions in tons of carbon dioxide equivalent produced by the agricultural sector in the Canadian province Quebec between 1995 and 2005 (o), fitted linear model to these data (–) and fitted model to the full data set from 1990–2021 (--). (b) Greenhouse gas emissions in tons of carbon dioxide equivalent produced by the agricultural sector in the Canadian province Quebec between 1990 and 2021 and fitted linear model to these data (--).

To illustrate some additional difficulties, let us consider an example. For this, we use the publicly available data about greenhouse gas emissions from the Canadian province of Quebec.[5] Among others, one can find the greenhouse gas emissions in tons of carbon dioxide equivalent produced by the agricultural sector.

Figure 10.1a plots the greenhouse emissions in tons of carbon dioxide equivalent produced by the agricultural sector between 1995 and 2005. The trend is essentially linear and seems qualitatively fairly well captured by a linear model. However, if we use the full available dataset (with data from 1990 to 2021 at the time of writing this book), we obtain a fairly different fitted linear model. This illustrates the problematic that model training depends on the used dataset. One may argue that, most likely, the larger dataset will lead to a better model. That may indeed be the case, but the real problem is different. Even when using the larger dataset, the larger dataset is still a subset of reality as we can never obtain all possible data we wish. In other words, this dependence on data will always exist, regardless of how large the dataset is. The reader may be familiar with this problem from statistics. Imagine we wish to measure the mass of an object. For this, we take 10 measurements and compute their mean. If we repeat the process with 10 other measurements, we will get a different mean. None of them will be the exact mass of the object. In this case, statistics provides a means to estimate the error, or confidence interval, associated with using the mean of 10 measurements as the value of our mass. This example points to another problematic in the machine learning field: how to come up with methodologies to estimate the prediction error that applies to more complex models.

Consider now Figure 10.1b displaying the greenhouse emissions in tons of carbon dioxide equivalent produced by the agricultural sector between 1995 and 2021. We observe that the model developed over the partial data (for the years from 1995 to 2005) can't generalize to years beyond 2005 because the trend is no longer linear. Starting in 2005, the greenhouse gas emissions continued to increase, but at a slower rate. This illustrates the problem that if the dataset isn't representative of the intended use, the predictions will be of poor quality. In our example, this is not surprising; greenhouse gas emissions are influenced by many factors beyond the year, such as governmental policies or behavior of humans. This highlights another challenge: it is crucial to develop a model based on data that contains relevant information to accurately describe and capture the effects on the outcomes we aim to predict.

10.2 HOW TO LEARN FROM DATA

The term *learning* in the statement "learning from data" may be somehow misleading, especially for the type of algorithms we discuss in the next chapters. There are several methods by which algorithms can "learn" from data. Learning refers to the concept that an algorithm can adjust itself to a given dataset to predict as good as possible new outcomes in the function of the chosen performance. Outcomes can be numerical values (e.g., the market value of a house) or a category (e.g., spam versus legitimate email). The first case is known as a regression problem, and the second one is a classification problem. Examples of both problems will be given in the coming chapters.

A typical example of the family of regression problems is least square linear regression. In linear regression, learning means selecting the linear model parameters by minimizing the least squared error. More generally, learning means adjusting model parameters by minimizing a loss function $\mathcal{L}$ (the squared error in the case of linear regression). However, recalling the discussion around the "No Free Lunch Theorem" and the example of greenhouse gas emissions in the Canadian province of Quebec, the problem is more complex. The model parameters adjusted using this approach, while they guarantee (at least in the case of linear regression) a minimum loss function, they do not guarantee however that new outcomes will also be predicted with a minimal error. Prior to addressing these challenges, let us delve into the origin of errors associated with data-trained models. We begin with the concept of variance-bias trade-off in the next section.

10.3 VARIANCE-BIAS TRADE-OFF

In this section, we discuss the nature of errors (and their meaning) in the process of model training using a given dataset $\mathcal{D}$. There are three types of errors. One error originates from the choice of the model we intend to train, and two errors emanate from the dataset. The bias error is associated with the nature of the model, whereas the data-dependent errors are known as the variance and noise. Let us illustrate these concepts through an example.

Let us imagine a process that produces an outcome $h(x)$ in the function of an input x. Let the outcome be defined as follows:

$$h(x) = 2\sin(2\pi x) - 1 \tag{10.1}$$

Furthermore, let us perform m observations of the process $h(x)$ to create a dataset $\mathcal{D} = \{(x_i, y_i)\}_{i=1..m}$. These m observations are accompanied by noise (e.g., measurement errors). For the purpose of our example, we model this noise with white noise, meaning an error e that follows a Gaussian distribution centered in zero and variance σ.

Figure 10.2a displays a graph with an example of $m = 50$ observations of the process. We can train a model $y(x)$ on this dataset $\mathcal{D}$. At this stage, it is not important which model is trained.[6] What is important to understand is that if we draw a different dataset $\mathcal{D}'$ from the process, the trained model will become slightly different (Figure 10.2b). We use the notation $y(x;\mathcal{D})$ to express that the model $y(x)$ was trained using the specific dataset $\mathcal{D}$.

The question we are interested in is to estimate the magnitude of the error $|y(x;\mathcal{D}) - h(x)|$ associated with predicting the outcome of a value x. As seen in Chapter 6, it is more convenient to measure the error of the model as the squared error $[y(x;\mathcal{D}) - h(x)]^2$.

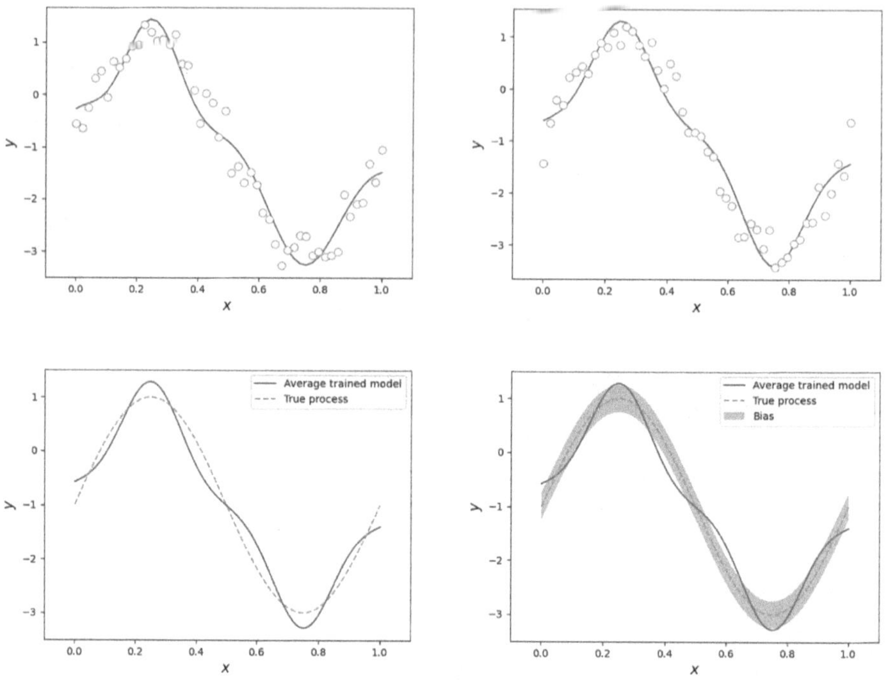

FIGURE 10.2 (a) An example of m=50 observations (o) of a process following Equation (10.1) with Gaussian white noise of variance $\sigma = 0.25$ together with a trained model (–). (b) Another example of m=50 observations (o) of the same process. As the dataset is different, it results in a slightly different trained model. (c) Average (–) of trained models from 100 drawn datasets and true process (–) according to Equation (10.1). (d) The shaded area represents the bias as computed according to Equation (10.3).

However, for every new dataset $\mathcal{D}$, the trained model changes, and its predictions change as well. It is, therefore, more relevant to speak of the expected squared error $\mathbb{E}_{\mathcal{D}}\left[\left\{y(x;\mathcal{D})-h(x)\right\}^2\right]$ obtained when averaging over all possible datasets.

To get a feeling about the expected squared error $\mathbb{E}_{\mathcal{D}}\left[\left\{y(x;\mathcal{D})-h(x)\right\}^2\right]$, let us repeat the process of model training on several datasets. As an example, let us draw 100 datasets and average all the trained models. Figure 10.1c compares the obtained average of the 100 trained models with the true process according to Equation (10.1). The average of trained models is not identical to the true process. This is not surprising as the model we used to train the datasets is not identical to Equation (10.1). This is generally the case. Unless we can model the underlying process and derive an equation that accurately represents the true process, the machine learning model will always be different than the actual process. The difference between the true process and the model is called the *bias* and is defined as follows:

Definition 10.2. BIAS

The difference between the average predictions of the trained model $y(x;\mathcal{D})$ and the true process $h(x)$, is called bias:

$$\text{bias}(x) = \mathbb{E}_{\mathcal{D}}\left[y(x;\mathcal{D})\right] - h(x) \tag{10.2}$$

where $\mathbb{E}_{\mathcal{D}}$ is the expected value computed over all datasets $\mathcal{D}$ drawn from the process and $y(x;\mathcal{D})$ is the prediction of the trained model using the dataset $\mathcal{D}$.

Instead of measuring the bias in a particular point x, one can as well define the bias by integrating (i.e., summing) the differences (10.2) over all values of x. This yields a measure analog to the mean squared error MSE:

$$\text{bias}^2 = \int \left[\mathbb{E}_{\mathcal{D}}\left[y(x;\mathcal{D})\right] - h(x)\right]^2 p(x)\,dx \tag{10.3}$$

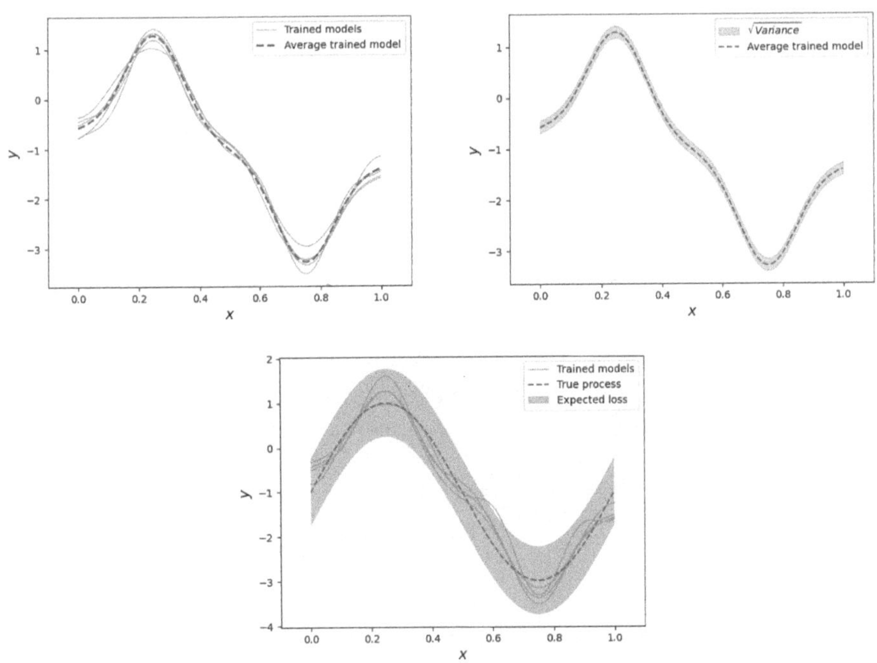

FIGURE 10.3 (a) Examples of five trained models (–) based on five different datasets and average trained model (––) from 100 datasets. (b) Average trained model (––) from 100 datasets and extension of variance around it (shaded area). (c) True process $h(x)$ (––) and some examples of trained models (–) form different datasets. The shaded area represents the extension of the to-be-expected error of the predictions of a trained model.

were $p(x)$ is the probability density function of x, i.e., $p(x)dx$ is the probability that x is within the interval $[x, x+dx]$. The reader should pay attention to the notation used herein. We use bias(x) to designate the bias in a particular point, as defined in Equation (10.2). We use bias to designate the root mean squared bias as defined in Equation (10.3). The context may help as well to understand if the bias in a particular point or the mean squared bias is used.

In Figure 10.2, the root mean squared bias is about 0.26, and the shaded area of Figure 10.2d illustrates its extension around the true process $h(x)$.

The bias measures the smallest root mean square error we achieve if we could train the model over an infinite number of noise-free datasets. Bias is the intrinsic error due to not using a model that correctly describes the true process $h(x)$. In practice, we do not have access to an infinite number of datasets, nor are these datasets noise-free. Using a limited number of noise-laden datasets generates models that vary from one another. These models obtained differ from the ideal trained model $\mathbb{E}_{\mathcal{D}}[y(x;\mathcal{D})]$ that we would obtain if averaging over an infinite number of datasets was possible (Figure 10.3a).

To measure the variation between these models, the concept of variance is defined as follows.

Definition 10.3. Variance

The variation in model predictions resulting from a variation in the datasets $\mathcal{D}$ is called variance and is defined as:

$$\text{variance}(\text{x}) = \mathbb{E}_{\mathcal{D}}\left[\left\{y(x;\mathcal{D}) - \mathbb{E}_{\mathcal{D}}[y(x;\mathcal{D})]\right\}^2\right] \tag{10.4}$$

In other words, the variance measures the variation from the ideal model that would be obtained if averaging over an infinite number of datasets was possible. A high variance means that models obtained with different datasets produce largely different outcomes. A low variance indicates the opposite, meaning the outcomes of models trained with different datasets are relatively similar. It is important to understand that a low variance does not imply a low bias. In fact, as we will learn, it is, in fact, often the opposite.

Similarly to the bias, the mean variance is defined as follows:

$$\text{variance} = \int \mathbb{E}_{\mathcal{D}}\left[\left\{y(x;\mathcal{D}) - \mathbb{E}_{\mathcal{D}}[y(x;\mathcal{D})]\right\}^2\right] p(x)\,dx \tag{10.5}$$

In Figure 10.2 and Figure 10.3, the mean variance is about 0.1, and the shaded area of Figure 10.3b illustrates its extension around the average model obtained from a large number of datasets (in our case, 100).

It is possible to prove that the expected squared error $\mathbb{E}_{\mathcal{D}}\left[\left\{y(x;\mathcal{D}) - h(x)\right\}^2\right]$ is the sum of the bias squared and variance as stated in Theorem 10.1.

Theorem 10.1. BIAS Variance Decomposition

The expected squared error $\mathbb{E}_{\mathcal{D}}\left[\left\{y(x;\mathcal{D})-h(x)\right\}^2\right]$ when training a model $y(x)$ on data generated from a process $y=h(x)$ can be decomposed as the sum of the bias squared and the variance:

$$\mathbb{E}_{\mathcal{D}}\left[\left\{y(x;\mathcal{D})-h(x)\right\}^2\right]=\left[\text{bias}(x)\right]^2+\text{variance}(\mathrm{x})=$$

$$\left\{\mathbb{E}_{\mathcal{D}}\left[y(x;\mathcal{D})\right]-h(x)\right\}^2+\mathbb{E}_{\mathcal{D}}\left[\left\{y(x;\mathcal{D})-\mathbb{E}_{\mathcal{D}}\left[y(x;\mathcal{D})\right]\right\}^2\right] \quad (10.6)$$

In practice, we will often be interested in a single measure for the error rather than a measure for each individual prediction (similar to the mean squared error MSE being a single measure of the error associated with linear regression). This measure is called the expected loss in the context of machine learning and is defined as:

$$\mathbb{E}[\mathcal{L}]=\iint\left\{y(x)-t\right\}^2 p(x,t)\,dxdt \quad (10.7)$$

where $p(x,t)dxdt$ is the probability of making an observation t for the value x (an element of one of the datasets). Practically, it represents the average, over a large number of datasets, of the sum of the squared errors.

To obtain the expected loss, we average Equation (10.6). We can prove that this results in the following expression:

$$\text{Expected loss}=\left[\text{bias}\right]^2+\text{variance}+\text{noise} \quad (10.8)$$

where the bias is defined in Equation (10.3) and the variance in Equation (10.5).

The noise is the error measurement associated with one observation and is defined as

$$\text{noise}=\iint\left\{h(x)-t\right\}^2 p(x,t)\,dxdt \quad (10.9)$$

In the present example, noise is the Gaussian white noise we added to the process $h(x)$. Figure 10.3c gives a graphical illustration of Equation (10.8). Some trained models, based on randomly drawn datasets, are shown. Their predictions are within the shaded region that is constructed by adding the expected loss around the true process.

The bias-variance decomposition and expression (10.8) for the expected loss show that the error in the prediction of a trained model results from three contributors.

First, the choice of the model. The difference between the model and the actual process results in the bias term. Second, the dataset used to train the model. Since we can only use a single (or a few) randomly drawn datasets extracted from an actual process, the variance becomes a contributing term to the error. Finally, the noise emanates from measurement error when collecting observations.

We obviously aim to minimize the expected loss to maximize the prediction accuracy of the trained model. In machine learning, we have essentially the freedom to choose a model. The size and range of the dataset and the quality of the observations (noise) can not be influenced by the choice of the model to be trained. But, both contributions can significantly reduce the overall quality of the predictions of the trained model (even if the model is very close to the actual process). The main challenge we face is to choose a model that minimizes the expected loss. This is indeed a very challenging task, as we will learn later in this book because we have limited data available, and, as such, we cannot evaluate the expected loss. We can only compute the loss over a limited dataset. Therefore, the challenge is to anticipate the loss associated with model predictions depending on the dataset used. This is known as the model generalization problem presented in Section 10.1.

It is tempting to choose models that result in the lowest loss possible when evaluated on the dataset. However, such models will generalize very poorly. This is known as the *bias-variance trade-off* problem. To illustrate this behavior, we revisit the example from process (10.1) using three different models. The three models are polynomial regression models, and they are trained (i.e., fitted), as we learned in Chapter 6.

The first model uses a polynomial of degree two and is depicted on the first row of Figure 10.4. Its variance is low (models trained from different datasets are similar), but its bias is very high (the average of all trained models differs significantly from the true process). Contrarily, the model on row three of Figure 10.4 uses a polynomial of degree 15. Its variance is very high (models trained from different

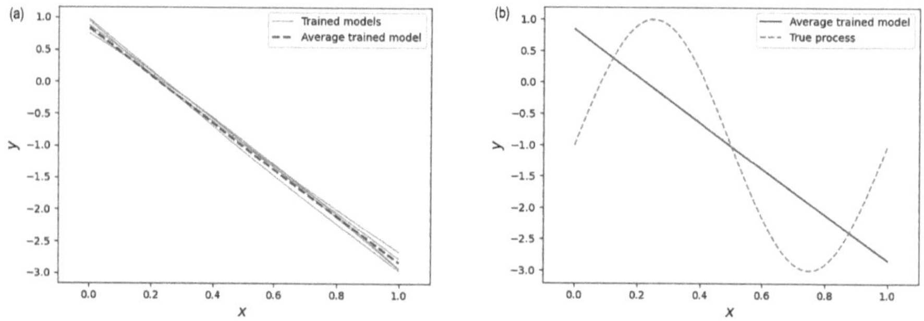

FIGURE 10.4 Three polynomial models trained on datasets generated from the process $h(x) = 2\sin(2\pi x) - 1$ with a Gaussian white noise of variance $\sigma = 0.5$. The left column shows five examples of three trained models from five randomly drawn datasets containing 50 points. The right column shows the true process (−−) together with the average of models trained on 100 randomly drawn datasets. The first row uses a model with a polynomial of order two, the second row a model with a polynomial of order four, and the third row a model with a polynomial of order 15. For the second order model, the variance is low while the bias is high. In the fourth order model, both variance and bias are low.

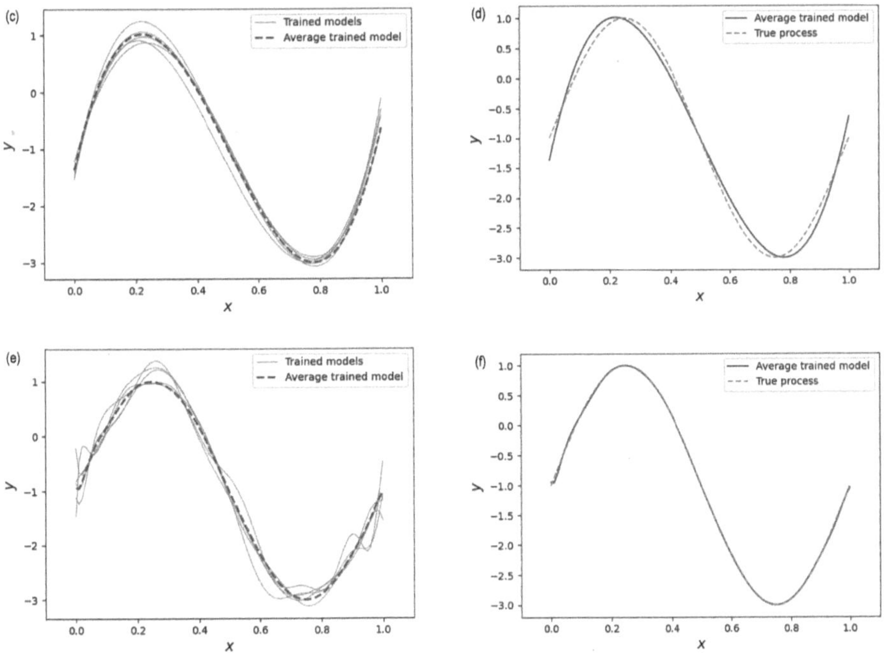

FIGURE 10.4 Continued.

datasets differ significantly), while its bias is very low (the average of all trained models is almost identical to the true process). An optimal solution seems to be for the model in the second row of Figure 10.4, which uses a polynomial of degree four. This time, both variance and bias are relatively low, even if the bias is not as low as in the third row or the variance is not as low as in the first row.

A trained model from the first or last row of Figure 10.4 will not generalize well. In both cases, the prediction will come with a large error either because the bias is too large (the model is too different from the true process) or because the variance is too high (the model is attempting to capture elements that are not typical of the true process, but rather come from the noise or the particular dataset used for the training). We use the terminology *under-fitting* for the first case and *over-fitting* for the second. An under-fitted model is a model that is too simple to capture the actual process, while an over-fitted model is too complex. In the next chapters, we learn how under- and over-fitting can be identified in real situations (situations where we do not know the actual process) and how the optimal trade-off between variance and bias can be found.

NOTES

1 Arthur Samuel (1901–1990), American pioneer in the field of computer gaming and artificial intelligence.
2 Tom Michael Mitchell (1951–), American computer scientist.

3 Recall that in the context of machine learning, the term training refers to the process of adjusting the model to a dataset.
4 David Wolpert (1961–), American physicist and computer scientist.
5 www.donneesquebec.ca/recherche/dataset/inventaire-quebecois-des-emissions-de-gaz-a-effet-de -serre
6 For the interested reader, the model trained is a general linear regression model using Gaussian basis functions, as will be discussed in Section 11.3.3.

11 Regression Models

Methods to perform linear and multilinear regression were presented in chapter 6. In the present chapter, we revisit the topic in order to introduce the vocabulary used in the field of machine learning and to illustrate the usage of the Python library SciKit-Learn. This library has become very popular in the field and comes with an excellent documentation. The aim here is not to expose the library in-depth, but rather to show important concepts that should allow the reader to explore further the documentation of SciKit-Learn.

11.1 FEATURES, LABELS, AND MODEL WEIGHTS

As explained in Chapter 6, the aim of linear regression is to adjust a model to a set of data points. A data point consists in n input variables $x_1^{(i)}, x_2^{(i)}, \ldots, x_n^{(i)}$ and one output variable $y^{(i)}$. In this notation, the index refers to the number of the input variable and the subscript (i) refers to the number of the data point. In the context of machine learning, a data point is often referred to as a *sample*. A collection of samples constitutes the dataset. Furthermore, the term *feature* is commonly used to designate the input variables. In our example, we have n features and we use the notation $x_1, x_2, \ldots, x_n$ to designate them. Note that the term feature refers to the mathematical variable and not to the specific value a feature takes in one sample of the dataset. Finally, the output variable y is referred to as the *label*. A linear model is then written as:

$$y(\underline{x}; \underline{w}) = w_o + w_1 x_1 + \ldots + w_n x_n \tag{11.1}$$

The coefficients $\underline{w} = [w_0, w_1, \ldots, w_n]^T$ of the model are, in the context of machine learning, referred as *model weights*. Model weights are normally coefficients of a model that will have to be adjusted to the dataset. In the case of linear models, the coefficient w_o has a special name and is referred to as the *bias term*.[1]

It is common in machine learning to regroup all the samples of a dataset into the so-called feature matrix X. This matrix is constructed by writing in each row one sample of the dataset and the columns corresponds to the features. For a dataset with m samples this would be:

$$X = \begin{bmatrix} x_1^{(1)} & \cdots & x_n^{(1)} \\ \vdots & & \vdots \\ x_1^{(m)} & \cdots & x_n^{(m)} \end{bmatrix} \tag{11.2}$$

DOI: 10.1201/9781003262121-13

The reader is invited to compare the feature matrix X with the coefficient matrix A used in the normal equations (Section 6.2.2 of Chapter 6). They differ by the first column of matrix A, which consists of a column of ones. Symbolically, one can write:

$$A = \begin{bmatrix} 1 & X \end{bmatrix} \tag{11.3}$$

This first column can become a source of confusion when using libraries implementing linear regression models. It is important to read the documentation to know if the library requires the matrix X or the matrix A. The SciKit-Learn library expects the matrix X and not the matrix A.

A further source of confusion comes from the fact that some authors and libraries will define the feature vector $\underline{x}$ as $\underline{x} = \begin{bmatrix} 1 & x_1 & \dots & x_n \end{bmatrix}^T$, adding a 1 in the first entry of the vector. In this case, the linear model (11.1) writes as

$$y = \underline{w}^T \cdot \underline{x} \tag{11.4}$$

with the usual inner product of the vector space $\mathbb{R}^{n+1}$. This notation is particularly popular in the field of neural networks. As with the matrix X, respectively A, it is important to understand well the definition used in the document one is consulting. In the frame of the present book, we will use the matrix X and we will define the feature vector $\underline{x}$ as $\underline{x} = \begin{bmatrix} x_1 & \dots & x_n \end{bmatrix}^T$. These are the conventions used by the SciKit-Learn library. In these notations, the linear model (11.1) writes as

$$y = w_0 + \underline{w}^T \cdot \underline{x} \tag{11.5}$$

We will learn in this chapter that features don't necessarily have to be numerical values. In fact, it is common to find categorical data where some features are categories. For example, the color could be such a categorical feature. We will discuss in Section 11.3.4 how such features are handled.

Example 11.1. A Small Dataset

To illustrate the concepts of this section, let us consider a small dataset regrouping some data about fish. The interested reader can find the full dataset with related explanations in the Journal of Statistics Education archive under https://jse.amstat.org/jse_data_archive.htm.

The data contains the following information on fish (the full set contains $m = 159$ samples):

- Fish species
- Weight of the fish in grams
- Lenght1: Length from the nose to the beginning of the tail (in cm)
- Length2: Length from the nose to the notch of the tail (in cm)
- Length3: Length from the nose to the end of the tail (in cm)

- Height%: Maximal height as percentage of Length3
- Width%: Maximal width as percentage of Length3
- Sex: Sex of the fish (1 = male; 0 = female)

The set contains eight features, some of them are categorical data (fish species and fish sex) while the others are numerical features.

As an example, let us build the feature matrix X for the particular case where the fish species is whitefish and using the three features x_1 = Length3, x_2 = Height% and x_3 = Width%. The dataset contains six samples of the fish species whitefish as shown in Table 11.1. The table contains further the information about the weight of the fish, which we will use as the label, i.e. the quantity that later will have to be predicted by the model.

TABLE 11.1
Some Data About Whitefish. Full Dataset Can Be Found Under https://jse.amstat.org/jse_data_archive.htm

y = Weight	x_1 = Length3	x_2 = Height%	x_3 = Width%
270	28.7	8.3804	4.2476
270	29.3	8.1454	4.2485
306	30.8	8.778	4.6816
540	34	10.744	6.562
800	39.6	11.7612	6.5736
1000	43.5	12.354	6.525

The features matrix for the dataset based on Table 11.1 writes as

$$X = \begin{bmatrix} 28.7 & 8.3804 & 4.2476 \\ 29.3 & 8.1454 & 4.2485 \\ 30.8 & 8.778 & 4.6816 \\ 34 & 10.744 & 6.562 \\ 39.6 & 11.7612 & 6.5736 \\ 43.5 & 12.354 & 6.525 \end{bmatrix} \tag{11.6}$$

In case we want to use the fish weight as a label, the label vector $\underline{y}$ becomes

$$\underline{y} = \begin{bmatrix} 270 & 270 & 306 & 540 & 800 & 1000 \end{bmatrix}^T \tag{11.7}$$

11.2 MODEL TRAINING AND VALIDATION

In Chapter 6, we learned how, using the technique of normal equations, a linear model can be adjusted to a set of data points. This approach could be used to develop a linear model for the dataset of Example 11.1, and the reader is invited to try it as an exercise. In this section, we will show how the Python library SciKit-Learn can achieve the same task.

11.2.1 Training a Linear Regression Model

In the context of machine learning, the process of adjusting model parameters according to a chosen procedure is called *model training*. In this section, we illustrate this for the case of linear regression models. As an example, we use the data from Example 11.1.

SciKit-Learn offers a variety of models. One of the strengths of the library is that all these models are handled with an identical interface. In a simplified way, they all derive from a common class that will expose a certain number of common methods. One of them is the method `fit`, which is a method used to train the model with a given dataset. In its simplest form, the method `fit(X, y)` asks for two parameters, `X` and `y`. The first one is the feature matrix X, and the second one is the label vector y. Technically speaking, the feature matrix must be provided as a `numpy` array of shape `(n_samples, n_features)` and the label vector as a `numpy` array of shape `(n_samples,)`. All these structures are required to import the Python library `numpy`.

Let us illustrate how to build these two quantities with the data from Example 11.1. The features matrix can be built like so:

```
import numpy as np
X = np.array([[28.7, 8.3804, 4.2476],
              [29.3, 8.1454, 4.2485],
              [30.8, 8.778, 4.6816],
              [34, 10.744, 6.562],
              [39.6, 11.7612, 6.5736],
              [43.5, 12.354, 6.525]])
```

The label vector is defined like this:

```
y = np.array([270, 270, 306, 540, 800, 1000])
```

We can verify that the shapes of the defined structures are indeed what they should be:

```
>>> X.shape
(6, 3)
>>> y.shape
(6,)
```

Having defined our data, we are ready to train the model. For a linear regression model, SciKit-Learn defines the class `LinearRegression` as one that implements least square linear regression. To use this class, we start by creating an instance of it and then train the model by calling the `fit` method:

```
from sklearn.linear_model import LinearRegression
reg = LinearRegression()
reg.fit(X, y)
```

This procedure is identical to other models SciKit-Learn implements. One will only have to replace the `LinearRegression` class with the desired model. The rest remains identical.

It is possible to inspect the model parameters. This part is, in general, model-specific, as every model will have different types of parameters to be adjusted. In the case of linear regression models, the trained parameters are the bias term w_o and the feature weights w_i. The bias term can be displayed by inspecting the model attribute `intercept_`:

```
>>> reg.intercept_
-1246.987738670462
```

whereas the features weights w_i are stored in the model attribute `coef_`:

```
>>> reg.coef_
array ([25.66001055, 132.37534855, -78.55363506])
```

As we have three features, the model training adjusted three features' weights. The order in which the numbers appear in the array is the same as the order of the features in the features matrix X. In our case, it is in the order of Length3, Height% and Width%.

11.2.2 Making Predictions

Now that we have trained our model, we want to use it to compute some predictions. Predictions are made with the method `predict` of the `LinearRegression` class. As the `fit` method, the `predict` method is a method common to the various models implemented by SciKit-Learn. The method takes one parameter, which is an array of features for which a prediction is desired. For example, let us predict what would be, according to our trained model, the weight of a whitefish with these characteristics: Length3 = 32 cm, Height%=9 and Width%=5. This is achieved like this:

```
>>> reg.predict([[32, 9, 5]])
array([372.74256045])
```

The weight predicted is about 372.7 grams. It is possible to make, in a single call of `predict`, predictions for different cases. For example, with the following call, we predict the weights of two fish:

```
>>> reg.predict([[32, 9, 5], [29, 10, 6]])
array([372.74256045, 349.5842423])
```

11.2.3 Validating the Model

In Chapter 6, we discussed some methods to evaluate the quality of the fit. The first indicator is to compute the residual errors of all samples in the dataset. As SciKit-Learn made the design choice to have the same format for the input parameter of the method `predict` than the feature matrix X, this calculation is straightforward:

```
>>> err = y - reg.predict(X)
>>> err
array ([ 4.85148529, 20.63438415, -31.57469782,
7.77558855, -9.66145288, 7.9746927])
```

A more compact metric, which results in a single number, is the root mean square error. SciKit-Learn implements the calculation of the root mean square error, among other metrics in their module `sklearn.metrics`. Its usage is as follows:

```
from sklearn.metrics import mean_squared_error
RMSE = mean_squared_error(y, reg.predict(X),
                          squared=False)
```

The first two parameters of the function `mean_squared_error` are the actual and predicted labels. The named argument `squared` allows one to choose between the calculation of the mean square error MSE (`squared=True`) and the root mean squared error RMSE (`squared=False`).

It is useful to represent this information with pertinent plots. The Python library Matplotlib is a popular one, and SciKit-Learn is designed to work hand in hand with Matplotlib. The following code snippet gives an example:

```
import matplotlib.pyplot as plt
plt.plot(err, ".", markersize=15,
         markerfacecolor='white')
plt.plot([0, len(y)], [RMSE, RMSE], '--',
         color='#ff7f0e')
plt.plot([0, len(y)], [-RMSE, -RMSE], '--',
         color='#ff7f0e')
plt.xlabel('Sample number')
plt.ylabel('Residual errors')
plt.show()
```

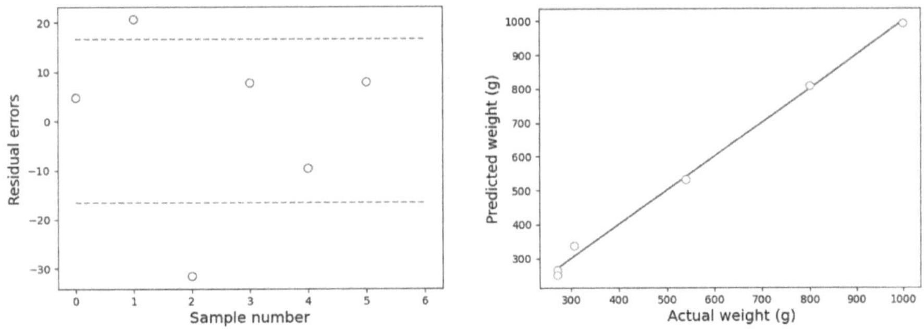

FIGURE 11.1 (a) Residual errors (o) and root mean squared error (--) of the samples in the dataset from Example 11.1. (b) Predicted versus actual weights of whitefish.

Figure 11.1a depicts the residual errors, together with the root mean squared error for all samples in the dataset from Example 11.1.

Another pertinent graph is a plot of the predicted labels versus the actual labels of the dataset as produced with the following code snippet (Figure 11.1b):

```
plt.plot(y, y, "-").
plt.plot(y, reg.predict(X), ".", markersize=15,
         markerfacecolor='white')
plt.xlabel('Actual weight (g)')
plt.ylabel('Predicted weight (g)')
```

It is important to realize that the metrics discussed here and displayed in Figure 11.1 only evaluate how well the dataset is represented by the linear model. They do not state anything about the predictive value of the model. This key aspect of machine learning will be discussed in Chapter 12.

11.3 FEATURES TRANSFORMATIONS

In Section 6.2.3 of Chapter 6, we discussed the idea of general linear regression. This approach can be extended to linear models (11.1) written in terms of features:

$$y(\underline{x};\ \underline{w}) = w_o + w_1 g_1(x_1) + \ldots + w_n g_n(x_n) \tag{11.8}$$

where we apply a function g_i to each feature. Such functions are called feature transformations. An example is the case of polynomial regression. In this case, an initial feature x is used to compute n, new features such as to form a polynomial of degree n. In this case, the functions g_i are $g_i(x) = x^i$. Note that such transformations can be applied to one or several original features. Let us first illustrate the case where the initial feature is a single one.

11.3.1 Polynomial Transformation

We start by generating a synthetic dataset. In machine learning, synthetic datasets are sets of known properties generated for the purpose of illustrating aspects of models or exploring their characteristics. The idea is similar to using problems of known solutions in traditional numerical algorithms. In this example, we generate a dataset with points following the equation:

$$y(x) = -2.5x^2 + 2x + 2 \tag{11.9}$$

To add some randomness, we draw $m = 30$ points from the interval $[-3,3]$ and add some Gaussian noise to the y values:

```
import numpy as np
m = 30
X = 6*np.random.rand(m, 1) - 3
y = -2.5*X*X + 2*X + 2 + 2*np.random.randn(m, 1)
```

This way, we build a feature matrix X and a label vector y representing our dataset. From a technical point of view, the built structures `X` and `y` are `numpy` arrays, as required later by the `fit` method.

Our aim is to adjust a quadratic model to this dataset. For this, we need first to compute the following two features:

$$\begin{cases} x_1 = x \\ x_2 = x^2 \end{cases} \tag{11.10}$$

and afterward adjust the linear model

$$y(x; \underline{w}) = w_o + w_1 x_1 + w_2 x_2 \tag{11.11}$$

to the dataset. SciKit-Learn offers a feature transformer to perform this task out of the box. Feature transformers are classes that implement transformations on features. They have a `fit_transform` method that performs the desired transformation. For polynomial feature transformation, i.e., a transformation applying $g_i(x) = x^i$ to each input feature, the usage is as follows:

```
from sklearn.preprocessing import PolynomialFeatures
poly_features = PolynomialFeatures(degree=2,
                                   include_bias=False)
x_poly = poly_features.fit_transform(X)
```

The class constructor has the named parameter `degree` that allows for the specification of the degree of the polynomial. We further add the option `include_bias=False` to avoid adding the bias term. We will come back to this aspect later.

The `fit_transform` method takes as input parameter exactly the same type of data as the `fit` method from the model, i.e., a `numpy` array of shape `(n_samples, n_features)`.

To understand better which transformation the `PolynomialFeatures` applies to its input, let us consider this code:

```
>>> poly_features.fit_transform([[1],
                                  [2],
                                  [3]])

array([[1., 1.],
       [2., 4.],
       [3., 9.]])
```

The `fit_transform` computed the features $x_1 = x$ and $x_2 = x^2$, and afterward created the new transformed feature matrix X_t, by placing x_1 into the first column and x_2 into the second column. In Octave, the same code would be written:

```
>> x = [1, 2, 3]'
>> Xt = [x x.^2]
```

Note that the transformed features matrix does not contain a first column of ones. This comes from the fact we used the option `include_bias=False` during the construction of the `PolynomialFeatures` transformer. This is important, as the `LinearRegression` model from SciKit-Learn expects a features matrix without this bias term, as discussed previously.

Once the transformation is performed, the training of the linear model is conducted the usual way:

```
from sklearn.linear_model import LinearRegression
reg = LinearRegression()
reg.fit(x_poly, y)
```

Figure 11.2 depicts the synthetic dataset together with the trained model. Let us spend some time explaining how predictions should be performed using transformed features.

To perform predictions with the trained model, it is important to apply the polynomial feature transformation prior to prediction. For example, to compute the value of y in $x = 2$, we proceed like this:

```
>>> reg.predict(poly_features.fit_transform([[2]]))
array([[-4.11590708]])
```

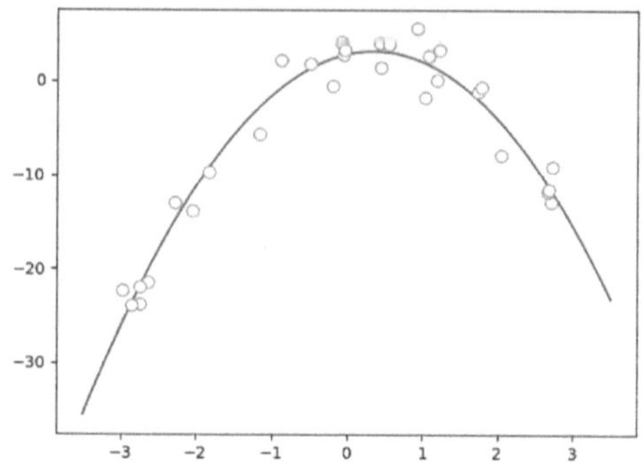

FIGURE 11.2 An example of a polynomial fit to a dataset. The set was generated from the equation $y(x) = -2.5x^2 + 2x + 2$ by adding some Gaussian noise.

The reader may obtain a slightly different result, as the synthetic dataset used to train the model has some randomness in it.

Similarly, we can predict the values of several points. For example, the following python code computes the prediction of 100 points distributed uniformly in the interval $[-3.5, 3.5]$:

```
reg.predict(poly_features.fit_transform(
    np.c_[np.linspace(-3.5, 3.5, 100)]))
```

This syntax looks rather complex. Unfortunately, the syntax in Python, when handling vectors, is not as intuitive as in Octave. This comes from the fact that `numpy` doesn't use the concept of vectors and matrices but instead uses samples, which are represented by `numpy` arrays, and datasets that are lists of samples. It is worth spending some time to explain this.

The `numpy` function `linspace` provides an array (in our case, an array of 100 points between -3.5 and 3.5). Compared with Octave, this could be described as a vector. However, the methods `fit_transform` and `fit` require as input a feature matrix and not a vector. To build this feature matrix column by column, the `numpy` indexing routine `c_` is helpful. Conceptually, even if technically not entirely correct, we can imagine it equivalent to the Octave construction:

```
>> A = [a b c]
```

which creates a matrix of three columns from the three column vectors a, b and c. In Python, this would write like:

```
>>> a = np.array([1, 2, 3])
>>> b = np.array([4, 5, 6])
>>> c = np.array([7, 8, 9])
>>> np.c_[a, b, c]
```

```
array([[1, 4, 7],
       [2, 5, 8],
       [3, 6, 9]])
```

With this in mind, we can understand that `np.c_[np.linspace(-3.5, 3.5, 100)]` will create the needed features matrix made of the single column from the vector `np.linspace(-3.5, 3.5, 100)`. This is the format required by the method `fit_transform` that will create a new features matrix with the transformed features (11.10). This feature matrix is then passed to the `predict` method of the trained model to perform the desired predictions.

As in the case of predictions, we apply the feature transformations prior to computing metrics such as the root mean squared error.

```
>>> X_poly = poly_features.fit_transform(X)
>>> mean_squared_error(y, reg.predict(X_poly),
                       squared=False)
1.952318781304255
```

If the reader chooses to run the code, they may find that the numerical value could differ slightly from the one here due to the random nature of the generation of the synthetic dataset used to train the model.

11.3.2 Pipelines

The explanations of the previous section, 11.3.1, showed that the notations used in Python become quickly cumbersome and difficult to follow. This situation becomes worse when applying not only polynomial feature transformations, but others, too. The developers of SciKit-Learn introduced an interesting tool to reduce the complexity of codes: the concept of pipeline. The idea of pipelines is to describe all the operations that need to be performed once, starting with the raw data and reaching the format desired for training a model. In the case of polynomial feature transformation, the chain is simple: initial feature -> polynomial feature transformation -> features matrix. But we will soon see more complex cases.

In SciKit-Learn, pipelines are instances of the class `pipeline`. A simple way to create pipelines is to use the function `make_pipeline`. In this function, classes are chained together to create the desired pipeline. Any class having a `transform` method can be chained. The pipeline will successively call the `transform` methods of each class within the pipeline, using the output of the previous class as an input to the next one. The classes representing models, such as the `LinearRegression` class, and feature transformers, such as the `PolynomialFeatures`, are both classes that implement a `transform` method.

For example, the following code creates a pipeline that will chain a `PolynomialFeatures` transformer (for a quadratic polynomial) and the `LinearRegression` model:

```
poly_model = make_pipeline(
    PolynomialFeatures(degree=2, include_bias=False),
    LinearRegression())
```

Once the pipeline is created, it can be used to train the model. Let us now consider a dataset represented by the feature matrix `X` and label `y`. The reader can use the synthetic dataset used in Section 11.3.1. To fit the model, we simply use

```
>>> poly_model.fit(X, y)
```

This automatically applies the `PolynomialFeatures` transformer to the feature matrix `X` and then uses the transformed matrix to train the `LinearRegression` model.

Predictions of the trained model are then performed like so:

```
>>> poly_model.predict(X_pred)
```

where `X_pred` is some feature matrix representing the features one wants to use for prediction. This could be a single sample like `[[2]]`, or a range of samples like `[[2], [3], [4]]`.

In case the previous code is executed in a Jupyter[2] Notebook, a graphical illustration of the pipeline is displayed with an image like this one:

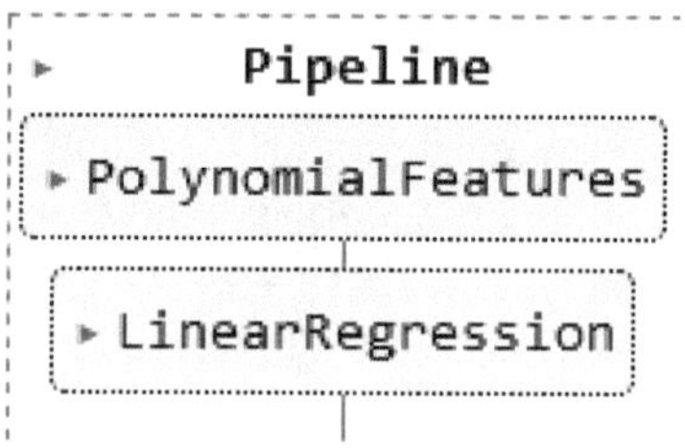

Comparing the few lines of code in this section with the serval ones from 11.3.1 should convince the reader of the usefulness of using pipelines.

11.3.3 Custom Transformers

Polynomial feature transformers are only a particular example. In fact, in principle, any kind of transformation can be applied to features. In this section, we show how a custom transformer can be implemented in Python. For this purpose, we consider general linear regression using Gaussian basis functions. The model we aim to train writes:

$$y(x; \underline{w}) = w_o + \sum_{i=1}^{n} w_i \phi_i(x) \tag{11.12}$$

where $\phi_i(x)$ are the so-called Gaussian basis functions:

$$\phi_i(x) = \exp\left[-\frac{(x-\mu_i)^2}{2\sigma^2}\right] \tag{11.13}$$

These functions have two parameters μ_i and σ that define the behavior of the Gaussian basis function. The parameters μ_i are the centers around which the Gaussian basis functions are built, and σ is the spatial extension of each of them. Parameters that need to be chosen prior to training, such as μ_i and σ, are, in machine learning, referred to as *hyper-parameters*. In fact, most models will have such hyper-parameters. One example we encountered already without realizing it is the case of the degree of the polynomial used in polynomial regression. Hyper-parameters can have complex effects on the training of a model, and they need to be tuned carefully. In fact, we will learn that they influence whether a model has acceptable or unacceptable predictive value. We postpone this important discussion until we have learned further concepts. For now, we will simply choose some arbitrary values for hyper-parameters and, in the coming examples, we choose $\sigma = 0.1$ and μ_i as uniformly distributed in the interval $[0,1]$.

Figure 11.3 plots five Gaussian basis functions where the hyper-parameters are $\sigma = 0.1$ and the locations μ_i of the basis functions are distributed uniformly over the interval $[0,1]$. Each function is centered at its location μ_i (for example ϕ_3 in $\mu_3 = 0.5$) and extends around it with a typical size of $\sigma = 0.1$. Conceptually, we can imagine Gaussian basis functions as fulfilling a similar role to the Lagrangian polynomials in Lagrangian interpolation.

To train model (11.12), we first need to transform the feature x using the Gaussian basis functions and create the transformed features matrix X:

$$X = [\phi_1(x) \quad \phi_2(x) \quad \ldots \quad \phi_n(x)] \tag{11.14}$$

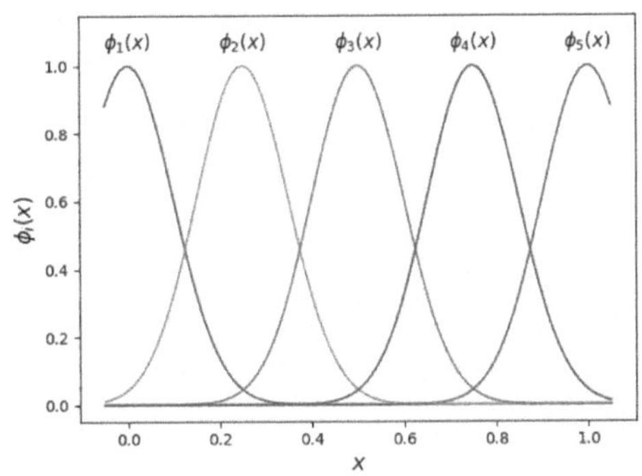

FIGURE 11.3 Example of five Gaussian basis functions ($\sigma = 0.1$ and the locations μ_i of the basis functions distributed uniformly over the interval $[0,1]$).

An elegant way to do so is to define a custom transformer and afterward use a pipeline. The writing of a custom transformer requires advanced skills in Python, including the understanding of Python-Mixin and derivation of Python classes. The documentation of SciKit-Learn explains the process to follow. The essential steps consist of creating a new class derived from the `BaseEstimator` class and using the `TransformerMixin`.

The constructor of the derived class can be used to assign parameters. In our case, we have two parameters: `n`, the number of Gaussian base functions and `s` the spatial extension σ of them. We further compute the location μ_i by distributing them uniformly in the interval $[0,1]$.

The derived class must further implement the two methods `fit` and `transform`. The `fit` method is rather simple. The key method is the `transform` method. It takes the features to be transformed as input and must return the transformed feature matrix, which in our case is Matrix (11.14).

The following code shows the result and can be used as a template for the creation of other custom transformers.

```
import numpy as np
from sklearn.base import BaseEstimator
from sklearn.base import TransformerMixin
class GaussianFeatures(BaseEstimator,
                       TransformerMixin):
    """
    Transformer for uniformly spaced Gaussian basis
    functions for one-dimensional input
    """
    def __init__(self, n, s):
        self.n = n
        self.s = s
        self.mu_i = np.linspace(0, 1, n)

    def _gauss_basis(self, x, i):
        returnnp.exp(-(x-self.mu_i[i])**2/(2*self.s**2))

    def fit(self, X, y=None):
        return self

    def transform(self, X):
        X_b = np.c_[self._gauss_basis(X, 0)]
        for i in range(1, self.n):
            X_b = np.c_[X_b, self._gauss_basis(X, i)]
        return X_b
```

With the help of this custom transformer, the fitting of a dataset follows the same approach we used in polynomial regression.

Let us first create a synthetic dataset:

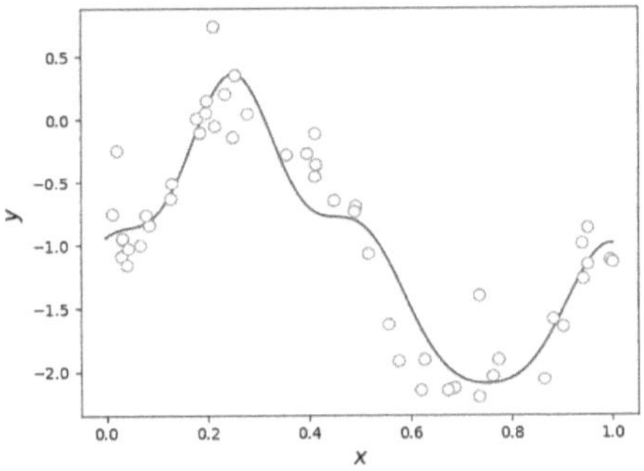

FIGURE 11.4 An example of a $n = 5$ Gaussian base function functions fit to a dataset (the spatial extension σ of the base function is equal to 0.1). The set was generated from the equation $y(x) = \sin(2\pi x) - 1$ by adding some Gaussian noise.

```
x = np.random.rand(50)
y = -1 + np.sin(2*np.pi*x) + np.random.normal(0, 0.25, 50)
X = np.c_[x]
```

The fitting of the dataset is then performed using a pipeline:

```
from sklearn.pipeline import make_pipeline
from sklearn.linear_model import LinearRegression
gauss_model = make_pipeline(
    GaussianFeatures(n=5, s=0.1),
    LinearRegression())
gauss_model.fit(X, y)
```

Figure 11.4 shows the synthetic dataset together with the trained model.

11.3.4 Categorical Data

As seen in Example 11.1, some features can be of a categorical nature, such as the fish species for example. Obviously, linear models, such as Model (11.1), cannot handle such types of data. A way to still include them in a model is to assign some arbitrary numerical values to them. This method is called ordinal encoding and is implemented in the `OrdinalEncoder` transformer in SciKit-Learn. The `OrdinalEncoder` operates by finding all categories in the dataset first and then assigning some arbitrary unique numbers to them. The following simplified example illustrates the `OrdinalEncoder`.

Example 11.2. Ordinal Encoder

To illustrate the usage of the `OrdinalEncoder` transformer from SciKit-Learn, let us consider a simplified dataset that contains only one entry: Fish species.

We start by defining a features matrix for this dataset:

```
>>> import numpy as np
>>> X = np.array([['Bream'], ['Roach'], ['Whitefish'],
                  ['Parkki'], ['Perch'], ['Pike'],
                  ['Smelt'], ['Whitefish'], ['Pike'],
                  ['Roach'], ['Bream'], ['Parkki']])
```

As mentioned previously, to simplify this discussion, we keep a single feature, which is a categorical feature. To use ordinal encoding, we use the `OrdinalEncoder` transformer from SciKit-Learn. Its usage follows the same principle as other transformers we have discussed so far:

```
>>> from sklearn.preprocessing import OrdinalEncoder
>>> ord_enc = OrdinalEncoder()
>>> ord_enc.fit(X)
```

By calling the `fit` method, the encoder will identify the categories in our features matrix and assign to each of them a number. We can check the categories that the encoder found by inspecting the `enc.categories _` attribute:

```
>>> ord_enc.categories_
[array(['Bream', 'Parkki', 'Perch', 'Pike',
'Roach', 'Smelt', 'Whitefish'], dtype='<U9')]
```

To obtain the encoded numbers for some elements, we use the `transform` method of our encoder:

```
>>> ord_enc.transform([['Roach'], ['Whitefish']])
array([[4.],
       [6.]])
```

In this case, the number 4 is attributed to the category Roach and the number 6 to the category Whitefish. We can list all categories with their numbers attributed like this:

```
>>> for cat in ord_enc.categories_[0]:
        print("{0:15} {1}".format(
            cat + ":",
            ord_enc.transform([[cat]])))

Bream:      [[0.]]
Parkki:     [[1.]]
Perch:      [[2.]]
Pike:       [[3.]]
Roach:      [[4.]]
Smelt:      [[5.]]
Whitefish:  [[6.]]
```

Note that the encoder is case-sensitive. If, for example, your dataset is as shown in the following, where the last entry uses a lower first letter instead of a capital first letter like the previous entries:

```
>>> X = np.array([['Bream'], ['Roach'], ['Whitefish'],
                  ['Parkki'], ['Perch'], ['Pike'],
                  ['Smelt'], ['Whitefish'], ['Pike'],
                  ['Roach'], ['Bream'], ['parkki']])
```

The encoder finds an additional category, "parkki", and encodes entries "parkki" and "Parkki" differently.

Ordinal encoding is very simple to apply; however, its usefulness is limited. Indeed, an arbitrary ordering takes place during the encoding. For instance, in Example 11.2, the fish species Perch and Pike are distant from each other by one unit, whereas the fish species Perch and Whitefish are distant by four units. If we use some linear regression model, then mathematically, the model considers that perch and pike are more similar to each other than perch and whitefish. This may not be the case in reality. This example shows the limitation of ordinal encoding. For that reason, another encoding methodology was developed: the one-hot encoding.

A one-hot encoder will assign a vector to each category. The length of the vector equals the number of categories in the dataset. Each category vector will be zero everywhere except in one entry where there will be a one. The position of the entry "1" is unique for the category. In this way, all categories are distant from one another equally and, contrary to the ordinal encoding, no arbitrary ordering is introduced.

Example 11.3. One-Hot Encoding

We use the same dataset as in Example 11.2. One-hot encoding is implemented with the `OneHotEncoder` class in SciKit-Learn. The encoding process is identical as with the `OrdinalEncoder`:

```
>>> from sklearn.preprocessing import OneHotEncoder
>>> hot_enc = OneHotEncoder()
       [0., 0., 0., 0., 0., 0., 1.]])
```

To obtain the encoded vectors for some elements, we use the `transform` method of our encoder. However, to facilitate the readability of the result, it is useful to use further the `toarray` method to transform the result into an array:

```
>>> hot_enc.transform([['Roach'],
                       ['Whitefish']]).toarray()
array([[0., 0., 0., 0., 1., 0., 0.],
       [0., 0., 0., 0., 0., 0., 1.]])
```

Note how each category is represented by a unique vector. We can verify this by displaying all category vectors:

```
>>> for cat in hot_enc.categories_[0]:
        print("{0:15} {1}".format(
            cat + ":",
            hot_enc.transform([[cat]]).toarray()))

Bream:          [[1. 0. 0. 0. 0. 0. 0.]]
Parkki:         [[0. 1. 0. 0. 0. 0. 0.]]
Perch:          [[0. 0. 1. 0. 0. 0. 0.]]
Pike:           [[0. 0. 0. 1. 0. 0. 0.]]
Roach:          [[0. 0. 0. 0. 1. 0. 0.]]
Smelt:          [[0. 0. 0. 0. 0. 1. 0.]]
Whitefish:      [[0. 0. 0. 0. 0. 0. 1.]]
```

All categories are equally different from each other. No arbitrary ordering is introduced.

11.3.5 Advanced Pipelines

Datasets often contain a mix of numerical and categorical data (see, for example, 11.1). This requires the first step of encoding the categorical data. In this section, we show step by step how, using pipelines, some data can be transformed with a chosen transformer (for example, a one-hot encoder) while others are transformed with other transformers or simply not transformed at all.

Building pipelines that apply different transformations to different features are more complex to setup then a pipeline that applies the same transformation to all features in the dataset. However, they offer more flexibility and better control on the way data is transformed.

To explain how to set up advanced pipelines, we work with an extended version of the dataset of Example 11.1. Here, we add the fish species in the first column and extend the set with four additional samples:

```
import numpy as np
X = np.array([['Whitefish', 28.7, 8.3804, 4.2476],
              ['Whitefish', 29.3, 8.1454, 4.2485],
              ['Whitefish', 30.8, 8.778, 4.6816],
              ['Whitefish', 34, 10.744, 6.562],
              ['Whitefish', 39.6, 11.7612, 6.5736],
              ['Whitefish', 43.5, 12.354, 6.525],
```

```
                  ['Perch', 8.8, 2.112, 1.408],
                  ['Perch', 14.7, 3.528, 1.9992],
                  ['Perch', 16, 3.824, 2.432],
                  ['Perch', 17.2, 4.5924, 2.6316]],
                  dtype=object)
y = np.array([270, 270, 306, 540, 800, 1000, 5.9, 32,
                  40, 51.5])
```

Note the `dtype=object` option added in the definition of the features matrix. This is needed as, otherwise, `numpy` will create a homogenous array composed exclusively of strings instead of a mixture between strings and numbers.

Let us now build a pipeline that applies a one-hot encoder to the feature fish species and keep the remaining features as they are. Figure 11.5 shows the pipeline we aim to build. The feature matrix is transformed by applying a one-hot-encoder to column number 0 (the column with the fish species) and a function transformer that simply copies the data as is to the remaining columns (the numerical data). Once transformed, we use the transformed features matrix as an input to a linear regression model.

To build this pipeline, two auxiliary pipelines need to be defined: one for the features that do not require one-hot encoding and another for the feature that requires one-hot encoding.

The first auxiliary pipeline keeps the features as they are. It is built using the `Pipeline` class. A `Pipeline` class is constructed by passing a named argument `steps`. Steps are defined as a list of tuples of names (names can be freely chosen) and classes of operations executed by the pipeline.

```
from sklearn.pipeline import Pipeline
from sklearn.preprocessing import FunctionTransformer
numeric_transformer = Pipeline(
    steps=[
         ("copy", FunctionTransformer(lambda X : X))
    ]
)
```

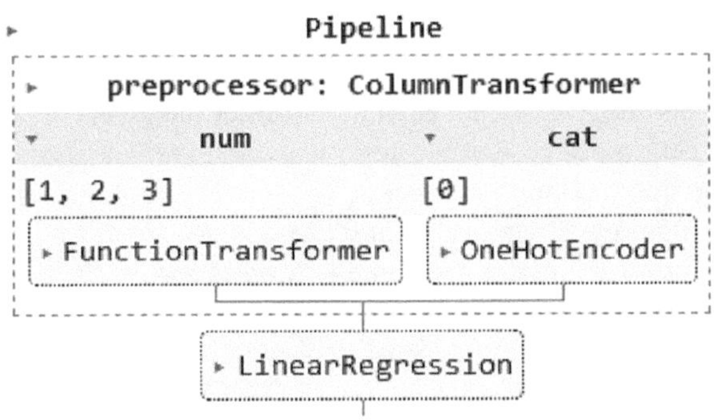

FIGURE 11.5 An example of a $n = 5$ Gaussian base function fit to a dataset (the spatial extension σ of the base function is equal to 0.1). The set was generated from the equation $y(x) = \sin(2\pi x) - 1$ by adding some Gaussian noise.

The defined Pipeline consists of a single step: a FunctionTransformer. The FunctionTransformer accepts one argument which consists of the function to be applied. This function takes the features matrix as input and must return the transformed features matrix. In our case, we simply return the features matrix as is since we do not want to apply any transformation.

The second auxiliary pipeline applies the OneHotEncoder:

```
from sklearn.preprocessing import OrdinalEncoder
categorical_transformer = Pipeline(
    steps=[
            ("encoder", OneHotEncoder())
    ]
)
```

We are now able to set up our pipeline which we call preprocessor:

```
from sklearn.compose import ColumnTransformer
preprocessor = ColumnTransformer(
    transformers=[
        ("num", numeric_transformer, [1, 2, 3]),
        ("cat", categorical_transformer, [0])
    ]
)
```

We use a ColumnTransformer to define it. The ColumnTransformer allows transformers to be applied on specific columns of the features matrix. It requires a list of transformers in the form of tuples with the format (name, transformer, columns). The name can be freely chosen. The transformer must be a Pipeline, and its columns consist of numbers on which the transformer will be applied.

In our case, we have two transformers. The first one, which we decided to name “num”, is the numeric_transformer pipeline we defined previously. The second one, which we chose to name “cat”, is the categorical_transformer pipeline we defined previously.

The column numbers on which each transformer has to act depends on the format of the dataset. In our case, the categorical data, the fish species, to undergo a one-hot encoder are in column number 0. The remainder of the columns (numbers 1, 2, and 3) are numerical data we want to leave untouched.

We can now use our pipeline the same way we did with the ones we presented so far. First, we call the fit method:

```
>>> preprocessor.fit(X)
```

This will set up the one-hot encoder. After the call of the fit method, the pipeline is ready to perform transformations:

```
>>> preprocessor.transform([['Perch', 29, 9, 3]])
array([[29, 9, 3, 1.0, 0.0]], dtype=object)
```

The features matrix `[['Perch', 29, 9, 3]]` is transformed as intended. The numerical features remain untouched. They appear first in the transformed matrix as in our `preprocessor` pipeline, which applies first to the `numeric_transformer`. The categorical data was one-hot encoded into the vector `[1.0  0.0]`. It appears last in our `preprocessor` pipeline, which applies the `categorical_transformer` after the `numeric_transformer`.

We can also transform several samples together:

```
>>> preprocessor.transform([['Perch', 29, 11, 3],
                            ['Whitefish', 35, 9, 5]])
array([[29, 11, 3, 1.0, 0.0],
       [35, 9, 5, 0.0, 1.0]], dtype=object)
```

Both samples were transformed as intended.

Now that the `preprocessor` pipeline is set up, we can use it in a pipeline that is employed to train a model. For example, we can define a pipeline as follows:

```
from sklearn.linear_model import LinearRegression
fish_weight = Pipeline(
    steps=[("preprocessor", preprocessor),
           ("regression", LinearRegression())]
    )
```

This pipeline first uses the `preprocessor` pipeline to transform the samples and then uses a linear regression model. To train the model, we call the `fit` method:

```
>>> fish_weight.fit(X, y)
```

After the model is trained, we can use it to predict the weights of some fish with the `predict` method:

```
>>> fish_weight.predict([['Perch', 15, 3, 2],
                         ['Whitefish', 35, 9, 5]])
array ([21.75799259, 486.48359951])
```

Note that we did not attempt to create a model of good quality in terms of prediction value. The sole purpose of the previous discussion is to demonstrate how we can build pipelines.

11.4 LAB: WORKING WITH REAL DATA

In this last section of the chapter, we show some typical steps involved when working with real data. Let us highlight that, for space limitations, we can only provide a high-level overview of data manipulation. Handling real data is an extensive topic and would require a book in itself. However, we aim herein to outline some major steps. The next example is a start point for the reader before delving deeper in the field of data processing.

11.4.1 Data and Preparation

The dataset we will use in this example can be found on the website Open Canada[3] by the government of Canada, where several data of general interest are made publicly available. Among them is the dataset "Wages by education level"[4] published by the province of Ontario, which presents information on wages across Canada. At the time of writing this textbook, the data is available under the link https://open.canada.ca/data/dataset/1f14addd-e4fc-4a07-9982-ad98db07ef86.

Detailed information on this dataset is available, and the data itself is published in the form of a CSV[5] file that can be downloaded under the link https://data.ontario.ca/dataset/1f14addd-e4fc-4a07-9982-ad98db07ef86/resource/7b325fa1-e9d6–4329-a501–08cdc22a79df/download/v0913_05.csv.

Once downloaded, the CSV file can be processed. The Python `pandas` library is powerful library to handle data. Several data structures are available; among them is the `dataframe` structure. Readers not familiar with `dataframes` are encouraged to follow some online tutorials. The few commands we will use in this section should help as a start-point for further explorations.

A CSV file can be loaded into a `dataframe` with the command `read_csv`:

```
>>> import pandas as pd
>>> raw_data = pd.read_csv('v0913_05.csv')
```

When running this command, it is important to ensure that the CSV file is downloaded into the same directory that contains the code we will execute. Alternatively, we can add the correct path using the `read_csv` method.

The method `head` allows obtaining a quick overview of the structure of the data in the form of a table (for space reasons the output was truncated):

```
>>> raw_data.head()
YEAR Geography                 Type of work  ...     Male    Female
1997    Canada  Both full- and part-time  ...   5954.5    5410.0
1997    Canada  Both full- and part-time  ...    983.1     894.7
1997    Canada  Both full- and part-time  ...   2244.3    2030.6
1997    Canada  Both full- and part-time  ...   4971.4    4515.3
1997    Canada  Both full- and part-time  ...   1602.2    1445.7
```

It is very common that the data is not in a format suitable for further processing, and typically, data transformation and cleaning steps are required. These steps are usually lengthy and require additional analysis to detect erroneous data entries or outliers. In our case, the data is already in relatively good shape, and only a few transformations and cleaning steps are required.

As a first step, we want to understand what type of "Education level" is stored in the dataset. For this, we extract all unique entries in the column "Education level" of the `dataframe`:

```
>>> raw_data['Education level'].unique()
array(['Total, all education levels', ' 0 - 8 years',
       ' Some high school', ' High school graduate',
       ' Some post-secondary',
       ' Post-secondary certificate or diploma',
       ' Trade certificate or diploma', ' Community
         college, CEGEP',
       ' University certificate below bachelors degree',
       ' University degree', " Bachelor's degree",
       " Above bachelor's degree", 'PSE (5,6,7,8,9))',
       'No PSE (0,1,2,3,4)'], dtype=object)
```

For our example, we will be interested in the wages of all types of education levels. These are the entries with "Total, all education levels" in the "Education level" column.

We repeat the same exercise on the other columns and inspect the type of data they contain. Based on the entries, we narrow down the data according to our interest. For the purpose of the present example, we extract the data as follows:

```
df = raw_data[
    (raw_data['Education level'] == 'Total, all educa-
     tion levels') &
    (raw_data['Age group'] == '15 years and over ') &
    (raw_data['Type of work'] == ' Full-time ') &
    (raw_data['Wages'] == 'Average hourly wage rate')
].copy()
```

In other words, we are interested in the average hourly wage rate, regardless of education level, for all full-time workers aged 15 or more.

Once the data is extracted, we can remove the columns that are no longer relevant to the `drop` method. We further rename some of them with the `rename` method to ease access later:

```
df.drop('Education level', axis=1, inplace=True)
df.drop('Age group', axis=1, inplace=True)
```

```
df.drop('Type of work', axis=1, inplace=True)
df.drop('Wages', axis=1, inplace=True)
df.rename(columns={'YEAR': 'Year',
                   'Both Sexes': 'Combined',
                   ' Male': 'Men',
                   ' Female': 'Women'}, inplace=True)
```

The new constructed `dataframe df` contains only the data we are interested in:

```
>>> df.head()
 Year       Geography                    Combined Men   Women
672 1997    Canada                          16.54 17.83 14.79
2352 1997   Newfoundland and Labrador 13.88 15.16 12.24
4032 1997   Prince Edward Island         12.32 12.99 11.56
5712 1997   Nova Scotia                  13.76 14.81 12.37
7392 1997   New Brunswick                13.63 14.77 12.08
```

Essentially, we have listed, per province and year, the average hourly wages of men, women, and men/women combined.

11.4.2 Defining the Labels and Features

Besides obtaining the data, it is essential to define clearly the objective of developing a model. In our case, we want to develop a model that reproduces the dataset. In particular, we would like to model the average hourly wage rates (combined, men and women) in function of the province and year. This means that the labels will be the average hourly wage rates:

```
y_combined = df['Combined']
y_men = df['Men']
y_women = df['Women']
```

and the features will be the province and the year:

```
import numpy as np
X = np.c_[df['Year'], df['Geography']]
```

The feature "Geography" is a categorical feature and, consequently, we need to encode it. We choose to use one-hot encoding. For this purpose, we define a transformer pipeline following Section 11.3.5:

```
from sklearn.preprocessing import OneHotEncoder
from sklearn.preprocessing import FunctionTransformer
```

```
from sklearn.pipeline import Pipeline
from sklearn.compose import ColumnTransformer
numeric_transformer = Pipeline(
   steps=[("copy", FunctionTransformer(lambda X : X))]
)
categorical_transformer = Pipeline(
   steps=[("encoder", OneHotEncoder())]
)
preprocessor = ColumnTransformer(
   transformers=[
        ("num", numeric_transformer, [0]),
        ("cat", categorical_transformer, [1])
   ]
)
```

We can verify that the transformer operates as intended:

```
>>> preprocessor.fit(X)
>>> preprocessor.transform([[1999, 'Quebec']]).toarray()
array([[1.999e+03, 0.000e+00, 0.000e+00, 0.000e+00,
0.000e+00, 0.000e+00,
        0.000e+00, 0.000e+00, 0.000e+00, 0.000e+00,
1.000e+00, 0.000e+00]])
```

The transformer indeed leaves the numerical feature "Year" unchanged, and one-hot encodes the province. Note that the one-hot-encoded vector is of length 11, which is the 11 categories we have: the 10 provinces of Canada and Canada itself.

11.4.3 Model Definition and Training

In the next step, we define the model pipeline. In this example, we want to implement linear regression. Since we have three labels to predict (the combined, men, and women average hourly wages), we need three models.

```
from sklearn.linear_model import LinearRegression
combined_wage = Pipeline(
   steps=[("preprocessor", preprocessor),
          ("regression", LinearRegression())]
)
men_wage = Pipeline(
   steps=[("preprocessor", preprocessor),
          ("regression", LinearRegression())]
)
```

```
women_wage = Pipeline(
   steps=[("preprocessor", preprocessor),
          ("regression", LinearRegression())]
)
```

All three model pipelines are identical. They start by preprocessing the data, followed by the linear regression model. They need to be trained on different datasets. For all of them, the features matrix is the same, but the labels change:

```
combined_wage.fit(X, y_combined)
men_wage.fit(X, y_men)
women_wage.fit(X, y_women)
```

11.4.4 Model Evaluation

We can now evaluate how well our models represent the dataset. As an example, we use the province of Quebec. For this, we start by creating a sub-dataset containing only the data from Quebec:

```
df_QC = df[df.Geography=='Quebec']
X_QC = np.c_[df_QC['Year'], df_QC['Geography']]
```

The model predictions can then be computed. For example, for the combined hourly wage rates:

```
y_pred_combined = combined_wage.predict(X_QC)
```

which can then be compared with the actual ones:

```
y_combined_QC = df_QC.Combined
```

Similarly, we can compare the prediction for the men's and women's average hourly wage rates. Figure 11.6a plots the predicted (straight line) and actual (points) average hourly wage rates. As the trends are linear, the model represents the data fairly well. The average hourly wage rates increase over the years. However, there is a clear difference between men and women, where women earn less than men. As a comparison, the data of the province of Prince Edward Island is shown, too, on Figure 11.6b. The rates are lower than in Quebec, but there is almost no difference between men and women.

Let us stress out that we only evaluate if the models reproduce the existing data. No attempt was made to evaluate the predictive value of the models. This important and difficult aspect is the topic of the next chapter.

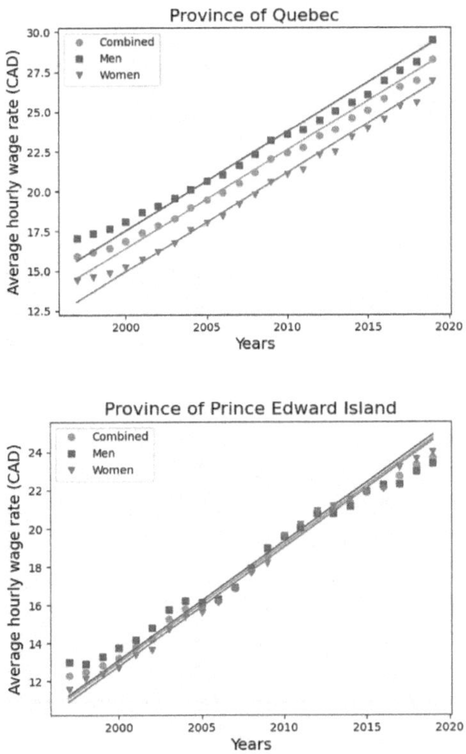

FIGURE 11.6 (a) Average predicted (–) and actual (data points) hourly wage rates (in Canadian dollars) in the province of Quebec from 1997 to 2019 for men (□), women (∇), and both combined (o). (b) Average predicted (–) and actual (data points) hourly wage rates (in Canadian dollars) in the province of Prince Edward Island from 1997 to 2019 for men (□), women (∇), and both combined (o).

NOTES

1 The term bias in *bias term* should not be confused with the concept of bias as defined in Chapter 10.
2 https://jupyter.org
3 https://open.canada.ca
4 Dataset ID: 1f14addd-e4fc-4a07–9982-ad98db07ef86; License: www.ontario.ca/page/open-government-licence-ontario
5 Comma-separated values.

12 Model Selection

The process of model selection consists of choosing, among a list of models, the one that will have the highest predictive value. As we know from Chapter 10, the predictive value is conditioned by the bias-variance trade-off. A model can have a very low bias, meaning its prediction on average will be close to the actual process, but it will, in general, present a high variance (i.e., be very sensitive to the particular data set used to train it). Such an over-fitted model will have limited predictive value due to its high variance. On the other hand, a model optimized for low variance will, in general, have a high bias (i.e., its predictions, on average, will be far from the actual process). Such a model is an under-fitted model, and it has, as an over-fitted model, little predictive value. The question is how one can optimize the training of a model in order to balance bias and variance and minimize its prediction error. This chapter will discuss this problematic. But, in the first step, we will formalize the types of errors one encounters in machine learning. As was the case with round-off and truncation errors, a deeper understanding of the nature of the errors allows us to obtain improved results. In the case of machine learning, a deeper understanding of the nature of errors will allow us to come up with a methodology, cross-validation, to optimize the training of models with the aim of maximizing their predictive value.

12.1 TYPES OF ERRORS

In Chapter 10, we discussed the bias-variance decomposition. In this context, we introduced, without defining them explicitly, some error types specific to the context of machine learning. In the present section, we will define them more formally.

12.1.1 Reference Process

To help us understand the various abstract definitions that will be introduced, a concrete example based on a reference process will be used. We use the process $h(x)$ defined by

$$h(x) = 2\sin(2\pi x) - 1 \tag{12.1}$$

with some Gaussian white noise added to it. We can define a Python function that will allow us to generate datasets based on this process.

 DOI: 10.1201/9781003262121-14

```
import numpy as np

def h(x):
    return 2*np.sin(2*np.pi*x) - 1

def generateData(m, sigma):
    x = np.linspace(0, 2, m)
    y = h(x) + np.random.normal(0, sigma, m)
    return np.c_[x], y
```

The first function defines our reference process. The second function, `generateData`, takes two parameters, which are `m`, the number of samples in the dataset, and `sigma`, the variance of the white noise added to the process. The function returns a dataset in the form of a feature matrix and corresponding labels. The dataset draws `m` samples uniformly distributed in the interval $[0, 2]$ and adds the random noise on the labels.[1]

12.1.2 Generalization Error

The first type of error we introduce is the *generalization error*, and it is defined as follows:

Definition 12.1. Generalization Error

The generalization error $\text{Err}_{\mathcal{T}}$ is the expected error of independent examples (x, y) when a particular training set $\mathcal{T}$ is used to train a model $y = y(x)$:

$$\text{Err}_{\mathcal{T}} = \mathbb{E}_x\left[\mathcal{L}\left(h(x), y(x|\mathcal{T})\right)\right] \tag{12.2}$$

where $\mathcal{L}(y, y(x))$ is the loss function used to measure the error of the predicted values $y(x)$ by the model of the true values $y = h(x)$ of the underlying process. The symbol $\mathbb{E}_x$ stands for expected value over all x.

In practice, the generalization error can not be computed as the underlying process is not known. In that sense, the generalization error can be compared with the concept of true solution discussed in the first part of this book.

The expected value should, rigorously speaking, be computed over all possible examples $(x, y = h(x))$. Practically, this isn't possible. But, we can obtain an approximation of $\text{Err}_{\mathcal{T}}$ by computing the mean over a large number of such samples. The following Python function achieves this by computing an average of over 100 samples distributed uniformly in the interval `[0,2]`.

```
def generalization_error(L, h, model):
    x = np.linspace(0, 2, 100)
    return L(h(x), model.predict(np.c_[x])).mean()
```

The function takes three arguments. Besides the underlying process h and the trained model, it needs a loss function L, which must be able to handle numpy arrays and have the signature `L(y_true, y_pred)`. Examples of such loss functions are the mean squared error MSE and the root mean squared error RMSE. They can be defined by the following two Python functions:

```
def MSE(y, y_pred):
    SE = (y-y_pred)**2
    return SE.mean()

def RMSE(y, y_pred):
    SE = (y-y_pred)**2
    return np.sqrt(SE.mean())
```

To illustrate how to compute a generalization error, we use our reference process $h(x)$ and generate a training dataset from it. As a model, we use a polynomial model of degree 5.

```
from sklearn.preprocessing import PolynomialFeatures
from sklearn.linear_model import LinearRegression
from sklearn.pipeline import make_pipeline

X_train, y_train = generateData(m=50, sigma=0.5)
polyreg = make_pipeline(
    PolynomialFeatures(degree=5,
                       include_bias=False),
    LinearRegression())
polyreg.fit(X_train, y_train)
```

We can now compute the generalization errors. If we use the mean squared error as an error measure, we obtain:

```
>>> generalization_error(MSE, h, polyreg)
0.04963113412925867
```

And for the root mean squared error:

```
>>> generalization_error(RMSE, h, polyreg)
0.2227804617314064
```

When executing these codes, it is likely to obtain different values as the training set will not be the same as the one we generated in our example. This is important to understand as the generalization error is defined for a model trained on a *specific* training set.

Figure 12.1 summarizes the different concepts. A generated training set is used to train a polynomial model of degree 5. The shaded area depicts the generalization error computed based on the root mean squared error loss function.

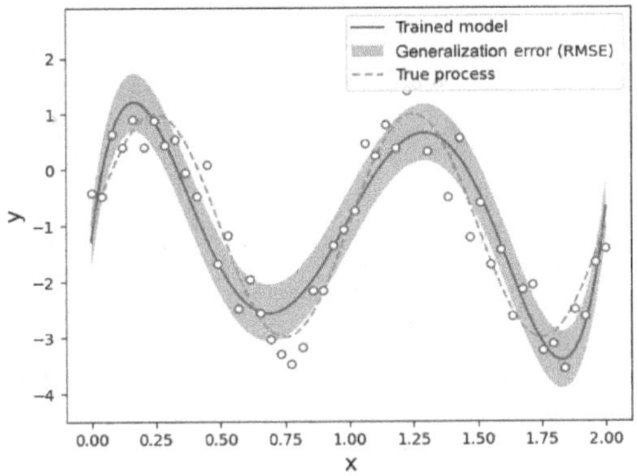

FIGURE 12.1 Polynomial model of degree 5 (–) trained with the generated dataset (o) from the process $h(x) = 2\sin(2\pi x) - 1$ (––) with Gaussian white noise with $\sigma = 0.5$. The shaded area is the generalization error computed based on the root mean squared error loss function.

12.1.3 Training Error

As the true process is not known, the generalization error can not be computed in real situations. One can, however, compute the training error, which is defined as follows:

Definition 12.2. Training Error

The training error $\overline{\text{err}}_{\mathcal{T}}$ is the average loss over all m training samples (x, y) in the set $\mathcal{T}$ used to train a model $y = y(x)$:

$$\overline{\text{err}}_{\mathcal{T}} = \frac{1}{m}\sum_{i=1}^{m}\mathcal{L}\left(y_i, y\left(x_i \mid \mathcal{T}\right)\right) \tag{12.3}$$

where $\mathcal{L}(y, y(x))$ is the loss function used to measure the error of the predicted values $y(x)$ by the model.

Even though the training error seems to be, at first sight, an approximation of the generalization error, this is unfortunately not true. There is no general relation between training and generalization error. This means that the training error alone cannot be used to estimate the test error. To obtain an intuitive understanding of this important statement, we will compute the training error for our particular example defined in Section 12.1.1 and compare it with the generalization error. The following Python function allows to compute the training error:

```
def training_error(X_train, y_train, L, model):
 return L(y_train, model.predict(X_train)).mean()
```

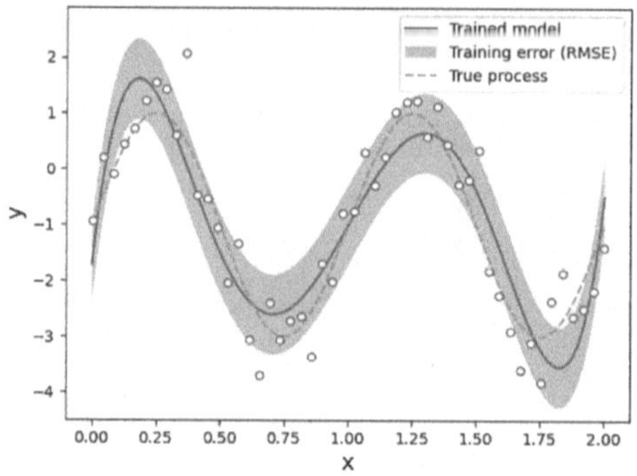

FIGURE 12.2 Polynomial model of degree 7 (–) trained with a generated dataset (o) from the process $h(x) = 2\sin(2\pi x) - 1$ (––) with Gaussian white noise with $\sigma = 0.5$. The shaded area is the training error computed based on the root mean squared error loss function.

Using our example, we can compute the training errors using the mean squared error as a loss function

```
>>> training_error(X_train, y_train, MSE, polyreg)
0.17779806714174626
```

and the root mean squared error as a loss function

```
>>> training_error(X_train, y_train, RMSE, polyreg)
0.4216610808952449
```

Note that despite the training set used to compute the training error being the same as in Section 12.1.2, the values of the training and generalization error differ.

Figure 12.2 summarizes the different concepts. A generated training set is used to train the polynomial model of degree 5. The shaded area depicts the training error computed based on the root mean squared error loss function. Note how the training error differs from the generalization error of Figure 12.1.

12.1.4 Generalization and Training Errors in Function of Model Complexity

In Section 10.3, we introduced the variance-bias trade-off. While discussing the expected loss, we learned that this quantity is minimal when both variance and bias are kept small. However, when attempting to minimize either the bias or the variance, the expected loss will grow. This important concept can now be revisited in the light of generalization errors. For this, we use our reference process from Section 12.1.1 to train polynomial models of increasing degrees and compute the generalization and training errors associated with them (Figure 12.3).

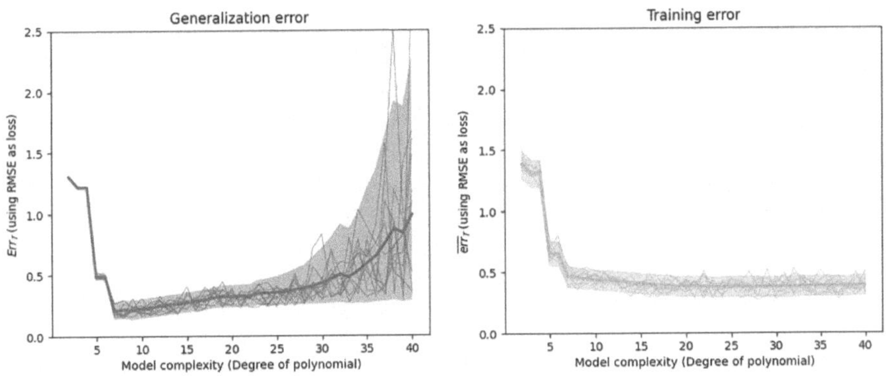

FIGURE 12.3 Generalization (left) and training errors (right) of polynomial models of increasing degrees 7 trained with datasets generated from the process $h(x) = 2\sin(2\pi x) - 1$ with Gaussian white noise with $\sigma = 0.5$. For each polynomial degree, 10'000 datasets were randomly drawn. The thin lines show the results of the first 10, and the thick line is the average over all 10'000 sets. The shaded area represents 90% of the results.

As the complexity of the trained model increases (i.e., higher polynomial degree), the generalization error decreases to a minimum for a polynormal of the seventh degree. Afterward, the generalization error increases, accelerating its growth progressively. By observing the shaded areas in Figure 12.3, where the majority of the results lie, we note that the variation of the results increases with model complexity. This observation aligns with our discussion on the bias-variance trade-off. For models with polynomials of low degrees (less than seven), the bias is high, while the variance is low, as illustrated in Figure 10.4 of Chapter 10. For degree seven, both are kept reasonably low. When increasing the order of the polynomial model further, the bias drops but the variance increases, which is why the computed generalization errors become very sensitive to the particular training set used and start to show high variability.

The behavior of the training error is very different. When increasing the order of the polynomial model, the training error decreases. In fact, as the degree of the polynomial model is increased, the fitted polynomial function represents better the particular training set, and the root mean squared error evaluated using the training set decreases. But this doesn't imply that the model generalizes well. On the contrary, as shown in the generalization error, the models are actually of lower quality. The reader is encouraged to compare this behavior with the Runge phenomenon discussed in Section 7.6. The model tends to pass by the points of the dataset, but for points not belonging to the dataset, the polynomial will not adequately represent the correct behavior. We say that the model is over-fitted. In machine learning, one says the model learned the behavior of a particular dataset (including the noise) but not the common behavior of the underlying process. As a result, the bias decreases, but the variance increases. On the other hand, when the degree of the fitted polynomial is low, the model is under-fitting as it is too simple to reproduce

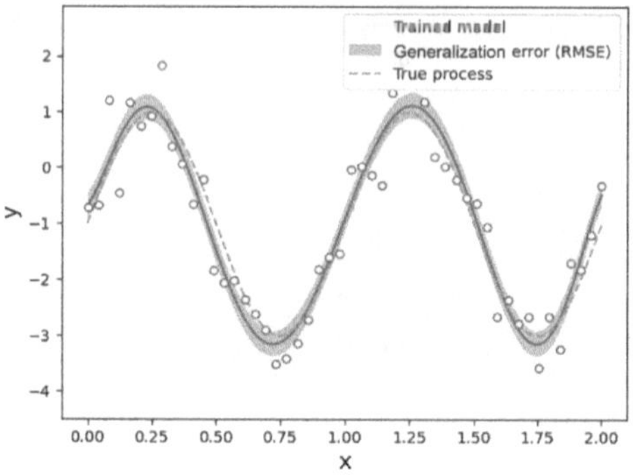

FIGURE 12.4 Example of a trained model (–) using a polynomial of degree 7 for a dataset (o) generated from the process $h(x)=2\sin(2\pi x)-1$ (––) with Gaussian white noise with $\sigma=0.5$. The actual value of $h(x)$ are essentially within the generalization error (shaded area) of the polynomial model.

the behavior of the underlying process. This results in large bias and low variance. In summary, for our example, the model using a polynomial of degree 7 results in the lowest generalization error and is the preferred model. An example of a trained model using a polynomial of degree 7 is shown in Figure 12.4.

The fact that generalization and training errors exhibit different behaviors presents a serious challenge in real situations. In real situations where the underlying process $h(x)$ is not known, only the training error can be computed. Considering our example, if we have only access to the training error, we cannot understand that for polynomial models of a degree higher than seven, the generalization error increases. There is also no way for us to tell that the optimal model is the one with a polynomial of degree seven. The tools presented in the next section, 12.2, will address this problematic.

12.2 ESTIMATING THE GENERALIZATION ERROR

As discussed in Section 12.1, the generalization error, which is the error of interest when evaluating the performance of a model, cannot be accurately estimated using the training error, which is the error we can calculate in practice. In other words, a model that performs well on the training dataset will not necessarily perform well on unseen data. In fact, as seen in Section 12.1.4, the model can even perform very poorly on new data. In this section, we will discuss some methods used in machine learning to estimate the generalization error of models.

12.2.1 Training and Test Sets

To estimate the generalization error, one needs to know the underlying process. A solution would be to develop methods to estimate the underlying process, but this is exactly what machine learning attempts to do. The problem becomes a circular process: to estimate the underlying process, we need the generalization error, but to estimate the latter, we would need the underlying process. To break this endless loop, the idea is to estimate the underlying process experimentally. A common way is to introduce two datasets: the training set and the test set.

The idea is simple. Before starting any analysis of the dataset, one first splits the set into two distinct sets: a training and a test set. The test set is put aside and not used during the development of the model, which uses only the training set. Once the model is developed, its performance is evaluated on the test set. As the data from the test set was never seen by the training process, it can be used to experimentally approximate the underlying process and, consequently, the generalization error.

Example 12.1. Training and Test Sets

Let us first generate a synthetic dataset based on the process

$$h(x) = -x^2 + x + 30 \tag{12.4}$$

we add a Gaussian white noise to $h(x)$ for the dataset generation:

```
import numpy as np

def generateData(m, sigma):
    x = np.linspace(-8, 8, m)
    y = -x**2 + x + 30 + np.random.normal(0, sigma, m)
    return np.c_[x], y

X, y = generateData(100, 10)
```

In the first step, we split the dataset into a training and a test set. SciKit Learn offers a standard function `train_test_split` to perform this task:

```
from sklearn.model_selection import train_test_split
X_train, X_test, y_train, y_test = train_test_split(
    X, y, test_size=0.2)
```

It requires two parameters to define the dataset. It offers several named parameters to control its behavior. In our example, we use `test_size=0.2`, which means 20% of the dataset will be used to generate the test set. The function `train_test_split` will generate the two sets by randomly choosing points to be attributed to each set.

Now, we use the training set to train a model. As an example, we choose a polynomial model of degree 10.

```
from sklearn.pipeline import make_pipeline
from sklearn.preprocessing import PolynomialFeatures
```

```
from sklearn.linear_model import LinearRegression
reg = make_pipeline(PolynomialFeatures(degree=10),
                    LinearRegression())
reg.fit(X_train, y_train)
```

To evaluate the performance of the model, we use the mean squared error MSE. To obtain the training error, we compute, using the SciKit learn function mean_squared_error, which calculates the MSE:

```
>>> from sklearn.metrics import mean_squared_error
>>> mean_squared_error(y_train, reg.predict(X_train))
175.99190159491442
```

When repeating the example, it is likely to obtain a different value considering the various random aspects in the various functions.

To obtain an estimation of the generalization error, we compute the MSE using the test set:

```
>>> mean_squared_error(y_test, reg.predict(X_test))
343.1419711078114
```

We can repeat this exercise for other models using polynomials of degrees 1 and 2 as an example. Table 12.1 summarizes the results.

According to the results of Table 12.1, when considering the training set only, the model with the lowest MSE is the model using a polynomial of degree 10. However, as we know from the discussion in Section 12.1, the training error is not representative of the capability of the model to generalize. A better indication is found by considering the MSE computed using the test set. In our example, the model with a polynomial of degree 2 has the lowest MSE and is, among the three models, the one which is likely to generalize best. This can be well illustrated by plotting the fitted models together with the test set (Figure 12.5). The linear model (degree 1) is too simple to reproduce the behavior of the underlying process and results in under-fitting. On the other hand, the polynomial model of degree 10 is too complex and results in over-fitting. The optimal model is the model using a quadratic polynomial as it should, considering that the underlying process is actually a quadratic polynomial.

TABLE 12.1
Metrics on Training and Test Sets of Various Models

Degree of Polynomial	MSE on Training Set	MSE of Test Set
1	593.5579284742174	770.0681608096289
2	192.1729567481866	277.21206611104884
10	175.99190159491442	343.1419711078114

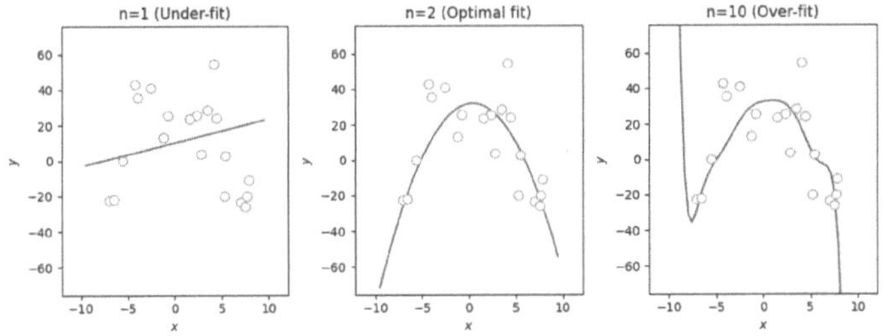

FIGURE 12.5 Example of three trained polynomial models with degree n=1, 2 and 10 (–) using a particular training set generated from the process $h(x) = -x^2 + x + 30$ with Gaussian white noise with $\sigma = 15$. The models are evaluated against the same test set (o).

Even Example 12.1 seems to suggest that evaluating the performance of a trained model on the test set will allow correct estimation of the generalization error; this is unfortunately not the case. While it is certainly significantly better to use the test set rather than the training set to evaluate the performance of a model, one must be aware that the performance evaluation still depends on the specific test set chosen. To understand this important aspect, the reader is invited to repeat Example 12.1 several times by generating different training/test sets splits and exploring various degrees for the models. The reader will discover instances where the MSE evaluated on the test set is lowest, not for the model of degree 2, but for a model of a different degree than 2. Such a result would suggest the model using a polynomial of a different degree than 2 is the best.[2] However, considering the used process (12.4), this conclusion is wrong. To avoid this difficulty, the approach of cross-validation can be used, as discussed in the next section.

12.2.2 Cross-Validation

The approach of splitting the dataset into a training and a test set, in which the data in the test set is never used during the model development and training, is one of the key methods in developing models and validating their performance. However, we know that the generalization error contains several contributions, among them the variance. The variance measures the sensitivity of the model to the set used to train it. This sensitivity is not measured by using a single training set as we did in Section 12.2.1. Cross-validation,[3] an approach well-known in statistics, addresses this problematic by further splitting the training set and training the model multiple times using different data.

There exist several variations of cross-validation, but the general idea consists of partitioning the training set into n folds, which are n distinct sub-sets of the training set of equal size (Figure 12.6). The model is trained using $n-1$ folds, and its performance is evaluated using the left-out fold by computing the error (e.g., the mean squared error) on that fold. The left-out fold is referred to as the validation set, and the $n-1$ folds used for model training are, unfortunately, often termed training set, which can lead to some confusion with the training set as

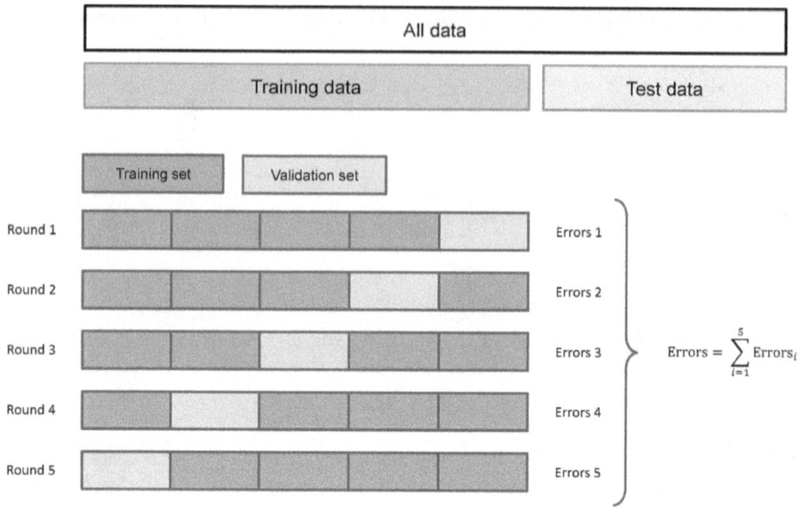

FIGURE 12.6 Principle of cross-validation. The training set is partitioned into n distinct equal-sized sub-sets called folds ($n = 5$ in the figure). The model is trained using $n-1$ folds and its performance is evaluated using the left-out fold by computing the error (e.g., the mean squared error) on that fold. After repeating the process n times, the average validation error is computed by commuting the average over the n errors.

defined in Section 12.2.1. Once the training is completed, the performance of the model is evaluated on the validation (the left-out fold) and the training set (the $n-1$ folds used for training). The performance is determined by evaluating some chosen error measure, such as the mean squared error. The performance obtained by computing the error over the training set is called the *cross-validation training error* (or simply *training error* if no confusion with Definition 12.2 can occur), and the one obtained using the validation set is called the validation error.

The calculation of one training and one validation error is called a round. In cross-validation, the process is repeated over all possible rounds, i.e., n times. After repeating the process n times, the average training and validation errors are computed by calculating the average over the n training and validation errors obtained in each round. Both the average validation error and distribution of the individual validation errors obtained in each round are valuable measures to estimate the generalization error.

The main variations in cross-validation are in the methods used to generate the folds. A simple approach consists of partitioning the training set into n folds. For example, for $n = 10$ the first 10% of the data is used for the first fold, the next 10% for the second, and so on. Alternatively, the various splits are constructed by drawing random samples from the data set instead of using consecutive drawing. This second method allows to repeat cross-validation multiple times over the same set in order to add further variations with the aim of improving the estimation of the generalization error.

Example 12.2. Cross-Validation

To illustrate cross-validation, we use a synthetic dataset generated from process (12.1), as explained in Section 12.1.1. We start by generating a dataset and splitting it into a training and test set:

```
from sklearn.model_selection import train_test_split
X, y = generateData(m=100, sigma=0.5)
X_train, X_test, y_train, y_test = train_test_split(
    X, y, test_size=0.2)
```

As a model, we use a polynomial model of degree 3:

```
from sklearn.preprocessing import PolynomialFeatures
from sklearn.linear_model import LinearRegression
from sklearn.pipeline import make_pipeline
poly_mod = make_pipeline(
    PolynomialFeatures(degree=3, include_bias=False),
    LinearRegression())
poly_mod.fit(X_train, y_train)
```

To perform cross-validation, we can use the SciKit Learn function `cross_validate`. We have to provide the model (`poly_mod`) and the data set (the training set `X_train`, `y_train`) as inputs. Several named parameters further control the behavior of the cross-validation. The parameter `cv` specifies the number of folds we want to use, and the parameter `scoring` allows us to choose which performance measure is used to evaluate the training and test errors. ScitKit Learn requires this error measure to be provided as a loss function, i.e., a function that is minimal if the performance is best. In our case, we want to use the root mean scared error RMSE as an error measure, which means that we can use the negative value of RMSE as a loss function to be provided to the parameter `scoring`. SciKit Learn provides several standard loss functions to choose from, among them `neg_root_mean_squared_error`, which is the opposite of the RMSE.

```
from sklearn.model_selection import cross_validate
scores = cross_validate(poly_mod, X_train, y_train,
                        cv=5,
                        scoring='neg_root_mean_
                        squared_error',
                        return_train_score=True)
```

The function `cross_validate` will return the computed errors if the parameter `return_train_score=True` is used. These errors are stored in a Python dictionary and can be accessed like so for the training error (when repeating the calculations, the results will differ due to the random nature of the process):

```
>>> scores['train_score']
array([-1.27362018, -1.24133025, -1.17955883,
-1.26819942, -1.27316566])
```

and for the validation error

```
>>> scores['test_score']
array([-1.22250871, -1.35740079, -1.55037426,
-1.25981604, -1.22312944])
```

As explained, these errors are loss functions and consequently negative. To obtain the RMSE, one can simply take their opposite values:

```
>>> val_err = -scores['test_score']
```

The validation error (called `test_score` in SciKit Learn), is the one that gives a rough estimation of the generalization error. In our case we have a mean validation error of about 1.3:

```
>>> val_err.mean()
1.3226458470760203
```

with a standard deviation of about 0.1:

```
>>> val_err.std()
0.12405376347323349
```

We can repeat these calculations for another model, for example, for a polynomial of degree 15. In our case, we obtained a mean validation error of about 0.58 with a standard deviation of around 0.05. The reader is invited to repeat these calculations several times. Every time a new dataset is generated and a new training/test split is performed, the results will differ. It is interesting to compare the results with Figure 12.3a, which displays the generalization error computed using the actual process (12.1). The reader will note that the validation errors, indeed, are of similar values to the generalization errors. The estimation is far from perfect, but the overall trend seems correct.

The usefulness of cross-validation is not so much in trying to estimate the generalization error, even though it can be used for that task, but rather to compare the performance of various models and select the one that is most likely to generalize best (i.e., result in the lowest generalization error). A typical use case is when a hyper-parameter has to be tuned in order to select the best model among a family of models.

Example 12.3. Hyper-Parameter Tuning With Cross-Validation

To illustrate how hyper-parameters can be tuned with cross-validation, we build on Example 12.2. Our aim is to determine which degree of the polynomial used in regression will likely result in the lowest generalization error. The idea is to compute the mean training and validation errors using cross-validation for models with polynomials of increasing degrees. Considering the random nature of the process, we will use a slightly improved version of cross-validation. Instead of building our folds by choosing consecutively the data from the training set, we will draw them randomly and repeat the process several times before taking the average of all results. In SciKit Learn, this is achieved by providing the `RepeatedKFold` function for the named parameter `cv` of `cross_validate`. The following code snipped shows how the calculations can be performed:

```
from sklearn.model_selection import RepeatedKFold
RMSE_train = []
RMSE_valid = []
for n in range(1, 18):
poly_model = make_pipeline(
    PolynomialFeatures(n),
    LinearRegression())
scores = cross_validate(
    poly_model, X_train,
    y_train,
    cv=RepeatedKFold(n_splits=10, n_repeats=50),
    scoring='neg_root_mean_squared_error',
    return_train_score=True)
RMSE_valid = np.append(
    RMSE_valid,
    -scores['test_score'].mean())
RMSE_train = np.append(
    RMSE_train,
    -scores['train_score'].mean())
```

Figure 12.7 plots the computed validation and training errors (RMSE) in the function of the degree of the trained polynomial model. The training error decreases constantly as the degree of the polynomial increases. This is expected: the higher the complexity of the model, the better it can represent the training set. However, considering the evolution of the validation error, even if the model explains better the training set, its generalization capability decreases. The validation error at first decreases with increasing degree of the model, but for degrees higher than about 8, the validation error increases. This shows that models with higher degrees are less likely to generalize well. They are, in fact, over-fitting the dataset, while polynomial models of

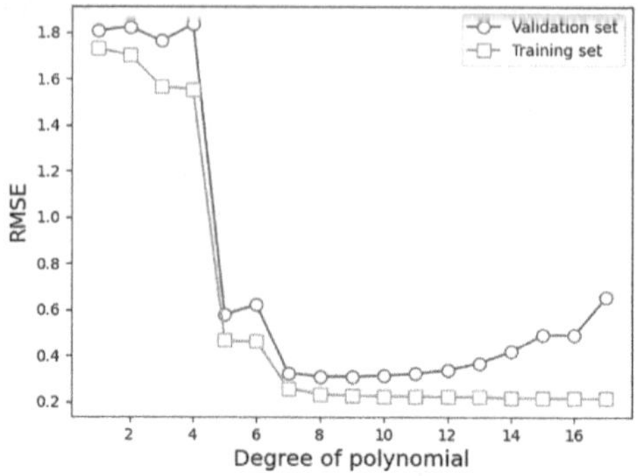

FIGURE 12.7 Validation and training errors (RMSE) in the function of the degree of the trained polynomial model. A generated dataset from the process $h(x) = 2\sin(2\pi x) - 1$ with Gaussian white noise with $\sigma = 0.5$ and 100 samples is used. Cross-validation is performed with 10 folds randomly generated 50 times.

lower degrees under-fit the dataset. The optimal value is for models of degree around 8.

The reader is invited to compare Figure 12.3 and Figure 12.7 and observe how the trend of the validation error follows the trend of the generalization error.

12.2.3 Regularization

In machine learning, hyper-parameters that are able, when tuned properly, to minimize the prediction error are said to *regularize* the model. Cross-validation is an example methodology that helps to *regularize* a model by finding the optimal values for hyper-parameters. There exist other regulation approaches. One of them consists in modifying the loss functions used in training. Famous examples are the family of elastic net regression. Elastic net regularization is used in linear regression models to modify the cost function in the training process. The regularization term is

$$\alpha\lambda\|\underline{w}\|_1 + \frac{1-\alpha}{2}\lambda\|\underline{w}\|_2^2 = \alpha\lambda\sum_{i=1}^{m}|w_i| + \frac{1-\alpha}{2}\lambda\sum_{i=1}^{m}w_i^2 \tag{12.5}$$

where α and λ are two hyper-parameters that will have to be tuned and w_i are the features weight in the linear model $y = w_0 + \underline{w}^T \cdot \underline{x}$. The regularization term is, in fact, the sum, with some coefficients, of the 1-norm and the 2-norm of the feature weights. For this reason, it is also often termed L1 or L2 regularisation, depending on the term of the 1-norm or the 2-norm is used. The linear model is then trained with the following modified cost function:

$$\mathcal{L}(\underline{w}) = \text{MSE}(\underline{w}) + \alpha\lambda\sum_{i=1}^{m}|w_i| + \frac{1-\alpha}{2}\lambda\sum_{i=1}^{m}w_i^2 \quad (12.6)$$

In other words, instead of minimizing the means squared error, we minimize a modified cost function on which the regularization term is added. Intuitively speaking, this cost function will penalize feature weights that are large. Two particular cases of elastic net regularization are the LASSO[4] regularization (for $\alpha = 1$) and the ridge regression (for $\alpha = 0$). Note that when using elastic net regularization, or any particular case of it, it is important to normalize the features first in order the regularization method to be efficient.

Finding the feature weights that minimize the cost function (12.6) can no longer be achieved using the normal equations method introduced in Chapter 6. Other algorithms that search for the minimum of a function are usually used. SciKit learn already implements specialized algorithms for LASSO regression, with the `Lasso` class, for the ridge regression, with the `Ridge` class, and for the elastic net regression with the `ElasticNet` class. Their usage is identical to other model classes we discussed so far. Example 12.4 gives an illustration of the LASSO regression.

Example 12.4. Lasso Regression

Let us consider a synthetic dataset generated based on a quadratic polynomial with some Gaussian noise added to it. The following code snippet generates the dataset with $m = 15$ samples, where the single feature `x` is randomly distributed in the interval $[-8, 8]$ and provides the feature matrix `X` and label vector `y`:

```
import numpy as np
np.random.seed(10)
m = 10
x = np.random.rand(m, 1)*20.0 - 8.0
X = np.c_[x]
y = 30 + x - x**2 + np.random.normal(0, 15, size=(m, 1))
```

Note the call to `np.random.seed`, which initializes the random number generator in order to generate the synthetic dataset in a deterministic way. This is added to allow the reader to reproduce the example.

If we attempt to train polynomial models on this dataset, we risk ending up with either an underfitted or overfitted model unless we choose the correct order for the polynomial (Figure 12.8). Cross-validation could be used, as in Example 12.3, to fine-tune the order of the model. In this example, we present the alternative approach of using LASSO regularization applied to a polynomial model of order 7.

LASSO regularization is implemented in SciKit learn with the `Lasso` class. As mentioned previously, it is important to normalize the features prior to using LASSO regression. The following pipeline implements these steps:

```
from sklearn.pipeline import make_pipeline
from sklearn.preprocessing import StandardScaler
from sklearn.preprocessing import PolynomialFeatures
from sklearn.linear_model import Lasso

reg = make_pipeline(StandardScaler(),
   PolynomialFeatures(7, include_bias=False),
   Lasso(alpha=0.1))
```

First, the features are normalized with a `StandardScaler`. Next, a `PolynomialFeatures` transformer is applied in order to create a feature matrix suitable for a polynomial model of order 7, and finally, the `Lasso` regression model is called. In the previous example, the hyper-parameter α

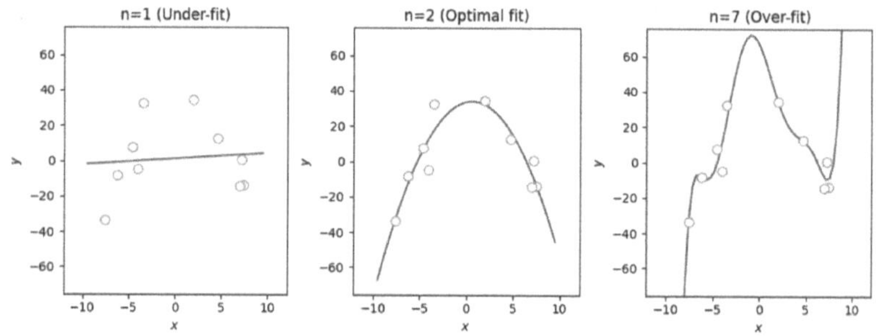

FIGURE 12.8 Three trained polynomial models (–) on a synthetic dataset (o). Depending on the choice of the order of the polynomial model, the trained model is under- or over-fitted.

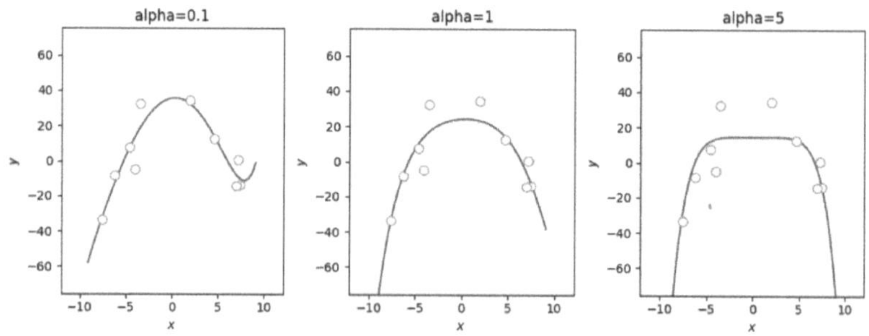

FIGURE 12.9 Three trained polynomial models of order 7 using LASSO regularisation (–) on a synthetic dataset (o). Three examples of the parameter α are shown.

of the LASSO regression is chosen to be equal to 0.1. Figure 12.9 displays two more examples for $\alpha = 1$ and $\alpha = 5$. In all cases, tshe model, contrary to the case of Figure 12.8, does not over-fit (at least not in an obvious way) the dataset.

Example 12.4 should not mislead the reader to believe that LASSO regularization does not require fine-tuning of the hyper-parameter α. Cross-validation can be used as a tool for this tuning. Note that since LASSO regression employs a numerical algorithm to minimize the cost function, there is a possibility that the algorithm may not converge. The parameter `max_iter`, for example, controls the number of iterations used in this minimizing algorithm. It may require some adjustment depending on the dataset used.

NOTES

1 The reader is invited to improve the dataset generation by further drawing the x values randomly from the interval [0, 2].

2 An example can be obtained by using the random seed `np.random.seed(10)` (prior to generating the dataset) and adding the named parameter `random_state=5` to the `train_test_split` function. In this case, the MSE evaluated on the test set with the quadratic model is about 184, whereas the MSE using a model with a polynomial of degree 7 is 183, lower than the quadratic model. This case would suggest that a polynomial model of degree 7 is better than the quadratic one.

3 Sometimes, cross-validation is referred to as rotation estimation or out-of-sample testing.

4 LASSO stands for least absolute shrinkage and selection operator.

13 Classification

In this chapter, the classification problem is discussed. Contrary to regression problems that aim to predict quantitative values, classification problems aim to predict if an entity belongs to a certain class. A typical example consists of classifying emails into spam and legitimate emails or assigning a label to an image (e.g., recognizing handwritten digits). As algorithms typically work with numbers (at least all the ones we discussed so far), it is common practice to encode classes numerically. However, some algorithms do not require this kind of encoding.

Another aspect that warrants discussion is how the performance of classification algorithms is measured. A straightforward approach is to count the fraction of correctly classified samples. However, as will be explained, this method is often too simplistic, and a more detailed analysis is required.

13.1 K-NEAREST NEIGHBORS CLASSIFICATION

As a first example of classification algorithms, we introduce the K-nearest neighbors classifier, which we will designate KNN in the rest of the book. This algorithm operates fundamentally differently than all the ones we discussed so far. There is no mathematical model with freely adjustable parameters underlying the algorithm. In fact, the algorithm uses a dataset consisting of samples and corresponding labels. A new, unseen sample is classified by comparing it with the ones from the labeled dataset. The principle is simple. The algorithm finds the K nearest neighbors of the sample to be classified in the labeled dataset.[1] It assigns the class to the unseen sample by majority voting; that is, if the majority of the K neighbors of the sample belong to class A, the sample is classified as belonging to class A. To define what constitutes a neighbor of a sample, a metric is used, such as the Euclidian distance between points. The parameter K, the number of neighbors used in the voting process, is a hyperparameter that must be adjusted to avoid over- or under-fitting.

In SciKit Learn, KNN classification is implemented with the class `KNeighborsClassifier`. Its usage follows the same pattern as all other models from SciKit Learn. To illustrate its usage and explore how the choice of the hyperparameter K affects the results, let us use an example based on a synthetic dataset. We will use the moon dataset, a collection of points in two dimensions, each of them labeled as either 0 or 1. The two classes are two half moons in the x-y plane. SciKit Learn has the method `make_moons` that allows the generating of such a dataset:

```
from sklearn.datasets import make_moons
X, y = make_moons(200, noise=0.3, random_state=1)
```


DOI: 10.1201/9781003262121-15

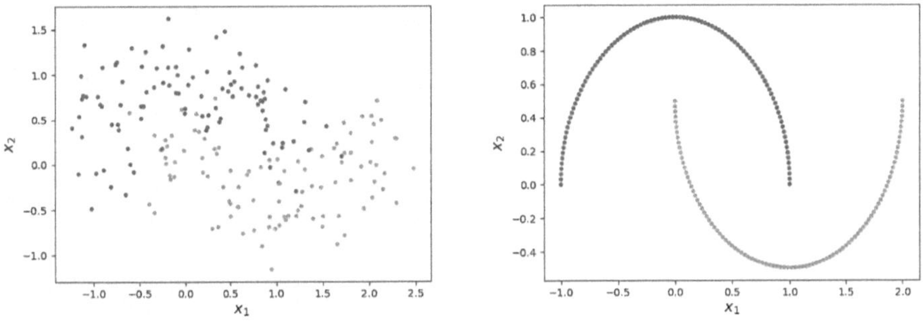

FIGURE 13.1 Moon dataset with two classes: (a) With noise level of 0.3 and (b) noise-free.

In this example, we generate 200 points that are stored in the features matrix `X` with their corresponding labels `y`. The parameter `noise` allows choosing the noise level added to the dataset. We add the seed `random_state=1`, which is useful when generating identical results as in the present example.

It is useful to plot the dataset (Figure 13.1a) by assigning different colors to each class (for reference, Figure 13.1b shows the ideal dataset when it is noise-free). The following code snippet achieves this.

```
import matplotlib.pyplot as plt
import numpy as np
colors = np.array(['tab:blue', 'tab:orange'])
plt.scatter(X[:, 0], X[:, 1], s=10, color=colors[y])
plt.xlabel('$x_1$')
plt.ylabel('$x_2$')
plt.show()
```

To deploy a KNN classifier, we use the `KNeighborsClassifier` class and fit it to the dataset the same way we did with previous algorithms discussed:

```
from sklearn.neighbors import KNeighborsClassifier
clf = KNeighborsClassifier(n_neighbors=5)
clf.fit(X, y)
```

The parameter `n_neighbors` is the value of K, i.e., the number of neighbors used in the voting process during classification. In our case, we choose to use five neighbors. We will soon discuss how the value of K has to be chosen.

To classify a sample, we use the `predict` method of the `KNeighborsClassifier` class. For example, to classify the sample $X_1 = 2$, $X_2 = -0.5$:

```
>>> clf.predict(np.c_[2, -0.5])
array([1])
```

The sample is classified as belonging to the class "1", which is the orange class in Figure 13.1. Indeed, the majority of samples in the vicinity of $X_1 = 2, X_2 = -0.5$ belong to the orange class. It is useful to define the boundary zone, which is the line in the (X_1, X_2) plane that delimits the regions of the classes defined by the KKN algorithm. Figure 13.2 delimits two zones. Any sample in the shaded zone (light green) is classified as belonging to the blue class, whereas any sample in the white zone is classified as belonging to the orange class. From the figure, we conclude qualitatively that the classification is rather good.

Obviously, the question arises: how should we choose the hyperparameter K? In fact, depending on the choice of K, the KNN classifier may under- or over-fit the dataset. Figure 13.3 displays three examples of the choice of K. For low values of K (K=1 in the figure), the dataset will be over-fitted. This is evident by observing that the boundary zones attempt to follow individual points, including the outliners. In fact, for K=1, the algorithm simply reproduces the dataset and will

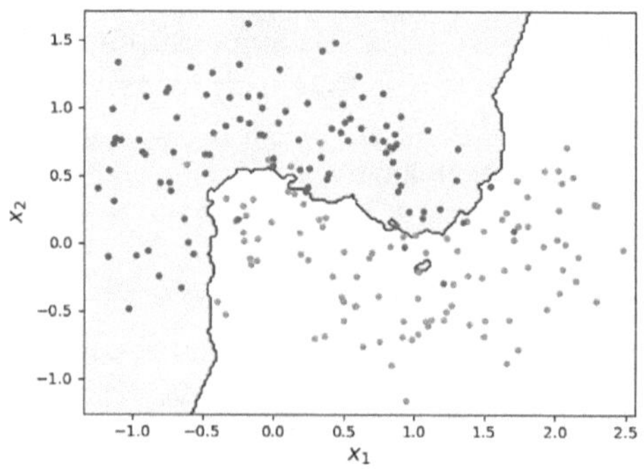

FIGURE 13.2 KNN classification of the moonset dataset from Figure 13.1 using a KKN classifier with K=5. Any sample in the shaded zone (light green) is classified as belonging to the blue class, whereas any sample in the white zone is classified as belonging to the orange class.

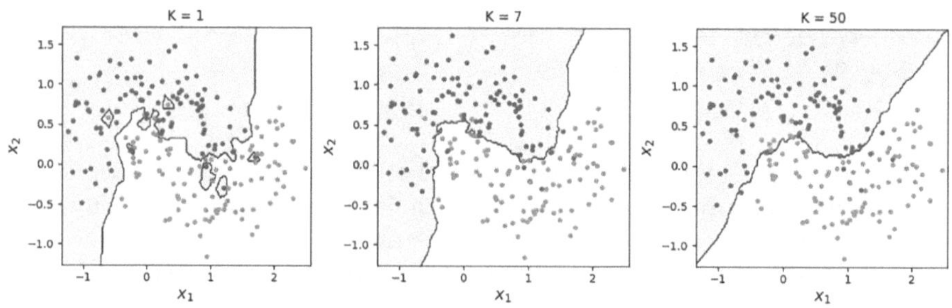

FIGURE 13.3 KNN classification of the moonset dataset from Figure 13.1 using a KKN classifier with K=1, 7, and 15 neighbors. Any sample in the shaded zone (light green) is classified as belonging to the blue class, whereas any sample in the white zone is classified as belonging to the orange class.

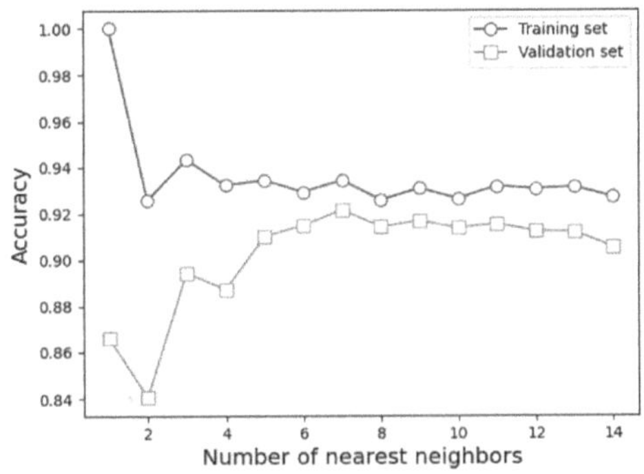

FIGURE 13.4 Cross-validation of KNN classification of the moonset dataset from Figure 13.1 using a KKN classifier with K ranging from 1 to 15. Ten random splits were used, and the process was repeated 50 times for each value of K. According to this plot, the highest predictive value of the KNN classifier is achieved for K=7.

always over-fit it. On the other hand, for large values of K (K=50 in Figure 13.3), the algorithm will under-fit the dataset. The boundary value simplifies the situation too much and tends to lose the actual behavior of the process underlying the data. For intermediate values (K=7 in the figure), the fit is optimal.

To fine tune the hyperparameter K, cross-validation can be used. For this, we define a performance measure. A common measure is accuracy, which is defined as the fraction of samples that are correctly classified. As we will learn later, this is not the only measure and not necessarily the most appropriate one. The optimal measure depends on the problem at hand. Figure 13.4 shows the results of cross-validation performed over 10 random splits repeated 50 times for each value of K. For K=1, the accuracy of the training set is equal to 1. This is normal, as KNN assigns each point in the set to the correct class. The accuracy of the validation set, on the other hand, will be lowest for K=1 as, in this case, we over-fit the data. As K increases, the validation-set accuracy improves, showing that the predictive value of the model becomes better. In our example, it reaches a maximum of around K=7 before decreasing again when the data becomes under-fitted.

13.2 CLASSIFICATION PERFORMANCE MEASURES

Accuracy, used in Section 13.1, is not necessarily an appropriate measure for the goodness of a classification algorithm. Suppose we aim to develop an algorithm that classifies people based on whether they are healthy or not. The algorithm can use vital signs such as body temperature, blood pressure, heart rate, and others to perform the classification. To test the classifier, we will randomly choose 100 people and compute the accuracy of our classifier. However, even if our classifier is not effective at determining the health status of a person, we may still obtain a high accuracy on our test set. For instance, imagine we have a very poor classifier that essentially classifies every person as healthy. When evaluating this classifier on randomly chosen people, the accuracy will be very high. This is because most people are generally healthy, and it is likely that the accuracy on the test set will be close to 100%, as most, if not all, individuals will be healthy in the test set. For this reason, better measures need to be used than the overall accuracy. One of these measures is the confusion matrix.

13.2.1 Confusion Matrix for Binary Classifiers

The confusion matrix computes not only the overall accuracy of a classifier but also the accuracy on a class-by-class basis. In the previous example, where we build a classifier that identifies health status, the confusion matrix would compute four numbers: true positive, false negative, false positive, and true negative. This is an example of a binary classifier. Binary classifiers are classifiers that have only two classes.

Table 13.1 helps us understanding the definition of the four cases that can happen in the case of a binary classifier.[2] The true positive (TP) value is the fraction of people who are in good health and who are predicted to be in good health. The false negative (FN) value is the fraction of people who are in good health but who are predicted to be not in good health. Symmetrically, the false positive (FP) value is the fraction of people who are not in good health but are predicted to be in good health. Finally, the true negative (TN) value is the fraction of people who are not in good health and who are predicted to be in not good health.

The confusion matrix provides a better measure of the goodness of the classifier compared to the overall accuracy. In the previous example, where we used a poor classifier that predicts almost everyone to be in good health, the confusion matrix would look similar to Table 13.2. Even if TP and FN are

TABLE 13.1
Confusion Matrix in Case of a Binary Classifier

	Classifier Predicts the Person to Be Healthy	Classifier Predicts the Person to Not Be Healthy
Test person is healthy	True positive (TP)	False negative (FN)
Test person is not healthy	False positive (FP)	True negative (TN)

TABLE 13.2
Confusion Matrix in Case of a Binary Classifier which Predicts Essentially Everyone to Be In Good Health

	Classifier Predicts the Person to Be Healthy	Classifier Predicts the Person to Not Be Healthy
Test person is healthy	$TP \cong 1$	$FN \cong 0$
Test person is not healthy	$FP \cong 1$	$TN \cong 0$

excellent (very high true positive predictions and nearly no false negative), the FP and TN are extremely poor (very high false positive and nearly no true negative).

One way to avoid being misled by the accuracy of the test set is to select the test set carefully. In the previous example, the selection of 100 people should not be random. On the contrary, it is best to carefully choose a balanced set that consists of about 50% healthy individuals and 50% unhealthy ones. In this way, the accuracy of the poor classifier would drop from close to 1 to close to 0.5, indicating that, indeed, the classifier isn't that accurate.

Scikit Learn has the method `confusion_matrix` to compute the confusion matrix. To illustrate its usage, we use the moon dataset and KKN classifier from Section 13.1.

First, we split our dataset into a test and a training set:

```
from sklearn.model_selection import train_test_split
X_train, X_test, y_train, y_test = train_test_split(
    X, y, test_size=0.2)
```

We then train a KNN classifier on our training dataset. We use K=7 as an example:

```
clf = KNeighborsClassifier(n_neighbors=7)
clf.fit(X_train, y_train)
```

We can now compute the confusion matrix using the predictions of the trained classifier on the test set:

```
>>> from sklearn.metrics import confusion_matrix
>>> confusion_matrix(y_test, clf.predict(X_test))
array([[18, 3],
       [1, 18]])
```

SciKit Learn computes the confusion matrix according to the definition of Figure 13.5. The question is: which class is the negative and which is the positive? For this, we can inspect the `classes_` attribute of the trained classifier. In our case, this is:

```
>>> clf.classes_
array([0, 1])
```

The first class displayed in the attribute, in our case, the "0" class (i.e., the blue class of Figure 13.1a), is the negative class. The second one, in our case, the "1" class (i.e., the orange class of Figure 13.1a), is the positive class. In other words, `confusion_matrix` will build the confusion matrix in the same order as the classes appear in the `classes_` attribute.

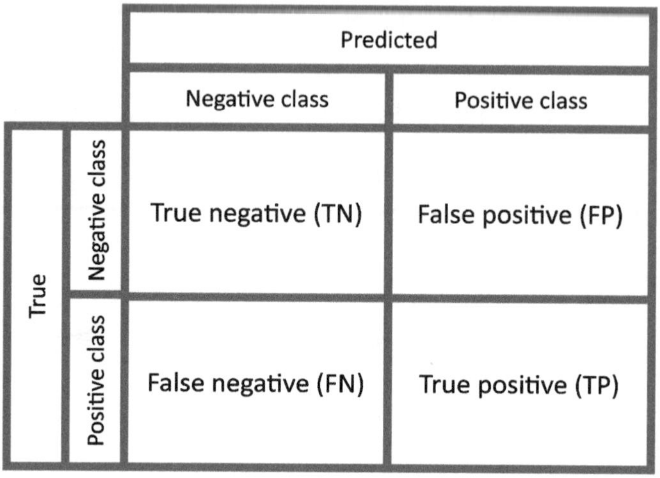

FIGURE 13.5 Definition of the confusion matrix in SciKit Learn.

For our example, this means that we had $18+3=21$ samples in the “0” class and $1+18=19$ samples in the “1” class. This is indeed the case:

```
>>> (y_test==0).sum()
21
>>> (y_test==1).sum()
19
```

Out of the 21 samples that correspond to class “0”, 18 were truly classified as “0” (i.e., about 86%), and three were wrongly classified as “1” (i.e., about 14%). Similarly, out of the 19 samples belonging to class “1”, one was falsely classified as “0” (i.e., about 5%) and 18 correctly as “1” (i.e., about 95%).

If we wish to have fractions rather than numbers, we can use the `normalize` parameter. It is possible to choose over which population the numbers are normalized. For example, if we want to use the true samples (as we did in the previous discussion), we can apply the following code snippet:

```
>>> confusion_matrix(y_test, clf.predict(X_test),
normalize='true')
array([[0.85714286, 0.14285714],
       [0.05263158, 0.94736842]])
```

13.2.2 General Confusion Matrix

The confusion matrix can be computed not only for binary classifiers but also for classifiers involving more than two classes. As an example, we consider the

FIGURE 13.6 Three types of Iris flowers: Setosa, Versicolor, and Virginica. The petals and sepals are further identified in the figure.

Image source: Wikipedia.

classification of iris flowers. The dataset we use contains three types of irises: Setosa, Versicolor, and Virginica. The aim is to classify an iris flower according to some simple features that consist of the length and the width of their petal and sepals.

The dataset can be loaded in Scikit Learn in a single command:

```
>>> from sklearn.datasets import load_iris
>>> iris = load_iris(as_frame=True)
```

The loaded Python structure contains various fields, among them a `frame` attribute that provides access to a dataframe containing the data. From this dataframe, we can build the features matrix and the labels:

```
>>> X = np.c_[iris.frame['sepal length (cm)'],
              iris.frame['sepal width (cm)'],
              iris.frame['petal length (cm)'],
              iris.frame['petal width (cm)']]
>>> y = iris.frame['target']
```

The types of irises are encoded in the form of numbers ranging from 0 to 2:

```
>>> iris.frame.target.unique()
array([0, 1, 2])
```

The corresponding labels are available in the attribute `target_names`:

```
>>> iris.target_names
array(['setosa', 'versicolor', 'virginica'],
dtype='<U10')
```

This means the Setosa iris is ladled as "0", the Versicolor as "1" and the Virginica iris as "2".

We will train a KKN classifier, and as an example, we arbitrarily choose K=5. The reader is invited to fine-tune the hyperparameter using cross-validation.

To train the model, we start by splitting the dataset into a training and a test set:

```
from sklearn.model_selection import train_test_split
X_train, X_test, y_train, y_test = train_test_split(
    X, y)
```

Even though the various features are of the same order of magnitude (they all range typically between approximately 0.1 and 8), it is in general safer to begin by normalising them to ensure all features are treated the same during the search of the neighbors. In our case, we choose a standard scalar for this and include it in the model pipeline.

```
>>> from sklearn.pipeline import make_pipeline
>>> from sklearn.preprocessing import StandardScaler
>>> from sklearn.neighbors import KNeighborsClassifier
>>> clf = make_pipeline(StandardScaler(n_neighbors=5),
                        KNeighborsClassifier())
```

Once the pipeline is defined, we can fit the model using the training dataset:

```
>>> clf.fit(X_train, y_train)
```

To evaluate the model performance, we compute the confusion matrix:

```
>>> from sklearn.metrics import confusion_matrix
>>> confusion_matrix(y_test, clf.predict(X_test))
array([[13, 0, 0],
       [0, 15, 1],
       [0, 0, 9]])
```

As we have three classes, the confusion matrix is not a 3x3 matrix. SciKit Learn has a second method, `ConfusionMatrixDisplay`, to compute the confusion matrix:

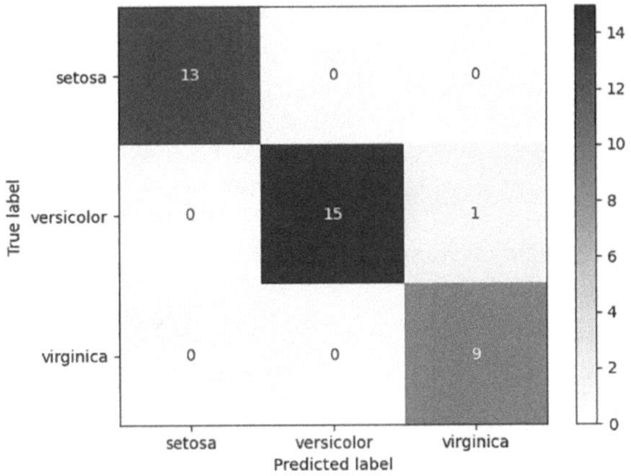

FIGURE 13.7 Confusion matrix of a KKN classifier trained on the Iris dataset.

```
>>> from sklearn.metrics import ConfusionMatrixDisplay
>>> disp = ConfusionMatrixDisplay.from_estimator(
        clf,
        X_test,
        y_test,
        display_labels=iris.target_names,
        cmap=plt.cm.Blues
    )
```

The calculations themselves are the same, but the result is displayed in a human-friendly way in the form of an image of the matrix with the relevant labels (Figure 13.7). As shown in the confusion matrix, the model performs very well on the particular test set we used. All three classes are predicted correctly, except in one case where a Versicolor iris is wrongly classified as a Virginica iris.

13.3 LOGISTIC REGRESSION

The logistic regression model, despite its name, is not a regression model but, indeed, a classifier. The classifier operates in a different way than the approaches we have learned so far. It attempts to predict the probability that a sample belongs to a certain class or not. Generically, the model can be summarized as follows for a binary classifier (two classes, which are class "0" and class "1"):

$$y = \begin{cases} 0, \text{ if } p = f\left(\underline{x};\underline{\theta}\right) < 0.5 \\ 1, \text{ if } p = f\left(\underline{x};\underline{\theta}\right) \geq 0.5 \end{cases} \tag{13.1}$$

where the function $p = f\left(\underline{x};\underline{\theta}\right)$ is the probability a particular sample $\underline{x} = \left[x_1 \quad \dots \quad x_n\right]^T$ belongs to class "1" and $\underline{\theta}$ are some parameters that can be adjusted during the training process. From the used notations, it follows that the dataset has n features.

In the case of logistic regression, the function $p = f(\underline{x};\underline{\theta})$ is written as:

$$p = \sigma\left(\underline{\theta}^T \underline{x}\right) \tag{13.2}$$

where $\sigma(t)$ is the sigmoid function

$$\sigma(t) = \frac{1}{1+e^{-t}} \tag{13.3}$$

and $\underline{\theta}^T \underline{x}$ is

$$\underline{\theta}^T \underline{x} = \theta_0 + \theta_1 x_1 + \theta_2 x_2 + \ldots + \theta_n x_n \tag{13.4}$$

which means that we have $n+1$ freely adjustable parameters. It is common to name the θ_i, for $i \geq 1$, the feature weights, and θ_0 the bias term.

Figure 13.8 displays the sigmoid function. The function presents the shape of a tilted letter “s”, which gives its name (the Greek version σ of the letter “s”). From the shape of the function, we can understand that high values of t are associated with high probabilities, whereas very negative values of t are associated with probabilities nearly equal to zero. The idea of logistic regression is to adjust the parameters $\underline{\theta}$ such that if one computes the value t, using Equation (13.4), a high value is obtained if the instance belongs to class “1”, and a very negative value is obtained if the instance belongs to class “0”.

To adjust the model parameters $\underline{\theta}$, a cost function is used such that the function is large in situations where the wrong probability $p = \sigma\left(\underline{\theta}^T \underline{x}\right)$ is predicted and low in case the probability is predicted correctly. Without entering into the

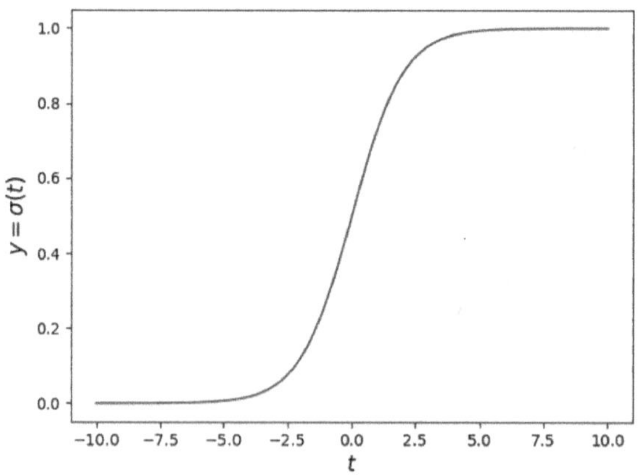

FIGURE 13.8 Sigmoid function.

mathematical justification, which would require the development of Bayesian theory, the cost function is based on the function $c(p)$ defined as follows:

$$c(p) = \begin{cases} -\log(p), \text{ if } y = 1 \\ -\log(1-p), \text{ if } y = 0 \end{cases} \tag{13.5}$$

The function $c(p)$ (Figure 13.9) indeed behaves as desired. In case the instance belongs to class "1", $c(p)$ is high for low values p (which corresponds to wrong predictions) and low for high values of p (which corresponds to the correct predictions). Symmetrically, for the case where the instance belongs to class "0", the cost is high for high values of p, which are wrong predictions and $c(p)$ is low when p is low, which are correct predictions.

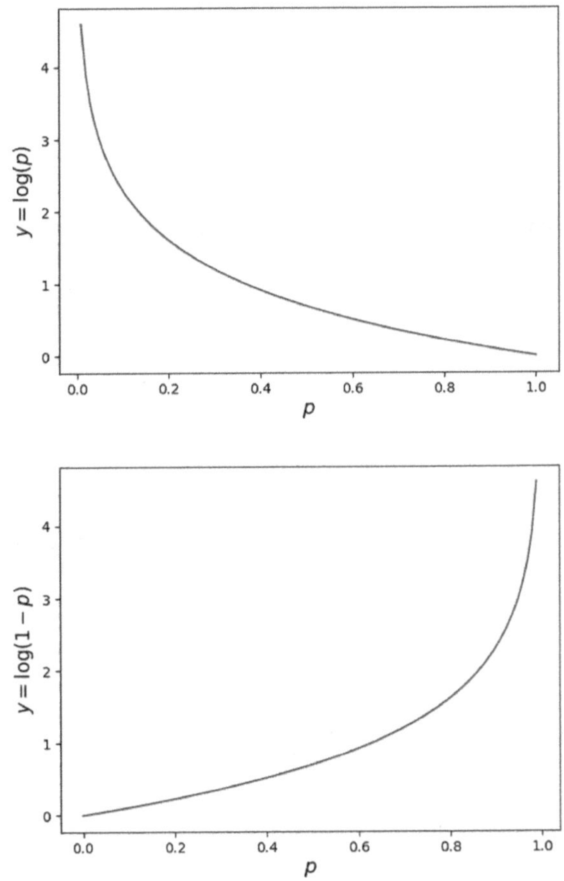

FIGURE 13.9 Cost function $c(p)$ as defined in Equation (13.5). (a) Case when the instance belongs to the class "1". (b) Case when the instance belongs to the class "0".

To train the model, the cost function needs to be evaluated for all m instances $\underline{x}^{(i)}$ in the training set. This means summing all contributions $c\left(p^{(i)}\right)$ originating from individual instances:

$$\mathcal{L}(\underline{\theta}) = -\frac{1}{m}\sum_{i=1}^{m}\left[y^{(i)}\log\left(p^{(i)}\right)+\left(1-y^{(i)}\right)\log\left(1-p^{(i)}\right)\right] \tag{13.6}$$

The resulting cost function $\mathcal{L}(\underline{\theta})$ is called the *log-loss* function. The probabilities $p^{(i)}$ for each instance $\underline{x}^{(i)}$ in the training set are computed using Equation (13.2), i.e., $p^{(i)} = \sigma\left(\underline{\theta}^T \underline{x}^{(i)}\right)$. The optimal values for the parameters $\underline{\theta}$ minimize the log-loss function $\mathcal{L}(\underline{\theta})$.

13.3.1 Binary Classifier

SciKit Learn has the class `LogisticRegression`, which implements the logistic regression classifier. As a first illustration of the class, let us attempt to develop a model that predicts the probability of passing or failing a course based on the midterm exam grade. This is an example of a binary logistic regression classifier.

Figure 13.10 displays the midterm exam grades out of a maximum of 100 points collected from a class of 90 students. If a student, at the end of the semester, did pass the course, the instance is assigned to the class "pass", and the number 1 is assigned to the corresponding label y. On the other hand, if a student fails the course, the instance is assigned to the "fail" class, and the number 0 is assigned to the corresponding label y.

We begin by defining the dataset in the form of a features matrix `X` and a target vector `y`:

```
>>> X = np.array([[54.28571429], [65.71428571],
[71.42857143], . . .])
>>> y = [1, 1, 1, . . .]
```

Only the first three instances are shown here for space limitations. Once the dataset is defined, we can split it into a training and test set:

```
from sklearn.model_selection import train_test_split
X_train, X_test, y_train, y_test = train_test_split(
    X, y)
```

and train a logistic regression classifier on the training set:

```
>>> from sklearn.linear_model import LogisticRegression
>>> clf = LogisticRegression()
>>> clf.fit(X_train, y_train)
```

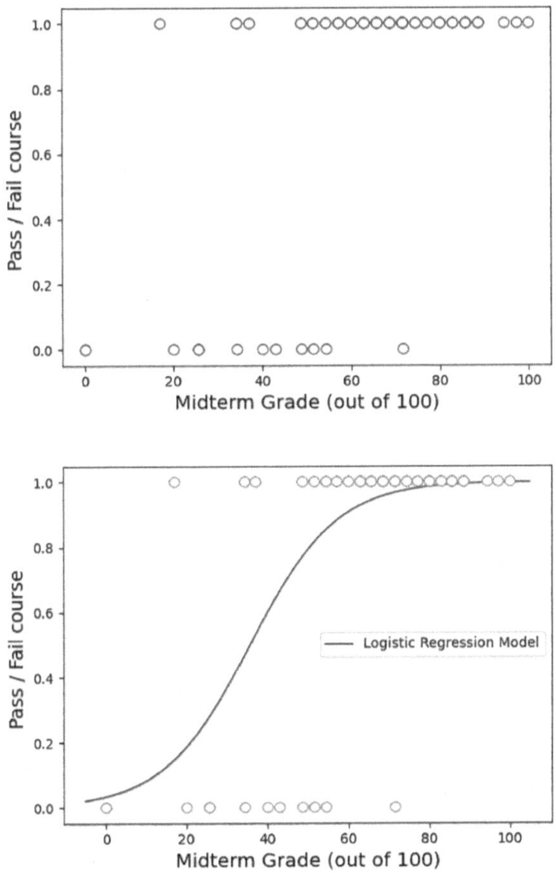

FIGURE 13.10 Probability of passing/failing a course based on the results of a midterm exam. (a) Dataset of grades of a midterm exam (out of 100) in a certain class with 90 students. (b) Probability to pass/fail the course based on a trained logistic regressor on the dataset.

Classifying a particular instance is achieved the usual way. For example, we can predict if a student passes or fails the course when the midterm exam grade is 70/100 or higher:

```
>>> clf.predict([[70]])
array([1])
```

which implies that the student will pass the course, as the model classifies the instance to the "pass" class (class "1"). Besides classifying the instance, logistic regression computes the probability of belonging to the class according to Equation (13.2). This result is available with the `predict_proba` method:

```
>>> clf.predict_proba([[70]])
array([[0.03828799, 0.96171201]])
```

which returns the probability for each class to belong to it. To know which class is which, the `classes_` attribute can be inspected:

```
>>> clf.classes_
array([0, 1])
```

The probabilities returned by `predict_proba` are in the same order as the `classes_` attribute. This means that in our example, there is a probability of about 4% to fail the course and a probability of about 96% to pass the course. Figure 13.10b shows the predicted probabilities over the full range of possible outcomes of the midterm grade.

As for the other classifiers we discussed so far, the confusion matrix computed from the test set can be used to obtain a rough idea of the predictive value of the classifier. In our case:

```
>>> from sklearn.metrics import confusion_matrix
>>> confusion_matrix(y_test, clf.predict(X_test))
array ([[1, 1],
        [1, 20]])
```

This means that, for our example, out of 21 students who passed the course, 20 are predicted correctly to pass the course, whereas one is predicted wrongly to fail the course. However, the test set we used had only two students who failed the course. This shows that the test set we used is not well-balanced, and the results should be interpreted with caution. In fact, the problem is already in the original dataset. As the majority of students pass the course (only 10 students out of 90 failed the course), the dataset is biased towards passing the course. The trained classifier will likely be biased towards predicting a pass. If a more reliable model is to be developed, more data should be collected, and a more balanced dataset should be used containing a similar number of examples with passing and failing cases.

13.3.2 Multiple Classes

Logistic regression can be extended to several classes, which is often termed *softmax regression* or *multinomial logistic regression*. Imagine a situation where we have K classes. In multinomial logistic regression, the classes are one-hot encoded, as explained in Section 11.3.4. This means the labels will be of vector $\underline{y}^{(i)}$ for the instance $\underline{x}^{(i)}$ in the dataset such that the component k will be $y_k^{(i)} = 1$ and all others will be equal to zero.

In the next step, for each class $k = 1,..,K$, we define a set of model parameters $\underline{\theta}^{(k)}$. For each instance $\underline{x}$ in the training set, we assign a score s_k according to:

$$s_k(\underline{x}) = \left[\underline{\theta}^{(k)}\right]^T \underline{x} \tag{13.7}$$

This is a generalization of $t = \underline{\theta}^T \underline{x}$ that we used so far. The probability p_k of the instance $\underline{x}$ to belong to the class k is then computed as:

$$p_k(\underline{x}) = \frac{e^{s_k(\underline{x})}}{\sum_{i=1}^{K} e^{s_k(\underline{x})}} \tag{13.8}$$

which is called the *softmax* function.

To train a multinomial logistic regression classifier, a cost function is required. A popular choice is the *cross-entropy loss function,* defined as

$$\mathcal{L}\left(\underline{\theta}^{(1)}, \underline{\theta}^{(2)}, \ldots, \underline{\theta}^{(K)}\right) = -\frac{1}{m}\sum_{i=1}^{m}\sum_{k=1}^{K} y_k^{(i)} \log\left(p_k^{(i)}\right) \tag{13.9}$$

where m is the size of the training set. Note that in the case of $K = 2$ classes, the cross-entropy loss function (13.9) is the same as the log-loss function (13.6). The optimal model parameters $\underline{\theta}^{(1)}, \underline{\theta}^{(2)}, \ldots, \underline{\theta}^{(K)}$ minimize the cross-entropy loss function.

The `LogisticRegression` class of SciKit Learn implements the multinomial logistic regression. To illustrate its utilization, we use the Iris dataset presented in Section 13.2.2. After splitting the dataset into a training and a test set, a multinomial logistic regression classifier can be trained:

```
>>> clf = make_pipeline(StandardScaler(),
                        LogisticRegression())
>>> clf.fit(X_train, y_train)
```

Similarly to the approach taken for a KNN classifier, we first scale all features. Logistic regression is, like the KNN algorithm, sensitive to differences in the scales of the features, and a normalization of the features should always be implemented. Note as well that the `LogisticRegression` class can work on labels that are not one-hot encoded (as is the case in the example with the Iris dataset). Predictions can be performed the usual way:

```
>>> clf.predict([[7.5, 3.2, 4.1, 1.4]])
array([1])
```

This classifies the example as part of class "1", which is the Versicolor Iris. Contrary to the KKN classifier, we can also associate the probabilities to the classification:

```
>>> clf.predict_proba([[7.5, 3.2, 4.1, 1.4]])
array([[0.00553332, 0.89955661, 0.09491007]])
```

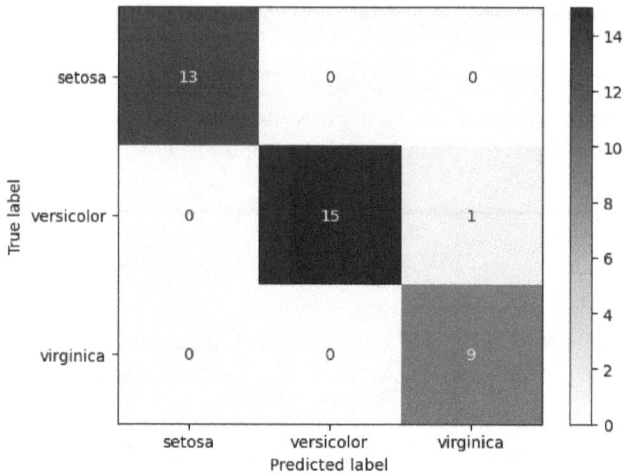

FIGURE 13.11 Confusion matrix of a trained multinomial logistic regression classifier on the Iris dataset.

The trained classifier is quite confident that the example is a Versicolor Iris, with a probability of about 90%, compared to about 0.5% for the Iris Setosa and 9.5% for the Iris Virginica. For other examples, the classifier may be less confident, like in this case:

```
>>> clf.predict_proba([[6.2, 2.8, 5, 1.5]])
array([[0.00417052, 0.57164668, 0.4241828]])
```

where the model hesitates between Iris Setosa (with about 57% probability) and Iris Virginica (with about 42% probability).

The confusion matrix computed with the test set allows us to have an idea about the predictive value of the model (Figure 13.11). In our example, it is identical to the one obtained with KNN regression (see Figure 13.7).

NOTES

1 The reader may have noted that there is a potential problem in this approach. What if the different features involved have very different scales? For example, one feature would be the number of rooms in a house, and another feature would be the total surface of the land on which the house is built. Of course, the numbers of rooms never differ too much (by a few rooms only, as most houses have a similar number of rooms). But, the surface of the land may differ significantly. The two features wouldn't be treated the same way by the algorithm when looking for the nearest neighbors. We will come back to this problem in Section 0.

2 As it is arbitrary which class is considered as the "positive" and which one is the "negative" class, the definition of these quantities is perfectly symmetric. Usually, one defines as the "positive" class the class that is the most important from a prediction point of view.

14 Tree-Based Algorithms

In this chapter, we explore some algorithms that can handle non-linear data more efficiently than the algorithms discussed previously. These algorithms, based on decision trees, can be used for regression and classification problems equally. An interesting aspect we will learn in this chapter, besides these new methods, is how algorithms with relatively low predictive value can be, under certain assumptions, modified to increase their predictive values. This technique, known in general as boosting, is based on the idea that variance can be partially removed from the error source by averaging a large number of predictions coming from algorithms trained under slightly different conditions.

14.1 REGRESSION TREES

To start our journey on tree algorithms, we present the idea behind regression trees. These algorithms aim to provide quantitative predictions of unseen data, as does, for example, linear regression. Besides the quantitative prediction, a trained regression tree will also provide a decision tree like the one shown in Figure 14.1a. Reading a decision tree is simple. Each box provides a condition (first row), the number of samples in the set that was used to determine this condition (second row), and a numerical value, which is the predicted outcome that is associated with any sample in that set (last row). One commonly calls each box a *leaf* of the tree.

Let us walk through an example. Say we would like to predict the label associated with the sample $x = 0.3$. We start with the first box, the one at the top of the tree, in Figure 14.1a. This box means that, if no further operation is to be performed, the predicted value y is -1.08. If, however, we want to perform at least one step in the tree, then the condition, which is to decide if the sample is smaller than or equal to or larger than 0.51, is to be used. In our case, the sample is smaller than 0.51. The convention in plotted decision trees is that if the condition is satisfied, then the left branch is to be followed. If the condition is not satisfied, we proceed with the right branch. In our case, since the condition is satisfied, we follow the left branch, leading us to the leftmost box on the second level. This means that the label is now predicted as $y = -0.33$. To move to the next level, the condition displayed in the box is applied. As $x = 0.3$ is larger than 0.07, the right branch must be followed, which leads us to the prediction $y = -0.24$. Similarly, a label y can be predicted for any sample x. Figure 14.1b displays the predictions of the tree from Figure 14.1a for samples x ranging from 0 to 1.

DOI: 10.1201/9781003262121-16

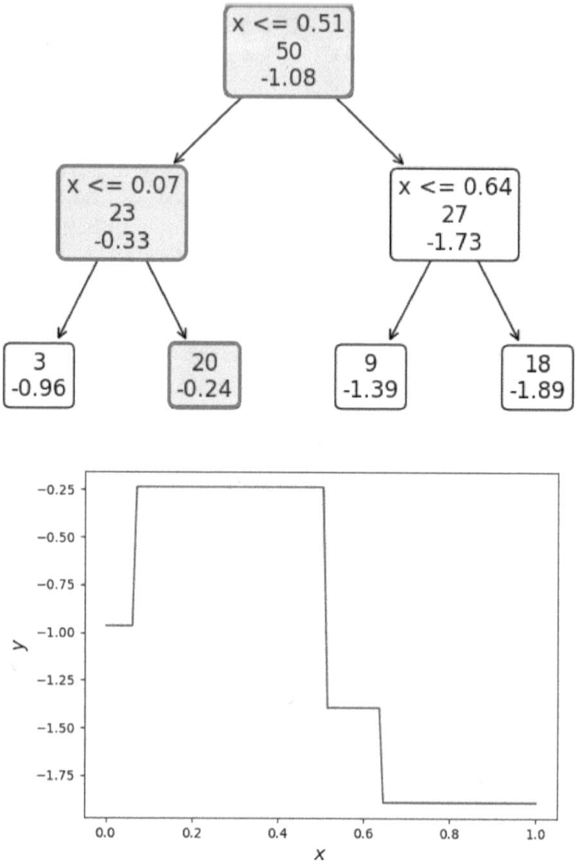

FIGURE 14.1 (a) Typical example of a trained regression tree. The highlighted boxes are the path to follow within the tree to predict the label of the sample $x = 0.3$. (b) Predictions of the tree for samples x ranging from 0 to 1.

14.1.1 CART Algorithm

Various algorithms exist to train regression trees. The one we present here is the Classification and Regression Trees (CART) algorithm. In this section, we present the idea of the algorithm for the case of regression.

The principle is relatively simple. First, a loss function must be chosen to allow measuring the quality of the training. As an example, we can think of the squared error SE. The algorithm progressively splits the training set into regions that correspond to the conditions used in the decision tree. At first, we have a single region R, the training set. The CART algorithm splits this region into two regions by choosing a cut value s within R. Samples lower than or equal to the cut value belong to the left region R_L, and samples larger than the cut value to the right region R_R. For each region, the algorithm assigns a label that consists of the average of the labels of all samples within that region. For example, for the left region:

$$y_L = \frac{1}{m_L} \sum_{i \in R_L} y_i \tag{14.1}$$

where the notation $i \in R_L$ designates the indexes of the samples that are in the left region, such that $x_i \leq s$, and m_L is the number of samples in the left region.

Then, the algorithm computes the loss function over each region. In case the SE is to be used as the overall metrics for the algorithm, the CART algorithm will compute the SE for the left region:

$$\mathrm{SE}_{\mathrm{L}} = \sum_{i \in R_L} (y_i - y_L)^2 \tag{14.2}$$

The overall loss function is then the weighted means of the loss over the left and the right regions:

$$\mathcal{L} = \frac{m_L \mathrm{SE}_L + m_R \mathrm{SE}_R}{m_L + m_R} \tag{14.3}$$

The splits are repeated for different cut values s. The cut value chosen is the one which results in the smallest total loss function. The process is then repeated for each new sub-region until some stopping criteria are reached (for example, stopping after n levels).

Let us illustrate the mechanism of the CART algorithm with an example. For this, we generate a synthetic dataset with $m = 50$ samples (a sine function with some noise added on top of it):[1]

```
import numpy as np
m = 50
sigma = 0.25
x = np.sort(np.random.rand(50))
y = -1 + np.sin(2*np.pi*x) + np.random.normal(0, sigma, m)
```

Note that we sorted the x values to ease the application of the CART algorithm. This will be shown further on. We will also see later that this approach is not needed when using decision trees in SciKit Learn.

Let us illustrate how the CART algorithm splits the dataset into two regions to minimize the loss function. The following code snippet computes the loss function over a series of splits of the training set.

```
L = []
for s in x[1:len(x)-1]:
```

```
# sets
R1 = np.where(x <= s)
R2 = np.where(x > s)
# sizes of sets
m1 = len(x[R1])
m2 = len(x[R2])
m = m1 + m2
# predicted target values of the sets
y1 = y[R1].mean()
y2 = y[R2].mean()
# squared errors and loss function
SE1 = ((y[R1]-y1)**2).sum()
SE2 = ((y[R2]-y2)**2).sum()
L = np.append(L, m1/m*SE1+m2/m*SE2)
```

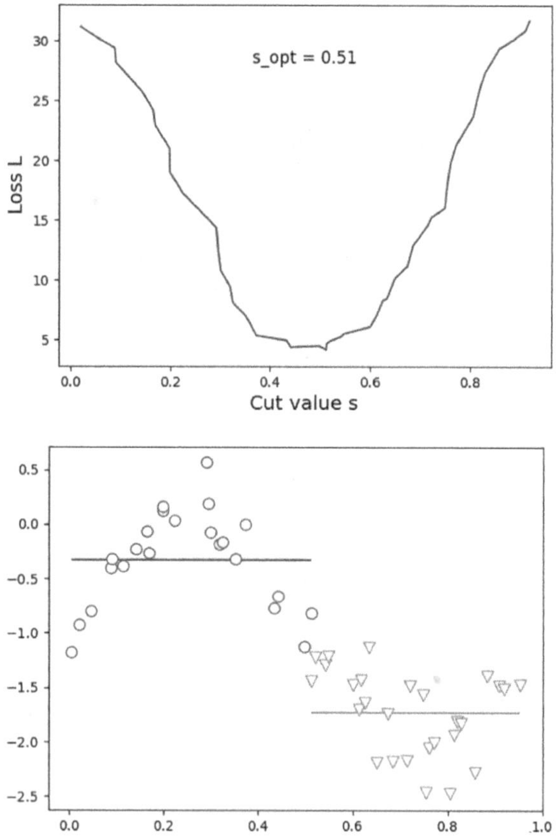

FIGURE 14.2 (a) Loss function computed over a range of cut values s. (b) Predictions (–) by the CART algorithm using the cut value $s_{opt} = 0.51$. The samples of the left region (o) and right region (∇) of the dataset are displayed too.

The optimal cut value s_{opt} is the cut value that results in the smallest loss function. From Figure 14.2a, this is $s_{opt} = 0.51$. Once the optimal cut value is found, the CART algorithm can make predictions using the condition $x \leq 0.51$ as illustrated in Figure 14.2b. For samples lower than 0.51, this results in the prediction of $y = -0.33$, which is the mean of all samples in the training set within the left region (the orange circles in Figure 14.2b). For a sample larger than 0.51, the prediction is y=-1.73, which is the mean of all samples from the training set within the right region (the blue triangles in Figure 14.2b).

The algorithm proceeds the same way for the next level, where, again, it splits each region into a left and a right region. The result of this process is the decision tree from Figure 14.1a.

SciKit Learn provides a pre-programmed CART algorithm with its `DecisionTreeRegressor` class. Its deployment follows the same pattern as other machine learning models we have discussed so far:

```
from sklearn.tree import DecisionTreeRegressor
X = np.c_[x]
reg = DecisionTreeRegressor(max_depth=2)
reg.fit(X, y)
```

As an input for the `fit` method, the usual features matrix is required, which is built with `X = np.c_[x]`. The `DecisionTreeRegressor` class has several arguments for controlling its behavior. The one we use here is `max_depth=2`, which tells SciKit Learn to construct a decision tree with two levels. Predictions of labels are achieved the usual way. For example, to predict the label of $x = 0.3$:

```
>>>> reg.predict([[0.3]])
array([-0.23529669])
```

SciKit Learn provides the function `plot_tree` to obtain a human-friendly graph of the decision tree. Various arguments can be used to control the output, and the reader is invited to consult the online documentation.[2] The following code snippet produces the tree from Figure 14.1a:

```
from sklearn.tree import plot_tree
plot_tree(reg, feature_names=['x'], label='none',
          impurity=False, rounded=True,
          precision=2, fontsize=12)
plt.show()
```

14.1.2 Regression Trees for Several Features

The CART algorithm, explained in Section 14.1.1, can be extended for datasets containing more than one feature. However, the number of splits to be scanned

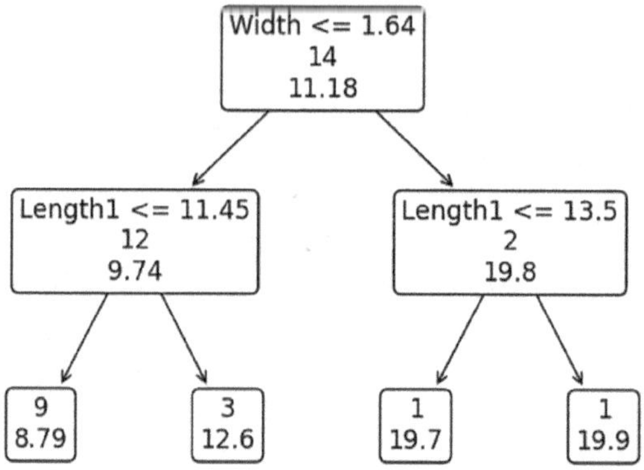

FIGURE 14.3 Regression tree trained on the fish dataset of Example 11.1 from Chapter 11.

increases exponentially with the number of features. This is due to the increasing number of possible combinations of splits on the various features. Computation time becomes excessively large and not practical. Generally, a simplified scanning approach is used, where the optimal cutting value for each feature is determined first. This approach splits regions according to one feature at a time rather than combinations of all of them. The split resulting in the lowest loss among all features is then kept.

As an illustration, let us use the fish dataset from Example 11.1 in Chapter 11. We consider the subset of smelt fishes and use three features: Length1, Height, and Width. As a label, we use the weight of the fish. Figure 14.3 shows the decision tree obtained when training the dataset down to a level of two. We note how, in each leaf, only one feature is used to create the split. For instance, at the first level, the feature used is Width. Moving to the second level, both leaves utilize the feature Level1. It's worth noting that this occurrence of both leaves using the same feature is incidental and not necessarily typical.

Figure 14.3 highlights how tree-based methods are simple and use easy-to-understand statements such as: "First consider the width of the fish. If it is lower than or equal to 1.64 % of Length3, the fish weight will be about 9.74 kg. If higher than 1.64%, the fish weight will be around 19.8 kg". Or a finer subdivision such as "A fish having a width higher than 1.64 % of Length3 and a Length1 lower than 13.5 cm will typically weigh 19.7 kg".

Furthermore, tree-based methods showcase the importance of the various features (an aspect we will come back to later in Section 14.3.2). In this example, the width of the fish appears to be the primary criterion, followed by Length1, in determining the fish's weight.

14.1.3 Model Selection

As with any machine learning algorithm, tree-based algorithms need to carefully choose the values of their hyper-parameters to avoid over- or under-fitting. As an example, consider the dataset from Section 14.1.1. Let us split the dataset into a training and test set the usual way, and let us train a regression tree by using the default hyper-parameters of SciKit Learn:

```
from sklearn.model_selection import train_test_split
X_train, X_test, y_train, y_test = train_test_split(
    X, y, test_size=0.1)
reg = DecisionTreeRegressor()
reg.fit(X_train, y_train)
```

Figure 14.4 shows that the model is overfitted. Indeed, we can see that the model follows almost perfectly the points from the training set, which means it learned the noise of the data and will most likely be sensitive to the particular training set (i.e., will present a high variance).

To avoid over-fitting (or under-fitting) the hyper-parameters have to be fine-tuned. Reading the documentation of the `DecisionTreeRegressor` class[3] will reveal that several parameters can be used to control the behavior of the CART algorithm. For example, the parameter `min_samples_leaf` defines the minimum number of samples required for a split region. The effect of this parameter is illustrated in Figure 14.5. When only a few samples per leaf are allowed, the tree will over-fit. It reproduces the dataset almost perfectly, and consequently learns the noise inside the dataset and will consequently be sensitive to the specific set used to train the tree (i.e., high variance, low bias). This is due to the fact that the algorithm grows a tree until reaching splits containing only a few numbers of samples. In the extreme case of `min_samples_leaf=1`, as in the first example of Figure 14.5, each split contains exactly one sample, and consequently, the label predicted will be the sample label of that specific sample. The dataset is reproduced exactly (as was the case with a KNN classifier when using K=1).

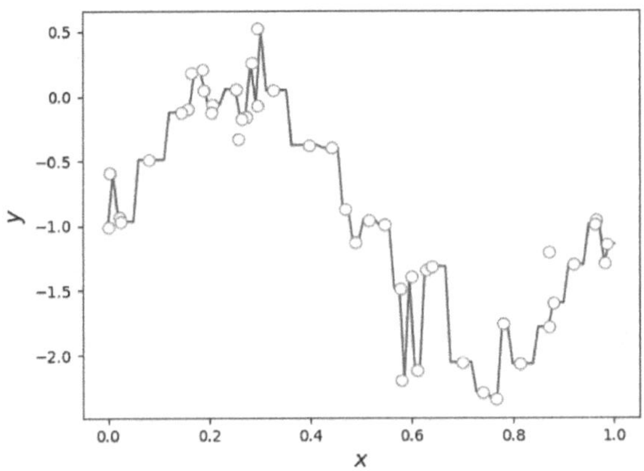

FIGURE 14.4 Predictions (–) of a regression tree trained on a dataset (o) using the default hyper-parameters from SciKit learn.

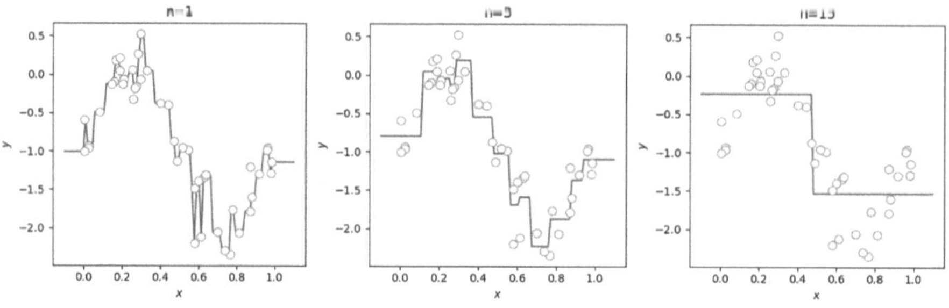

FIGURE 14.5 Three examples of trained regression trees on a dataset (o). The first one was trained with `min_samples_leaf=1`, the second one with `min_samples_leaf=3`, and the third one with `min_samples_leaf=15`. The first model is over-fitted, and the third one is under-fitted. The intermediate one achieves a trade-off between variance and bias and results in an optimal fit.

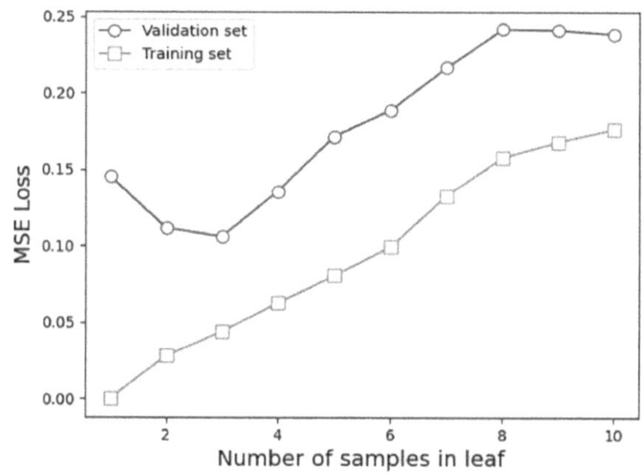

FIGURE 14.6 Cross-validation results for a decision tree regressor using the synthetic dataset of Section 14.1.1. The minimal number of samples in each leaf is used as a hyper-parameter to be tuned. In the present case, using at least three samples per leaf results in a model with the highest potential for generalization.

On the other hand, if `min_samples_leaf` is rather large, the tree will under-fit. Indeed, for a large number of samples in each split, the tree does not grow many levels of leaves. The model simply predicts a very rough, averaged trend of the data. An optimal fit is obtained for intermediate values. To determine that optimal value, cross-validation can be used, as we discussed in previous chapters. Figure 14.6 plots the validation and training error obtained by cross-validation (MSE is used as a metric). In the present case, using at least three samples per leaf results in a model with the highest potential for generalization.

As previously mentioned, other hyper-parameters are available besides the minimal number of samples in each leaf. Another example of a hyper-parameter sample is `max_depth`, which limits the number of levels the tree can grow. Letting the tree grow as deep as desired will result in over-fitting as, eventually, the tree creates splits containing only a single sample from the dataset and consequently reproduces exactly the set. On the other hand, limiting the number of levels excessively results in under-fitting as the tree does not have enough freedom to learn the trends from the dataset. In general, limiting the hyper-parameters of type `max_*` (as the `max_dep`) or increasing the parameters of type `min_*` (as the `min_samples_leaf`) avoids over-fitting. Proceeding the opposite way (i.e., increasing the `max_*` or limiting the `min_*` parameters) results in under-fitting.

14.2 CLASSIFICATION TREES

14.2.1 CART Algorithm for Classification

The approach used to develop decision trees for regression problems can be adapted to the case of classification. As in regression, the features space is split according to some cut values. In each region, the algorithm predicts the class of a sample to be classified as the class that is the most present in that region. The optimal values used to split the regions are determined by minimizing a loss function $\mathcal{L}$ as was the case for regression. The difference between regression and classification is the metrics used to build the loss function $\mathcal{L}$. The CART algorithm constructs the loss function in a general way according to:

$$\mathcal{L} = \frac{m_L}{m_L + m_R} H_L + \frac{m_R}{m_L + m_R} H_R \tag{14.4}$$

where, as in Section 14.1.1, the index L stands for left and the index R for right. The function H is a metric used to judge the error of the model. In regression, we used the function H as an example for the squared error SE. In classification, different metrics exist. We present two of them. The first one is the information entropy, defined as:

$$H = -\sum_k p_k \log_2 (p_k) \tag{14.5}$$

where p_k is the proportion of samples of class k within the region over which the information entropy is computed. The second metric is the Gini impurity, defined as:

$$G = 1 - \sum_k p_k^2 \tag{14.6}$$

It is worth noting that, for both metrics, they will be minimal (in fact, equal to zero) if all samples in the region are of the same class k (in which case $p_i = \delta_{ik}$).

On the other hand, if all samples are very diverse (i.e., all p_k are similar), the metrics will be large. The information entropy is equal to $\log_2(1/K)$ if K is the total number of classes. The Gini impurity, G, tends towards $1-\frac{\pi^2}{6}$ for a large number of classes. In other words, the CART algorithm tries to define regions that are as pure as possible, i.e., ideally containing only one class.

As an example, let us train a classification tree on the moon dataset introduced in Section 13.1 of Chapter 13. We generate $m = 200$ samples with a noise level of 0.1. About half of the samples belong to the orange class, whereas the other half belong to the blue class. Figure 14.7a shows the trained classification tree, and Figure 14.7b presents the corresponding predictions. For training, we use the default values of hyper-parameters in SciKit Learn.

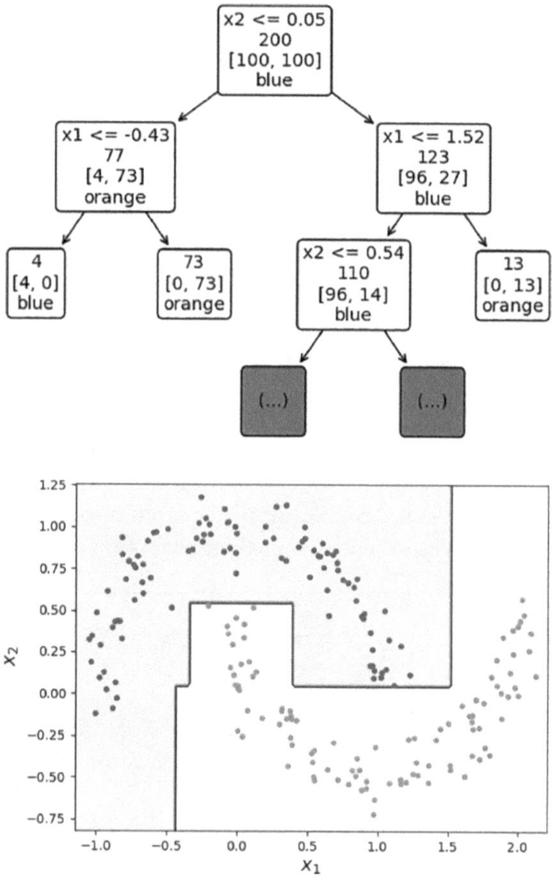

FIGURE 14.7 (a) Classification tree trained on the moon dataset. (b) Predictions by the trained classifier (samples in the shaded area are classified by the tree as belonging to the blue category) together with the dataset (o) used to train the tree.

14.2.2 Hyper-Parameters Fine-Tuning

As with regression trees, hyperparameters require tuning. The effective classification boundary produced by the tree in Figure 14.7 is accidental, as we were lucky with our choice of hyper-parameters. Figure 14.8, showing three classification trees trained with different hyper-parameters, illustrates this fact. As with regression trees, low values of `min_samples_leaf` result in over-fitted data, and large values result in under-fitted data. This is observed in the first figure, where the decision boundaries isolate almost every single data point. On the other hand, when `min_samples_leaf` is large, the decision boundary is too simple and does not capture the general trend of the moon dataset.

Cross-validation can help to estimate the optimal values of hyper-parameters. The validation and training loss can be measured using accuracy, which represents the fraction of samples correctly classified. Technically, this is not a loss function

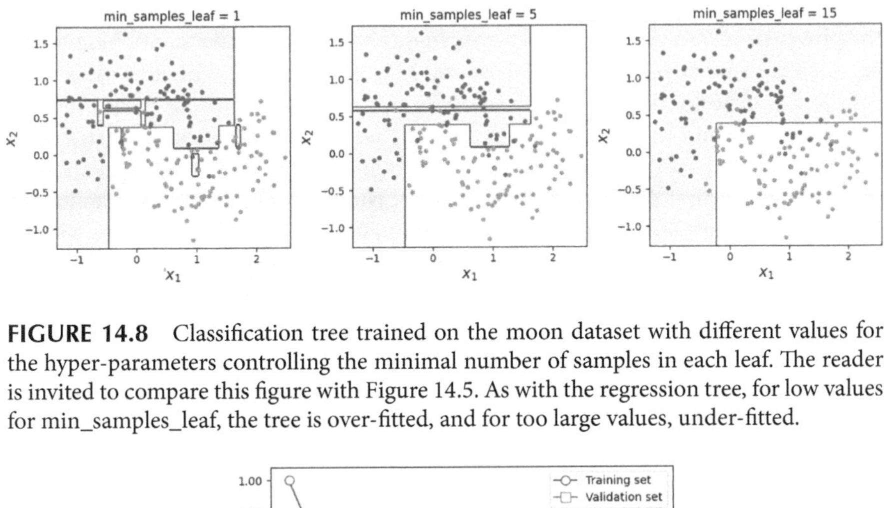

FIGURE 14.8 Classification tree trained on the moon dataset with different values for the hyper-parameters controlling the minimal number of samples in each leaf. The reader is invited to compare this figure with Figure 14.5. As with the regression tree, for low values for min_samples_leaf, the tree is over-fitted, and for too large values, under-fitted.

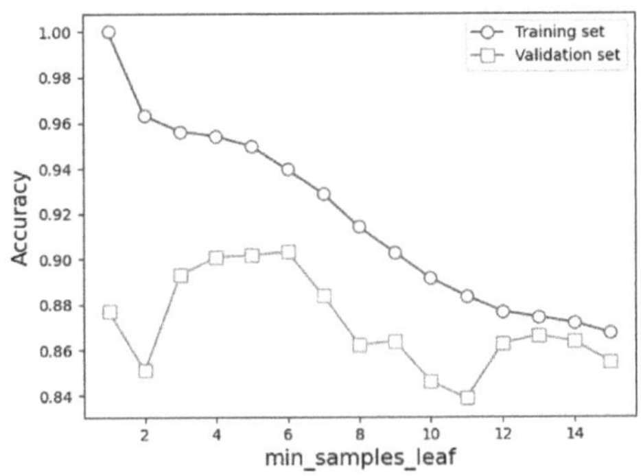

FIGURE 14.9 Cross-validation results for a decision tree classifier using the moon dataset ($m = 200$ samples and a noise level of 0.3). The minimal number of samples in each leaf is used as a hyper-parameter to be tuned. In the present case, using at least six samples per leaf results in a model with the highest potential for generalization.

as the model performs best when accuracy is highest, not smallest, as is the case with typical loss functions. Cross-validation can be performed as follows:

```
scores = cross_validate(clf, X, y,
   cv=RepeatedKFold(n_splits=10, n_repeats=100),
   scoring='accuracy', return_train_score=True)
```

where `clf` is the trained classifier used for cross-validation. Figure 14.9 presents the result of cross-validation for a decision tree classifier using the moon dataset ($m = 200$ samples and a noise level of 0.3) where the hyper-parameter `min_samples_leaf` is varied. The validation accuracy presents a maximum around `min_samples_leaf=6`. In other words, a classification tree trained with this value for the hyper-parameter will have the highest potential to generalize best.

14.2.3 Some Issues with Decision Trees

Despite their advantages, such as providing an intuitively understandable model, decision trees also present several issues. One significant problem is their tendency to overfit the dataset. Additionally, they are sensitive to the dataset due to their construction of decision boundaries aligned with feature axes. Consider, for example, the dataset from Figure 14.10a. This dataset presents two classes that are separated according to the axis $x_1 = 0$. A decision tree classifier can predict the decision boundary well as it is a vertical line. However, the same dataset rotated by −45° presents serious challenges for decision tree classifiers (Figure 14.10b). Indeed, this time, the two classes are separated by the boundary formed by the diagonal $x_1 = x_2$. Such boundary lines cannot be reproduced by a decision tree classifier (or only in an approximated way).

14.3 RANDOM FORESTS

14.3.1 Growing Forests

Decision trees tend to be very sensitive to the training dataset, and over-fitting is a common issue. This means decision trees present high variance. A way to reduce variance is to train several models and then take their average. Methods implementing this idea belong to the family of ensemble methods. A particular case is random forests, which are ensembles of decision trees. There are some conditions the models must satisfy, so the variance is indeed reduced when averaging their predictions. An important condition to satisfy is that the prediction of each model must be statistically independent. A simple way consists of training the same model over several sub-datasets randomly drawn with replacements from the training set. The prediction of the ensemble model is then computed as the average of the predictions of the models trained over the sub-datasets. This approach is known as *bagging,* which stands for bootstrap aggregating. Unfortunately, retraining the same model on randomly drawn sub-datasets often results

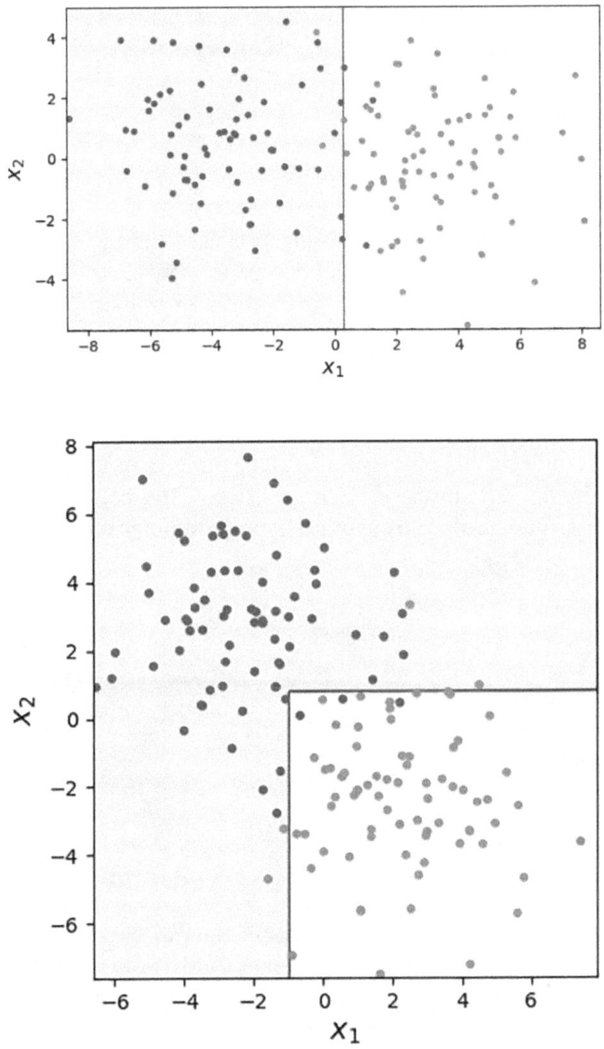

FIGURE 14.10 (a) Decision boundary of decision tree classifier for a dataset (o) separated according to the boundary line $x_1 = 0$. (b) Decision boundary of decision tree classifier for the same dataset but rotated by -45^0. The dataset is now separated by the boundary $x_1 = x_2$. The decision tree classifier cannot reproduce this boundary.

in correlated predictions and not in statistically independent ones, limiting the desired effect of variance reduction. This is why, in the case of random forests, one uses not only randomly drawn sub-datasets but also randomly drawn features (i.e., each decision tree will use a randomly drawn sub-set of features) to avoid too strong correlations between the trained trees.

SciKit Learn offers the class `RandomForestClassifier`, which implements random forest classifiers. Its usage follows the same principle as always:

```
from sklearn.tree import DecisionTreeClassifier
clf = RandomForestClassifier(n_estimators=50)
clf.fit(X, y)
```

Several parameters control the training process. Among the most important is `n_estimators`, which is the number of trained classifier trees (50 in the previous code snippet). Figure 14.11 gives three examples. When a single tree is trained, the result is essentially the same as a decision tree classifier. As we increase the number of trained trees, the forest predicts an increasingly better decision boundary, and it is likely that the forest from Figure 14.11 trained with 50 trees will generalize significantly better than the other two. Cross-validation (Figure 14.12) confirms this observation. As the number of trained trees increases, the accuracy of the validation set increases, too. This behavior makes random forests popular as they require little to no hyperparameter tuning.

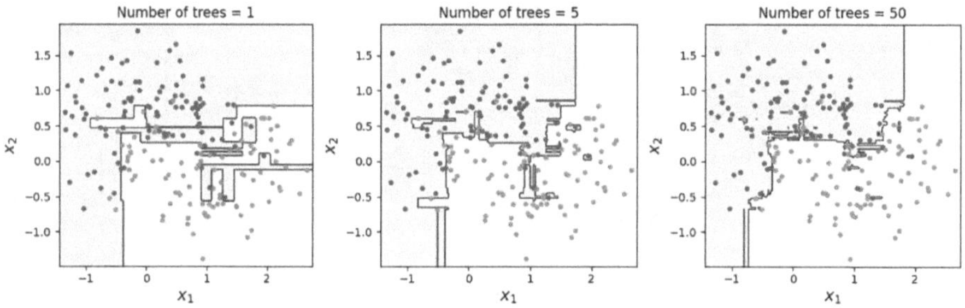

FIGURE 14.11 Example of tree-trained random forest classifiers with a variable number of classification trees. The dataset (o) is the moon dataset with a noise level of 0.4 and $m = 200$ samples.

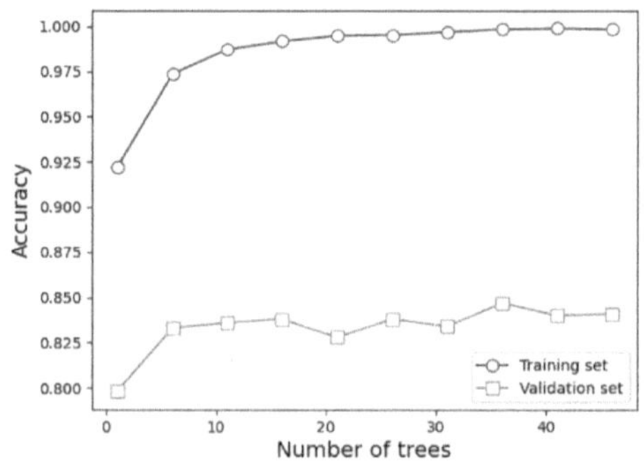

FIGURE 14.12 Cross-validation of a random forest model on the dataset from Figure 14.11. The number of grown trees is used as hyperparameter to be tuned.

14.3.2 Additional Properties of Random Forests

Random forests present some interesting properties that are lacking in the models previously discussed. They can classify a sample, but they can also associate a probability with the fact that a sample belongs to a given class. They are even able to order the features according to their importance in the model predictions.

Let us start by explaining how they can determine the probability of belonging to a certain class. Random forests proceed in two steps. First, for each trained tree, they compute the fraction of samples of the same class in the leaves. The fraction of samples in the leaf that is used to classify the sample (the leaves in the last level of the tree) is called the class probability. As an example, we consider the classification tree of Figure 14.7a and the sample $x = 0.2$. This sample is classified as belonging to the class "blue". The class probability of the class "blue" would be 96/110, and the class probability of class "orange" would be 14/110. Second, the forest determines the probability of a sample belonging to a given class by calculating the mean of all the class probabilities from the trees in the forest.

SciKit Learn implements the calculation of the probability of belonging to a given class using the method `predict_proba` of the `RandomForestClassifier`. As an illustration, we use the Iris dataset introduced in Section 13.2.2 from Chapter 13. We train a random forest classifier on the training dataset we constructed in Section 13.2.2:

```
>>> clf = RandomForestClassifier(n_estimators=20)
>>> clf.fit(X_train, y_train)
```

We used 20 trees based on cross-validation. Once the classifier is trained, we can use it to classify unseen samples. For example, a sample with a sepal length of 5.1 cm, a sepal width of 3.5 cm, a petal length of 4.9 cm, and a petal width of 1.5 cm:

```
>>> clf.predict([[5.1, 3.5, 4.9, 1.5]])
array([1])
```

which is class number 1. This class corresponds to the Versicolor iris:

```
>>> iris['target_names'][1]
'versicolor'
```

More relevant, however, are the probabilities:

```
>>> clf.predict_proba([[5.1, 3.5, 4.9, 1.5]])
array([[0., 0.65, 0.35]])
```

The order in which the probabilities are returned is the same as the classes. In this example, the classifier believes with a 65% probability that the sample is a Versicolor iris and with a 35% probability that it could as well be a Virginica iris. The classifier is very confident that the sample is not a Setosa iris.

Besides estimating the probability of belonging to a particular class, random forests can also rank the features according to their importance in making a decision. The importance of a given feature is computed as the (normalized) total reduction of the classification metrics across all trees in the forest. Without delving into a detailed explanation, it is worth stating that the forest determines for each trained tree how much the classification metrics are reduced for the feature of interest. It does this by considering all the leaves in the tree where the feature appears and determines how much the classification metrics (e.g., the Gini impurity) are improved. It then averages this quantity over all trees in the forest and further normalizes it such that the sum of all these numbers for all features adds to one.

SciKit Learn implements the ranking of the features in the method `feature_importances_` of the `RandomForestClassifier`. Considering again the Iris dataset we obtain:

```
>>> clf.feature_importances_
array([0.12094845, 0.01926502, 0.40787929,
0.45190724])
```

Presented in this way, the result is not very intuitive to read. It gives the relative importance of each feature (in the same order as they appear in the features matrix). A human-friendly way to present the result is given in the following code snippet, which prints the name of the feature together with the relative importance of it:

```
>>> for name, score in zip(iris['feature_names'],
                           clf.feature_importances_):
        print(name, score)
sepal length (cm) 0.12094845096157666
sepal width (cm) 0.019265018361694728
petal length (cm) 0.40787929253684674
petal width (cm) 0.4519072381398819
```

According to the trained forest, petal widths and petal lengths are the most important features to classify the sample. The sepal length and width seem significantly less important.

NOTES

1 If the reader would like to reproduce the same dataset as in our example, initialize the random number generator with `np.random.seed(10)`.

2 https://scikit-learn.org/stable/modules/generated/sklearn.tree.plot_tree.html

3 https://scikit-learn.org/stable/modules/generated/sklearn.tree.DecisionTreeRegressor.html

Index

Note: Page numbers in *italics* indicate a figure and page numbers in **bold** indicate a table on the corresponding page.

1-norm, 402–403
  calculation, 130
2-norm, 402–403
  induced norm, equivalence, 160
17th order, reaching, 42

A

absolute backward errors, evaluation, 132
absolute condition, 92
absolute error, 104
  approximations, 11
  estimation, convergence order basis, 85
  finding, Octave (usage), 35
  upper boundary, 55, 56
absolute forward error, 137–138
  approximation, 147
absolute value measurements, usage, 23
absolution condition number, 33
abstract derivation, Taylor series basis, 69
abstract normal equations, 207–212
abstract normal, matrix representation, 210
abstract vector spaces
  calculations, Octave (usage), 154–155
  normal spaces, usage, 207–210
abstract vectors, representation, 155–156
accelerometer, usage, 243
accumulated round-off error, 46
accuracy, 7–8
  precision, distinction, 7
actual mathematical number, difference, 16
adaptive IvP solvers, 338–343
  embedded adaptive IvP solver, 341
  operation, schematic, *340*
adaptive methods, 335–343, 346
  example, 336–338
advanced pipelines, 378–381
algebraic equations, solving, 50
algebraic systems, 323
algorithms, 54–55
  approximation, production, *4*
  behavior, understanding, 30
  bias, 7
  bias-free algorithms, development, 7
  bisection algorithm, 55–58
  convergence conditions, knowledge (requirement), 63
  convergence investigation, 79
  converging algorithm, defining, 6
  defining, 10
  description (Babylonian clay tablet YBC 7289), 64
  efficiency, 111
  Gauss elimination basis, 110
  generated approximations, error (decrease), 74–75
  iterative algorithm, 8
  mechanism, knowledge (requirement), 63
  progression, 64
  representation, *34*
  switching, 101
  usage, 44
approximation of a root (error estimation), backward error (basis), 93–94
approximations
  backward error, 90
  calculation/computation, 296–297, 315
  errors, estimation, 99
  estimated errors, 332–333
  estimated errors, plotting, 98
  forward/backward errors, computation, 131
  functions, approximations, 212–217
  general expression, 297
  generation, cessation, 65
  guarantees, 339
  improvement, computation (Richardson extrapolation allowance), 275
  incorrectness, 297
  polynomial function approximation, 212–214
  production, 8
  production, Euler's IvP solver/trapezoid IvP solver (usage), 330
  Richardson error estimations, *321*
  series (production), algorithm (usage), 6, 7
  series, generation, 84
  succession, 296
  true error, *313*, *316*, *319*
arguments, 390
asymptotic error constant, 76, 77
  bisection algorithm, convergence, 76
  computation, 86

root, 80
unknown value, 86
auxiliary pipeline, usage, 379–380
auxiliary variables, usage, 336
auxiliary vector (iterNumber), 64
average loss, 391

B

b
back substitution, 111–112
backward errors, 90–94, 137
basis, 93–94, 95
behavior, round-off error (impact), 91
calculation, 142
equations systems, 131
forward errors, relationship, *93*
function, consideration, 91
minimization, 196, 198
obtaining, 91, 194
small value, 148
squaring, 195
usage, 131–132
zero value, 142
BaseEstimator (class), derivation, 374
base results, usages, 210–211
base vectors, standard representation (usage), 208
Bayesian theory, development (requirement), 417
bias
concept, 7
generalization, 7
measurement, 356
representation, *354*
term, 361
term, addition (avoidance), 369
variance decomposition, 357–358
word, usage, 354–355
bias-free algorithms, development, 7
bias-variance trade-off problem, 358–359
big O notation, 28
intuitive interpretation, 29–30
bilinear mapping, 157
binary classifiers, 418–420
binary classifiers, confusion matrix, **410**, 410–412
binary logistic regression classifier, 418
bisection algorithm, 55–58
convergence, 76
root approximation, 56
usage, **57**
bisection method, 55–57
application, 56
iterations, number, 58
bits, number (usage), **17**
Boole's rule, 250
bracketing interval, movement, *62*
bracketing methods, 52–62, 102
algorithm, 54–55

C

calculus
radius of convergence, 27
theorems, 9–10
usage, 250
Canada, hourly wage rates (prediction/data points), *387*
capital first letter, avoidance, 377
categorical data, 375–378
Cauchy-Schwarz inequality, 159–160
consequence, 161
centimeters, units, change, 18
class
belonging, probability (estimation), 438
constructor, usage, 369
multiple classes, 420–422
classical gram-Schmidt orthonormalization algorithm, 176, 180
classical RK4 IvP solver, 317–319
classical RK4 method
Octave implementation, 331
usage, *332*, 334
classification, 406
algorithm, goodness, 409
CART algorithm, usage, 431–432
K-nearest neighbors (KNN) classification, 406–409
performance measures, 409–415
Classification and Regression Trees (CART) algorithm, 424–427
behavior, control, 429
mechanism, 425–427
usage, 431–432
classification trees, 431–434
moon dataset training, *432*, *433*
training, 432
classifier, 415. *see also* binary classifier; K-nearest neighbors classifier
trained multinomial logistic regression classifier, confusion matrix, *422*
closed-form expression, absence, 25
closed Newton-Cotes formulas, 279
interpolating points, usage, 249
weights, **249**
closed two-points Newton-Cotes formula, error, *279*

coefficient matrix, 107
building, 218, 240
calculation/computation, 194, 227, 247
conditioning number, computation, 146–147
conditioning number, high level, 148
consideration, 141, 196
defining, 146
identity matrix check, 234
obtaining, 196
QR decomposition, computation, 199–200
coefficients
calculation/computation, 216, 240
discovery, 225
coef_ model attribute, features weights storage, 365
ColumnTransformer, usage, 380
column vector
construction, 228
formation, 192
representation, 209
return, 295
unknowns grouping, 107
usage, 224–225
composite methods, convergence
order, estimation, 270–271
reduced order, 263–264
composite Newton-Cotes formulas, 255–259
examples, *256*
expressions, derivation, 258
Octave, usage, 258–259
order, study, 263–264
truncation errors, 259–265, **261**, 274
composite quadrature formulas
Newton-Cotes composite quadrature formula, consideration, 274–275
truncation errors, estimation, 265–271
composite quadrature methods, convergence order, 261–264
composite Simpson's-1/3 rule, *256*
application, 274
expressions, derivation, 258
implementation, 259
level, reduction, 264
composite trapezoidal method
truncation error, 271–274
usage, 278
composite trapezoidal rule, *256*, 256–257
application, 260, 264, 267, 273
application, performing, 261
leve, reduction, 264
panels, correspondence, 269
panels, number (usage), 262
precision, increase, 273–274
truncation error, 260
usage, 260–261, 267–269
computed digits, correction (estimation), 40
computing metrics, feature transformations (application), 371
conditioning, 129–146, 150–151
concept, *34*
usage, *34*, 92
condition number, 33
calculation/computation, 35, 147
estimation, derivation, 93
matrix condition number, 139
root-finding problem, 92
cond (Octave function), usage, 140
confusion matrix, **410**, 410–412
building, 411
computation, 411
definition (SciKit-Learn), *412*
general confusion matrix, 412–415
confusion_matrix, usage, 411
constants, existence, 75–76
continuous function, 9
continuously differentiable function, 9, 77
continuous random variables, probabilities, 243
continuous real-valued function, root, 52
contributions (sum), true error (equivalence), 46
convergence
bisection algorithm convergence, 76
plot, *65*
radius, 26, 27
radius equality, 28
reduced order, 263–264
speed, 74–75
convergence order, 74–89, 103–104
basis, 85, 86–87
composite quadrature methods, 261–263
cubic characteristic, *83*
determination, 301
estimation, 270–271, 322
finding, 262
fixed-point iteration method, 78–79
knowledge basis, 84–89
Newton-Raphson method, 82–84
order, estimation, 270
quadratic order, confirmation, 263
converging algorithm, 69
defining, 6
coordinate system, usage, 328
correct digits, loosing, 20
corrected trapezoidal rules, applications, 273
corrector, 311
cosine function
approximation, 25–26
closed-form expression, absence, 25

computation, 26
value, 25
cosine Taylor series, truncation error, 31
cosine x, replacement error, 31
cost function, 417
defining, *417*
Cotes numbers, 250
Cotes weights, 250
cross-entropy loss function, 421
cross_validate (function), usage, 399–400
cross-validation, 397–401, *409*
performing, 399, 434
principle, *398*
synthetic dataset, usage, 399–400
training error, 398
usage, 401–402
usefulness, 400–401
variations, 398
cubic polynomial, usage, 351
custom transformers, 372–375
creation, template, 374
cut point
computation, 58, 59
cutting, calculation, 60
cut values
optimal cut values, 427
usage, *426*

D

data
categorical data, 375–378
erroneous data entries, detection, 383
Gaussian base function, example, *379*
generalizing data, 350
learning process, 352–353
logarithmic display, *201*
ML component requirement, 350–352
original data, plotting, 206
quadratic model, adjustment, 368
structure, overview, 382–383
transformed data set, *203*
transformed data, usage, *203*
type, inspection, 383
usage, 382–387
weather data, modeling, 217–219
whitefish, data, **363**
dataframe df, construction, 384
data points
consideration, 187
defining, 199–200
model, application, 204–205
dataset
defining, parameter requirements, 395
example, 362–363
features matrix, defining, 2376
formation, *355*
generation, Gaussian white noise (addition), 395
moon dataset, 406, *407*
noise-laden datasets, impact, 356
over-fitting, 401–402
polynomial fit, example, *370*
randomly drawn datasets, *358–359*
samples, regrouping, 361–362
size, impact, 352
splitting, 395, 397, 411
sub-datasets, 434–435
synthetic dataset, creation, 374–375
synthetic dataset, depiction, 369, *370*
usage, 382, 406
decision tree classifier
cross-validation results, *433*
decision boundary, *435*
DecisionTreeRegressor (SciKit-Learn class), 427
decision trees
issues, 434
regressor, cross-validation results, *430*
definite integrals
additive property, usage, 255
approximation, 252, *262*
composite Newton-Cotes formulas, order (application), 263–264
composite trapezoidal formulas, order (application), 264–265
computation, 245
equality, 251
estimation, 247, 257, 279
Gauss quadrature, usage, 284, 285
interpolating polynomial, relationship, 281–282
mean value theorem, usage, 254
numerical approximations, computation, 261–262
numerical approximations, correction, 274
numerical approximations, obtaining, 275
quadrature method application, true errors, *273*
Simpson's-1/3 formula, order (application), 264–265
true value, 243
degree (parameter), class constructor naming, 369
Δy, plots, *40*, *41*
derivatives
divergence, 270
integration, 272
i-th order derivative, 326

Descartes, René, 153
determinant, computation, 123
differentiable function, 95
differential equation
basis, 302
dependence, 291
solutions, 305
digits
correct digits, loosing, 20
finite number, 16
infinite number, usage, 23
retention, 39
usage, requirement, 144–145
Dirac, Paul, 153
diverging Newton-Raphson method, 72–73
double precision floating point numbers, usage, *22*
double precision numbers, usage, 15
drop method, 384
DSV file, form, 382

E

Earth/Moon, average distance (estimation), 10
educational course, passing/failing (probability), *419*
eigenvalues, inequality, 132–133
eigenvectors, association, 132
Einstein, Albert (law of gravity reformulation), 335
elastic net regularization, 402
element-by-element division operator, usage, 94
element-by-element power operator, usage, 94
embedded adaptive IvP solver, 341
encoded numbers, obtaining, 2376
equal-sized panels, usage, 260
equations
inconsistent equations, 190, 191
inconsistent equations, least square solution, 191
normal equations, 191–194, 220–221
normal equations, QR decomposition (usage), 198–200
solution, approximation (finding), 51
swapping, 109
upper diagonal system, 108
equations, linear systems
conditioning, 141–146
error analysis, usage, 132
error magnification factor, 136–141
forward errors, estimation, 141
equation, solving
example, 2
specification, 50
equation solving algorithms, convergence speed, 74–75
equations systems
backward errors, 131
base results, usage, 210
errors, measurement, 129–135
forward errors, 131
graphical solution, *136*
involvement, 331–335
IvP, Euler's method (usage), 324–326
matrix equation form, 149
true errors, 129
erroneous data entries, detection, 383
error estimation
backward error, basis, 95
computation, 94, 267, 334
computation, Richardson's error formula (usage), 266
convergence order basis, 86–87
general error estimation formula, 94–96
method, usage requirement, 63
numerical IvP solver, involvement, 331–335
power law, fitting, 322
tools/methods, development, 30
unavoidable output error estimation, 35
error magnification factor, 92, 136–141
maximal relative error magnification factor, 140
terms, 142–143
error measurement, 4–6, 10–11
infinity norm, usage, 131
noise, association, 357
error of generated approximations (estimation), convergence order (knowledge basis), 84–89
error of numerical approximations, 291
error of the approximations, estimation, 99
error propagation, 32–42, 48
consideration, 35–36
cosine function, consideration, 36
errors, 5–6, 22–32
analysis, 90–96, 105
analysis, usage, 132
backward errors, 90–94
conditioning, 35–36
contributions, *44*
decrease, 42, 45, 100
forward errors, 90–94
function, numerical approximation (computation), 285–287
generalization error, 389–391
growth, 338
increase, 299
magnitude, estimation, 353

minimum, 213
order of convergence, determination, 301
power law, fitting, 264
reduction, *279*
residual errors, computation, 188
terms, simplification, 260
training error, 391–392
types, 388–394
estimated errors, 5–6, *337*
change, 58
numerical estimation, providing, 47
plotting, 98
upper boundary, providing, 47
estimated relative errors, 5
estimated truncation error, 26
Euclidean norm, computation, 130
Euler function (Octave), usage, 297, 299
Euler-Maclaurin formula, 272
Euler's algorithm, 293–301, 343–344
application, Octave function (usage), 324
basics, 293–296
usage, 293
Euler's IvP solver
approximation production, 330
usage, *327*, 330
Euler's method, 294–296
application, 294
convergence, 299–301
impact, 324
local truncation error, 303
schematic, *310*
stability, 296–299
usage, 324–326
Euler's number (*e*)
approximation, true error (usage), *43*
exact value, 42
numerical approximation, providing, 47
representation, true value (usage), 43
Euler's solver, usage, *329*
exact solution
numerical approximation, 3
words, usage, 3
existence theorem, 52
expansion order, 30
increase, *24*
expected loss, 357
minimization, 358
expected squared error (E_D), 354, 356
expected value, 7
computation, 389–390
explicit ordinary differential equation (ODE), 291
exponential function
equation, application, 44
first derivative, numerical approximation, *44*
Taylor series expansion, usage, 42
usage, requirement, 206
exponential function, series expansion, 23
exponential model, 200–204
expressions, evaluation, 23

F

false negative (FN) value, 410–411
false position algorithm, 58–62
application, *62*
graphical illustration, *59*
initiation, 60
root approximation, **61**
sub-intervals, production, 59
false position method
algorithm, 59–60
example, 60–61
features
defining, 384–385
regression trees, usage, 427–428
transformations, 367–381
weights, storage, 365
features matrix, 363
creation, 371
defining, 376
SciKit-Learn expectation, 369
feature, term (usage), 361
feature transformation, application, 371
feature transformer, SciKit-Learn offer, 368
finite number, 16
existence, 28
first derivative
approximation, analysis, 45
estimation, 44
first iteration, 57–58
first order approximation, 69
first order initial value problem, 292
approximation, Euler's algorithm (usage), 293
approximation, production, 311
solution, display, 301–302
solution, schematic representation, *301*
first order ordinary differential equation, 323
solutions, *305*, *306*
first root
approximation, 98–99
Newton-Raphson approximations (estimated errors), *100*
fit method, 380
calling, 365
fitted linear model, 352
fitted model
original data, *203*

plotting, 206
predictions, 202
residuals, presence, *219*
transformed data set, *203*
fit_transform method, 369
fixed-point algorithm
application, 64
initial guess, 74
usage, 63
fixed-point determination, 63
fixed-point finding problem, 68
fixed-point iteration
algorithm, application, *66*
scheme, 103–104
square root approximation, 63–66
fixed-point iteration method, 63–69
convergence, 77–80
convergence order, 77, 78–79
convergence plot, *65*
convergence theorem, 67
divergence, 68–69
non-converging fixed-point iteration method, 68–69
usage, 68, 78
fixed-point method, 102–103
iterations, 68
fixed-point, real number (equivalence), 52
fixed step sizes methods, problem, 338
fixed value, convergence, 64
floating-point numbers, 16
precision, differences (IEEE Standard), **17**
representation, 45
usage, 18, 99
floating-point representation, 16
forests. *see* random forests
growing, 434–436
for-loop
index usage, 89
nested for-loops, 113
forward elimination, 108–109
completion, 111
forward errors, 90–94, 137
approximation, 92
backward errors, *93*
equations systems, 131
equivalence, 140
estimation, 137, 141–142
observation, 93
usage, 131–132
forward relative error
lower limit, 138
upper limit, 139
forward substitution, 125
Fourier polynomials (continuous line), *216*
Fourier polynomials function approximation, 214–217
Fourier series, 215
fourth-level Newton differences, 232
fourth order Runge-Kutta methods, 316–319
functions
approximations, 212–217, 221–222, 325
approximations, Fourier polynomials function approximation, 214–217
arguments, 390
condition number, 33
definite integral, estimation, 247
deployment, 112
evaluation, floating-point number representation, 45
evaluation, usage, 37–42
first derivative (numerical estimation), algorithm (usage), 44
interpolating functions, 235–239
interpolating polynomial, calculation, 247–248
plotting, root display, 86
plot, usage, 56
replacement, *244*
root, approximation, 55–56
sine function, interpolating, 235–238
softmax function, 421
tangent, intersection (construction), *70*
uniform distribution, *373*
FunctionTransformer, argument acceptance, 379

G

Gauss elimination
basis, 110
coefficients, usage, 121
matrix, inverse, 121
naïve Gauss elimination, 109–114
packages, usage, 113
steps, 113–114
usage, 227–228
Gauss elimination algorithms, 108–119, 149–150
partial pivoting, usage, 115–119
Gaussian base function, example, *379*
Gaussian basis functions
example, *375*
usage, 372–374, *373*
Gaussian noise, 368
Gaussian white noise, *394*
addition, 388–389, 395, *397*
Gaussian white noise of variance, *354*
addition, 357

Gauss-Legrendre quadrature, 282
formula, 282, 289
Gauss quadrature, 279–286
idea, 281
usage, 284, 285
Gauss quadrature formula, 281–285
development, 283
interpolation points, **284**
two-point Gauss quadrature formula, 283
weights, **284**
general confusion matrix, 412–415
general error estimation formula, 94–96
general inner space, orthogonal projects, 163–164
generalization, 392–394
problem, 350–351
generalization errors, 389–392
computation, 390
computation, impossibility, 391
estimation, 394–405
polynomial models, *393*
values, 400
generalization of new instances, 350
generalized angle, 161
orthogonality, relationship, 160–162
generalized intermediate value theorem, 259
generalized orthogonal matrix, 180
generalized Pythagoras theorem, 162
generalizing data, 350
general linear models, 196
general *n*-order ordinary differential equation, 326
generated approximations estimation, convergence order (knowledge basis), 84–89
geometrical considerations, 58
Gini impurity, 431–432
global truncation errors, 301–303, 308
building, 307
estimation, 319–322, 339–340, *340*
estimation, Richardson's method (usage), 338
order, determination, 313, 315
order *q*, 339
set tolerance, contrast, 340
trapezoid method, 313
gram-Schmidt algorithm
implementation, 180
usage, 162, 178, 179
gram-Schmidt method, usage, 174
gram-Schmidt orthogonalization, 170–178, 184
initiation, 176
linearly independent vectors, consideration, 171–172
procedure, usage, 173
sub-space usage, 173
vectors, usage, 172
gram-Schmidt orthogonalization algorithm
iterative process, 170
matrix representation, 179
gram-Schmidt orthonormalization, 177
gram-Schmidt orthonormalization algorithm, 176, 177, 180
equation sequence, production, 178
gram-Schmidt's algorithm, usage, 280
greenhouse gas emissions, production data, *351*, 352
Gregory's algorithm, result, 272

H

Heun's IvP solver, 311
Heun's method, 311
high-condition number, avoidance, 210
higher order initial value problems, 323–335, 345–346
higher order ordinary differential equations, 326–330
high error magnification factor, *136*
linear equations systems, relationship, 135–136
h values, range, 39
h values range, approximation, *39*
hyper-parameters, 373
default values, usage, 432
fine-tuning, 433–434
tuning, cross-validation (usage), 401–402
usage, 429
values, 428–429

I

identity matrix, checking, 234
IEEE754 standard, 16
implemented decomposition, contrast, 126
implicit ordinary differential equation (ODE), 292
inbuilt variable pi, usage, 36
inconsistent equations, 190, 191
least square solution, 191
solving, Octave (usage), 200
indexes, attention, 268
induced norms, 158–159
2-norm, equivalence, 160
usage, 208
infinite loop, entry, 73
infinity norm, 134
usage, 131, 160

information entropy, 431–432
initial error estimation, writing, 58
initial guess, fixed-point, 77
initial value problem (IvP), 291
  adaptive initial value problem solvers, 338–343
  approximations, Euler's method (usage), **294**, *296*
  consideration, 306
  Euler's method, usage, 324–326
  existence and uniqueness theorem, 304
  first order initial value problem, 292
  higher order initial value problems, 323–335
  ill-conditioning, 305
  numerical approximations, depiction, *322*, *325*
  numerical approximations, Euler's method (usage), *298*, *300*
  numerical initial value problem solver, 292
  solution, 295
initial value problem (IvP) solvers
  analysis, 301–309
  classical RK4 IvP solver, 317–319
  numerical IvP solvers, local/global truncation errors, 301–303
inner products, 157–160
  calculation, 161, 174–175
  norm, inducing, 159
  norms, relationship, 158, 167–168
  standard inner product, 158
  usage, 160
inner spaces, 153–162, 182–183
  dimension n, 179
  general inner space, orthogonal projects, 163–164
  norms, 156
  notations, 153–156
  real-valued inner space, sub-space (consideration), 163
input
  Δx change (measurement), conditioning (usage), *34*
  error, propagation, 32–37
  feature, 368–369
integrable function, 9
integration interval, consecutive numbering, 256
intercept_ (model attribute), inspection, 365
intermediate value theorem, 9
  generalization, 259
  generalized intermediate value theorem, 259
interpolating functions, 235–239
interpolating points
  Legendre Polynomials, relationship, 283
  usage, *245*, *256*
interpolating polynomial
  calculation, 228
  coefficients, 238
  construction, 238
  defining, 236
  functions, replacement, *244*
  Newton interpolating polynomial, finding, 230
  obtaining, 283
  Vandermond interpolating polynomial, evaluation, 230
  writing, 229
interpolating surface, 239–241
interpolation
  advantages, 223
  error, 235, 242
  Lagrange interpolation, 233–235
  Vandermond interpolation, 223, 226–228, 241–242
interpolation points, *228*
  basis, 244
  display, 235
  distance, 258–259, 274, 275
  location, knowledge, 279
  number, 251
  number, increase, 255
  number, limitation, 252
  placement, *279*
  selection, 243
interpolation polynomial
  building, 238
  determination, Vandermond interpolation (usage), 236
  evaluation, 227
  representation, 226
  sine functions, *237*
  writing, 233
  yield, 224–225
interval
  finding, 59
  numerical distinction, avoidance, 21
  roots, presence, *53*, 60
  sub-intervals, dissection, 56–57
intuitive context, usage, 160–161
iris flowers
  dataset, 413
  labels, 414
  trained classifier, usage, 422
  trained multinomial logistic regression classifier, confusion matrix, *422*
  types, *413*

iteration number, 58
requirement, 58
true error, 65
true error, ratio, *79*
iterations, 57
computation, Newton-Raphson method (usage), 99
display, *62*
estimated errors, impact, 57
first iteration, 57–58
fixed-point iteration method, 63–69
minimal number, 58
Newton-Raphson algorithm, usage, 98
performed iterations, true error, *65*
performing, 64
results, **61**, 64
root approximation, 60–61
scheme, 63
storage, 64
iterative algorithm, 8
iterNumber (auxiliary vector), 64

J

Jupyter Notebook, code (execution), 372

K

Kepler motion, 328–330
Kepler's third law of planetary motion, 204–207
kilometers, unit (change), 18
K-nearest neighbors (KNN)
algorithm, impact, 408
classification, 406–409, *408*
classification, cross-validation, *409*
K-nearest neighbors (KNN) classifier
approach, 421–422
confusion matrix, *415*
deployment, 407–408
training, 411
usage, 429
KNeighborsClassifier (class), 406
k order expansion, 27
Kronecker symbol, 162, 233

L

labels, 361–363
correspondence, 406
defining, 384–385
usage, 385–386
vector, 364
"Lack of A Priori Distinctions Between Learning Algorithms" (Wolpert), 351
Lagrange interpolating polynomial, evaluation, 234–235
Lagrange interpolation, 233–235, 242
basis, 247–248, 288
polynomial, finding, 233–235
Lagrange polynomials
integration, 245–246
writing, 233
LASSO regression, 403–405
LASSO regularization, usage, *404*
LASSO regularization, 403–405
law of motion (Newton), 328–330
least square, 213
solution, finding, 192
least square linear regression, QR decomposition, 199
least square regression, 189
optimal model parameters, 195
usage, 195–198
left-out fold, usage, *398*
Legendre Polynomial of order five, *97*
first root, Newton-Raphson approximations, *97*
first root, Newton-Raphson approximations (estimated errors), *100*
Legendre Polynomials, 174–175, 279–281
construction, 280
determination, 280–281
family, 175
interpolating points, relationship, 283
n+1 Legendre polynomials, 282
orthogonal Legendre polynomials, usage, 282
properties, 281
roots, 96–99
set, defining, 280
linear combinations, basis, 113
linear damped oscillator, 336
linear equations
answers, arrival, 192
context, 153
linear equations systems, 107
absolute forward error, 137–138
construction, 141
equation, number, 190
high error magnification factor, association, 135–136
operations, 109
relative forward error, 139–141
true solution, 135
linear equations systems, solving, 123–125, 146–149

LU decomposition, usage, 124
PA=LU decomposition, usage, 127–128
round-off errors, usage, 143
linear interpolation polynomial
obtaining, 248
writing, 245
linearization, 200–207
linear least square regression, 186
data points, consideration, 195–196
definitions, 190–191
solution, 190–200
linearly converging algorithms
boundedness, 85
equation, solution, 84
ratio calculation, 88
linear models, 362
adjustment, 368
linearization, 200–207
linear regression, 423
implementation, 385–386
orthogonalization, 210–212
problem, formulation (geometrical problem), 210
problem, solving, 211
QR decomposition usage, 199–200
LinearRegression class, 371
SciKit-Learn defining, 364–365
Lipschitz condition, satisfaction, 304, 305
Lipschitz constant
approximations, 308
increase, 305
presence, 307
selection, 304
Lipschitz function, 304
appearance, 304
Lipschitz theorems, 303–309
local convergence, 67
local solution (t_i, w_i), 302
slopes, average (usage), *310*
slope, usage, 314
local truncation errors, 301–303, 308
accumulation, 307
estimation, 339
Euler's method, relationship, 303
extrapolation, *340*
order $q + 1$, 339
logarithmic function
application, 200, 202
approximation, 27–28
logistic regression, 415–422
binary logistic regression classifier, 418
classifier, training, 418–419
function, writing, 416
multinomial logistic regression, 420–421
trained multinomial logistic regression classifier, confusion matrix, *422*
LogisticRegression class (SciKit-Learn), 421
log-log plot, *205*
usage, 319, 334
loss
average loss, 391
expected loss, 357
expected loss, minimization, 358
loss function, 392
building, 431
computation, cut values (usage), *426*
cross-entropy loss function, 421
minimization problem, 192
selection, 424
usage, 389, 402
weighted means, 425
loss of significance, 19–20
lower diagonal matrix, 240
lower diagonal system (expression), Newton finite differences (usage), 229
lower first letter, usage, 377
lower triangular system, 229
L^p-norm, 157
LU decomposition, 119–123
computation, 121–122
usage, 123–125
lu (function), Octave implementation, 126

M

machine epsilon, 20–22, 145
magnitude, 147
number, size, 20
value, 21
machine learning (ML), 349
algorithms, components (requirement), 350–352
context, 357, 364
dataset samples, regrouping, 361–362
model, 354
make_pipeline (function), 371
Manahan norm, computation, 130
mass, orbitals, *329*
mathematical model, adjustment, 223
mathematical pendulum, *327*, 327–328
mathematical problem
convergence, 6
numerical solution, 1–2
replacement, 22
solution, precision (desire), 3
true solution, 5, 7

Matplotlib, SciKit-Learn interaction, 366
matrix (matrices). *see* coefficient matrix
  condition number, 139
  decompositions, 119–128, 150
  determinant, LU decomposition (usage), 123
  eigenvalues, inequality, 132–133
  equation, usage, 108
  features matrix, 363
  Gauss elimination matrix, inverse, 121
  infinity norm, 134
  LU decomposition, 119–123
  norm, computation, 133–135
  norms, submultiplicative property, 133
  orthogonal projection matrix, 164–170
  PA=LU decomposition, 125–127
  permutation matrices, usage, 125
  product, associative characteristics, 164–165
  reduced QR factorization, 180
  representation, finding, 209
  square matrix, decomposition, 120
  structure, 179
  transformation, naïve Gauss elimination (usage), 119
  triangular matrix, 230
  usage, 209
  Vandermond matrix, 147–149, 197
  version, usage, 226
  $Z\Lambda$ decomposition, 179
maximal relative error magnification factor, 140
mean squared error (MSE), 190
  analog, 355
  loss function, basis, *392*
  usage, 392, 396, 431
mean validation error, obtaining, 400
mean value theorem, 9
  usage, 254
mechanical energy, conversion, 328
midpoint IvP solver
  application, 314–316
  order, rank, 316
midpoint method, schematic method, *314*
midpoint rule, 252
  usage, 260–261
minimization problem, avoidance, 192
Mitchel, Tom, 350
model parameters, 206
  adjustment, 416–417
  choices, *189*
  finding, 201–202
model predictions, computation, 188
models
  complexity, function (generalization/ training errors), 392–394
  definition, 385–386
  evaluation, 386–387
  exponential model, 200–204
  fitted model, predictions, 202
  linearization, 200–207, 221
  mathematical model, adjustment, 223
  ML component requirement, 350–352
  over-fitting, 393–394
  performance, evaluation, 396, 414
  pipelines, inclusion, 414
  regularization, 402
  residuals, calculation, 206
  selection, 388, 429–431
  sensitivity (measurement), variance (usage), 397
  trained models, 354
  trained models, examples, *355*
  training, 353, 363–367, 372–373, 381, 385–386
  validation, 363–367
  weights, term (usage), 361
moon dataset, 406
  classes, inclusion, *407*
  KNN classification, *408*
  KNN classification, cross-validation, *409*
  training, *432*, *433*
multinomial logistic regression, 420–421
multiple classes, 420–422
multiple root
  approximation, error estimation, 95–96
  usage, 88

N

naïve Gauss elimination, 109–114
  application, 121–122
  calculations, involvement, 111
  impact, 119
  linear equations system, consideration, 110–111
  Octave, usage, 112
  operator form, 113–114
  overhead, addition, 126–127
naïve Gauss elimination algorithm
  execution, 120
  presentation, 113
  usage, 120
naive_gauss (Octave function), usage, 122
natural base, 226
  selection, 168

natural logarithmic function
approximation, 27
usage, 204
natural logarithmic, usage, 206
natural vector base, 154
nested for-loops, 113
Newton-Cotes composite quadrature formula, consideration, 274–275
Newton-Cotes formula, 288–289
closed Newton-Cotes formulas, 249–250
closed Newton-Cotes formulas, weights, **249**
composite Newton-Cotes formulas, 255–259
open 1-point Newton-Cotes formula, derivation, 252
open Newton-Cotes formula, 249
open Newton-Cotes formula, weights, **252**
usage, 256
weights, symmetry, 251
Newton-Cotes formula, truncation errors, 252–255, **253**
estimation, 254–255
expressions, derivation, 253
leveraging, 259
Newton-Cotes quadrature, 249–274
Newton differences, 231–232
Newton finite differences, usage, 229
Newton interpolating polynomial
coefficients, computation, 240
finding, 230
Newton interpolation, 229–232, 242
advantages, 229
benefits, 231
dataset, consideration, 229–232
usage, 239–240
Newton, Isaac (law of motion), 328–330
Newton polynomial, coefficients, 230
Newton-Raphson algorithm, 69
application, 69, *71*
cases, enumeration, 73
convergence, 73, 82
convergence plot, *88*
divergence, 73
execution, zero evaluation, 73
guaranteed local convergence, 74
implementation, 70
initial guess, 74
iterations, calculation, 86, 89
local convergence, 72
quadratical convergence, 82, *83*
running, 83
usage, 86, 98
usage, true value (knowledge, absence), 87
Newton-Raphson approximations
estimated errors, *97*, *100*
iteration errors, *83*
Newton-Raphson iteration scheme, 69
Newton-Raphson method, 69–74, 103
application, example, 70–72
convergence order, 80–84
diverging Newton-Raphson method, 72–73
examination, 104
iterations, 103
Newton-Raphson root-finding algorithm, graphical representation, *70*
"No Free Lunch Theorem," 351
noise-laden datasets, impact, 356
non-continuous function, *53*
non-converging fixed-point iteration method, 68–69
none zero constant, multiplication, 109
non-linear equations, solving, 96–101
non-linear expression, 196
n-order ordinary differential equation, 326
normal equations, 191–194, 220–221
least square regression, usage, 195–198
name, origin, 193
Octave, usage, 194
solving, QR decomposition (usage), 198–200
usage, 207–210
norm (function), Octave implementation, 134–135
norms, 156–157
2-norm, 160
induced norms, 158–159
infinity norm, 134
infinity norm, usage, 131, 160
inner products, relationship, 158, 167–168
L^p-norm, 157
minimization, 207
vector norms, 156–157
numbers
actual mathematical number, difference, 16
finite number, 16
succession, iterative production, 67
undistinguishable numbers, 20–22
numerical algorithms
Legendre Polynomials, role, 96–97
solutions, 368
numerical analysis, conditioning (concept), *34*
numerical answer (determination), schematic workflow (usage), 4f
numerical approximation, 250, *286*
change, 44
computation, 265, 285–287
display, *322*

error, 291
estimated errors, *337*
Euler's method, usage, *298*, *300*
finding, 3
plotting, 326
providing, 10
round-off error, determination, 36
true absolute error, 5
true errors, *332*
numerical calculations, 30
error-free assumptions, avoidance, 18
floating-point numbers, usage, 18
numerical definite integrals, estimation strategies, 243
numerical derivation, usage, 272
numerical errors, 15
impact, 337
numerical evaluation, 35–36
allowance, 25
double precision floating point numbers, usage, *22*
numerical function evaluations, display, 19
numerical integration, 243
Lagrange interpolation basis, 247–248, 288
polynomial interpolation basis, 243–246, 287–288
numerical IvP solver, 293
analysis, 304
error estimation, equations systems (involvement), 331–335
global truncation error, 308
global truncation error, estimation, 319–322
local truncation error, 303
problems, challenge, 307
Richardson's truncation error estimation, usage, 320
Runge-Kuta initial value problem solver, equivalence, 310
numerical IVP solvers, local/global truncation errors, 301–303
numerical methods
aim, 1–4, 10
definitions, relationship, 6–9
golden rule, 18
numeric_transformer pipeline, 380–381
numpy (Python library), importation, 364

O

observation, making (probability), 357
Octave
abstract vector spaces (calculations), Octave (usage), 154–155
addition, 155–156
code, consideration, 16
code, usage, 130
composite Newton-Cotes formula, usage, 258–259
construction, 370
cosine computation, assumption, 26
documentation, 101
function (naive_gauss), usage, 122
implementation, 94, 111
inbuilt variable pi, usage, 36
inconsistent equation solutions, 200
numerical approximation, 147
polynomial evaluation, 225–226
polynomial regression usage, 197
ratio convergence calculations, 89
square root computation, 38–39, 40
usage, 19, 35–36, 47, 112
vector index numbering, initiation, 65
OneHotEncoder (class), 377
one-hot encoder, vector assignment, 377
one-hot encoding, 377–378
open 1-point Newton-Cotes formula, derivation, 252
open methods, 62–74, 102–103
application, aspects (mastery), 63
open Newton-Cotes formulas, 249
weights, **252**
open trapezoidal rule, 248
usage, 251
operator form, 113–114
operator formulation, usage, 113
optimal cut values, 427
orbital computation, RK4 solver (usage), *335*
ordered orthonormal base, 280
order five, Legendre polynomial, *97*
order *k*
approximation, truncation error, 27
increase, error decrease, 42
order $O(x^8)$, term, 31
order $O(h)$, truncation error, *44*
OrdinalEncoder transformer
operation, 375
usage, 375–377
ordinal encoding, simplicity, 377
ordinary differential equation (ODE), 291–292
explicit ordinary differential equation (ODE), 291
first order ordinary differential equation, 323
first order ordinary differential equation, solutions, *305*
first order system, 336
general *n*-order ordinary differential equation, 326

higher order ordinary differential equations, 326–330
implicit ordinary differential equation (ODE), 292
theory, Lipschitz functions (relationship), 304
original data, plotting, 206
orthogonal base, formation, 170
orthogonality, 153
condition, translation, 208
definition, usage, 162
generalized angle, relationship, 160–162
idea, usage, 208
orthogonalization, linear regression, 210–212
orthogonal Legendre polynomials, usage, 282
orthogonal polynomials, construction, 280
orthogonal projection, 162–170, 183–184
matrix, 164–170
matrix, component selection, 166
orthogonal vector base
construction, 174–175
formation, 172
orthogonal vectors, 161
usage, 153
orthonormal base, 162
advantages, 169–170
selection, 209, 212
usage, 169
orthonormal, term (origin), 162
outcome, defining, 353
overfitted model, 403
over-fitting, 393–394, 401–402
avoidance, 428–429

P

PA=LU decomposition, 125–127
usage, 127–128, 145
performance (ML component requirement), 350–352
performed iterations, true error, *65*
permutation matrices, usage, 125
perturbated gravitational field, orbital motion, 333–335
phase space representation, 337
physical quantities, representation, 18, 132
Picard iteration algorithm, 311
pipeline (class), 371
Pipeline (single step), 379
pipelines, 371–372
advanced pipelines, 378–381
auxiliary pipeline, usage, 379–380
building, 379
model pipeline, inclusion, 414
numeric-transformer pipeline, 380
pivot equation, 110
planetary motion, Kepler's third law, 204–207
plane, vector spanning, *193*
PolynomialFeatures (class), 371
polynomial feature transformation, 368
polynomial fit, example, *370*
polynomial interpolation, 223
basis, 243–246, 287–288
definitions, 223–226, 241
error, expression, 252
problem, 224
polynomial model of degree, *391*, *392*
polynomial models
adjustment, 223
dataset training, relationship, *358–359*
generalization errors, *393*
training, attempt, 403–404
training errors, *393*
polynomial of degree
trained model, *394*
usage, 396
polynomial regression, 197–198
example, 367–368
polynomials
evaluation, 225–226
evaluation, Octave (usage), 225
Fourier polynomials function approximation, 214–217
function approximation, 212–214
interpolation verification, Octave (usages), 232
quadratic polynomial, finding, 213–214
representation, 224
roots, finding, 101
polynomial transformation, 368–371
positive definite function, 158
positive root, finding, 99–101
power law, 204–207, 319
fitting, 264, 322
usage, 269
power operator
display, 94
element-by-element power operator, usage, 94
regular power operator, 94
power series, 274
convergence, radius, 26
precision, 7–8
accuracy, distinction, 7
predictions
performing, trained model (usage), 369
predictions, making, 365–366
predict method
input parameter, 366
usage, 407

predictor, 311
predictor corrector solvers, 311
predict_proba method, 419–420
preprocessor pipeline, setup, 381
process
  expected value, 7
  variance, precision, 8
projection matrix, usage, 165
propagation, functions (considerations), 19
pseudo-code, usage, 176
Python classes, derivation, 374
Python function, defining, 388–389
Python-Mixin, understanding, 374
Python pandas library, usage, 382
Python structure, 413
Python syntax, issues, 370

Q

QR decomposition, 178–182, 184–185
  computation, 182
  least square linear regression, QR decomposition, 199
  reduced QR decomposition, 181–182
  usage, 198–200
QR factorization, 180
  reduced QR factorization, 181
quadratic convergence, 86
quadratic model, adjustment, 368
quadratic polynomial
  continuous line, *214*
  finding, 213–214
quadrature
  Gauss-Legrendre quadrature, 282
  Gauss quadrature, 279–286
  Gauss quadrature formulas, 281–285
  rule, 257
  weights, sum, 247
quadrature formula
  characterization, 246
  error, reduction, *279*
  names, 251
  obtaining, 245
  precision, 246, 254
  usage, 243
  weights, 247
  writing, 248
quadrature formula error, 244
  development, *245*
  surface, equivalence, *245*
quadrature methods, 320
  Romberg quadrature method, 275–278
  true errors, *273*
quantitative prediction, 423

R

R^3, orthogonal projections, 165–167
radius of convergence, 27
RandomForestClassifier (SciKit-Learn class), 436
random forests, 434–438
  model, cross-validation, *436*
  properties, 437–438
  tree-trained random forest classifiers, example, *436*
ratio convergence, Octave calculations, 89
real function
  approximation, problem, 213
  root, 51
real number
  consideration, 23
  fixed-point, equivalence, 52
  value, 21
real-valued continuous function, 52
  input, 59
  norms, defining, 156–157
  target accuracy, relationship, 54
real-valued inner space
  linearly independent vectors, consideration, 171–172
  sub-space, consideration, 163
real-valued random variable, process, 7
real-valued vector spaces, 153
  concept, exploitation, 224
  consideration, 155–156
  counterparts, 167
  geometrical problem, 210
  orthogonal vector base, construction, 174–175
  theory, 108
reduced QR decomposition, 181–184
reference process, 388–389
regression
  approach, usage, 223
  binary logistic regression classifier, 418
  goal, interpolation goal (comparison), 223
  LASSO regression, 403–405
  least square regression, 189
  linear least square regression, 186
  linear regression, QR decomposition usage, 199–200
  logistic regression, 415–422
  multinomial logistic regression, 420–421
  polynomial regression, 197–198
  softmax regression, 420–421
regression models, 361
  features, 361–363
  labels, 361–363
  weights, 361–3636

regression problems, 186–190, 219
  adjustments, requirement, 186
  parameters, tuning, *187*
regression trees, 423–431
  fish dataset training, *428*
  predictions, *429*
  trained regression tree, *424*
  usage, 428–429
Regula Falsi, 58
regularization, 402–405
  elastic net regularization, 402
  LASSO regularization, 403–405
regular power operator, 94
relative backward errors
  evaluation, 132
  low value, 135
relative condition number, 33
relative errors, 35
  approximations, usage, 11
relative forward error
  calculation, 147
  estimation, 145
  linear equations system, 139–141
  true relative forward error, 141
relative round-off error, 144
RepeatedKFold (SciKit-Learn function), 401
residual errors, *367*
  computation, 188, 198
residuals
  calculation, 206
  computation, 219
  computation, transformed data (usage), *203*
  display, 202
  presence, *218*
Richardson error estimations, computation, 321, *321*
Richardson error formula
  application, 334
  hypothesis, 266–267
  range, validity, 267–269
Richardson extrapolation, 274–275
  application, 276, 277
Richardson's error formula
  graphs, *268*, *271*
  usage, 266–268, 278, 321, 331, 334
  validity, 269
Richardson's formula
  usage, 269, 286
  validity, 269, 287
Richardson's method, usage, 266, 338
Richardson's truncation error estimation, 265
  usage, 320
right-hand side vector, 106–107
  calculation, 194
  relative round-off error, 144
RK2 IvP solvers, 316
RK2 methods, 316
RK4 function (Octave), 321
RK4 method, 319, 320, 322
  classical RK4 method, Octave implementation, 331
RK4 solver
  combination, 342
  usage, *335*
RK5 IvP solver, combination, 342
RKF45 IvP solver, 342–343
RKN IvP solver, expression, 316
R^m, representations (creation)(, 212
R^n
  column vectors, 212
  match, 179
  natural vector base, 154
  V, inner product, 178
Romberg integration, 274–278, 289
  basis, 277
Romberg quadrature method, 275–278
root(s)
  bracketing, 76
  first root, approximation, 98–99
  location, 60–61
  multiple root, usage, 88
  multiplicity, 80, 95–96
  multiplicity, estimation, 87–89
  positive root, finding, 99–101
  presence, *53*
  real number, 51
  root-solving problem, 70
  single roots, usage, 98
  triple root, 87
root approximation, 60–61, 98, 101
  bisection algorithm, usage, **57**
  computation, 55–56
  estimation, backward error basis, 93–94
  false position algorithm, **61**
root-finding problem
  condition number, 92
  equations, 102
  Newton-Raphson algorithm, application, *71*
  Newton-Raphson algorithm, convergence plot, *88*
  usage, 93–94
root mean squared error (RMSE), 190, *367*, 401–402
  bias measurement, 356
  calculation, 390
  calculation, SciKit-Learn implementation, 366

equations, 203, 206
usage, 399–400
validation/training errors, *402*
root mean square (RMS) value, 243
round-off errors, 15–22
absence, 18
avoidance, 23
determination, 36
doubling estimation, 40
effect, minimization, 202
error, creation, 147
example, 16, 39–42
formula sensitivity, 252
impact, 91, 148
increase, 331
indications, 39–40
introduction, 35
limits, imposition, 65
presence, assumption, 141, 143
propagation, 18–20
propagation, functions (consideration), 19
trade-off, 42–46, *45*, 48–49
trade-off, example, 42–43, 44–46
usage, 143
round-off errors, estimation, 65–66
ability, 41–42
function evaluation, usage, 37–42
R" representations, usage, 167–168
Runge-Kutta IvP solver
example, 309, 313
numerical ivP solver, equivalence, 310
schematic, *310*
usage, 342
Runge-Kutta methods, 309–323, 344–345
fourth order Runge-Kutta methods, 316–319
second order Runge-Kutta methods, 310–316
Runge-Kutta solver, order (rank), 316
Runge phenomenon, 238–239, 255
results, *239*

S

sample, term (usage), 361
satellite position, Richardson error estimates, *335*
SciKit-Learn
feature transformer offering, 368
library, conventions (usage), 362
models, implementation, 365–366
OrdinalEncoder, 375
pipeline introduction, 371
Python SciKit-Learn, 364
second-level Newton differences, 231
second order Runge-Kutta methods, 310–316
self-teaching computers, term (usage), 349
semi-log plot, usage, 74–75, *75*, 82
series of approximations, generation, 84
set tolerance, global truncation error (contrast), 340
sigmoid function, 416, *416*
signals, RMS value, 243
significance, loss, 19–20
significant digits, 18
cancellation, 39
concepts, understanding, 18
example, 18
Simpson's-1/3 composite method, usage, 268
Simpson's-1/3 composite rule
panel numbers, usage, 262
usage, 285, 286, *286*
Simpson's-1/3 rule, 250, 254
application, 264
application, performing, 261
composite Simpson's-1/3 rule, *256*
precision, 265
true absolute error, *262*, *263*, *264*
usage, 269, 286
sine function
interpolation polynomial, *237*
sine function, interpolating, 235–238
sine x, first derivative, 273
single roots, usage, 98
$sin(x) = x$ equation, solving (example), 2
softmax function, 421
softmax regression, 420–421
solution (accuracy), digits (requirement), 144–145
square error (SE), 187, 424–425
data points, consideration, 188–189
expected squared error (E_D), 354
minimization, 196
square root, 198
square matrix, decomposition, 120
square root approximation
fixed-point algorithm, usage, 63
fixed-point iteration, 63–69
square root function
round-off errors, 38–42
Taylor series expansion, 29–30
standard inner product, 158
standard normal distribution
interpolating surface, 239–241
Z score, *241*
starting interval, initial guess, 60
step size, 309
fixed step sizes methods, problem, 338
reduction, 299
usage, 319, *336*

sub-datasets, 434–435
sub-intervals
cutting, 60
production, 59
size, equivalence, 58
submultiplicative property, 133
sub-space
consideration, 165
dimension, 179
general vector, 208
gram-Schmidt orthogonalization, 173–174
orthogonal projection, 211
orthogonal projection matrix, 165
spanning, 171–172, 173
vector, 163
vector, construction, *171*
"sufficiently close" (meaning), 67
sum(f(x)) function (Octave), 259
summation, decrease, 25
sun/planet, mean distance (log-log plot), *205*
switching operations, usage, 125
symmetric function, 158
symmetric matrix, 194
synthetic dataset
creation, 374–375
depiction, 369, *370*
generation, 395
usage, 372, *430*

T

target accuracy, input, 54, 59
targeted diameter, average (measurement distance), 7
task, improvement (achievement), 350
Taylor polynomial of order, 24
Taylor remainder, basis, 25
Taylor remainder of order, 24
Taylor series
approximation, *27*
cosine Taylor series, truncation error, 31
increase, order *k* (decrease), 28
order, increase, 28
round-off error trade-off, example, 42–43
$T_k(1)$, numerical evaluation, 43
truncation error trade-off, example, 42–43
usage, 25
Taylor series expansion
construction, 47
increase, *24*
order $k = 6$, 31
orders, increase, *27*
theorem, 23–24
usage, 42, 47, 69, 214, 293
Taylor series truncation error
behavior, 29–30
usage, 23–28
Taylor theorem, graphical illustration, *24*
test_score (SciKit Learn), 400
test sets, 394–397
metrics, **396**
performance, evaluation, 395
third-level Newton differences, 231
time evolution, numerical approximations, *327*, *329*
tolerance (*TOL*), 340, 341
TOL target accuracy, 55–56, 58
total error, *45*
determination, truncation errors (usage), 269
dominance, 269
truncation error, contrast, 43
trained models, 354
examples, *355*
polynomial of degree, *394*
usage, 369
trained multinomial logistic regression classifier, confusion matrix, *422*
trained polynomial models
example, *397*, *404*
LASSO regularization, usage, *404*
trained regression tree, example, *424*, *430*
training errors, 391–394, 398, *402*
behavior, contrast, 393–394
calculation, 398
computation, MSE (usage), 392
polynomial models, *393*
training sets, 394–397
metrics, **396**
partitioning, 397–398
usage, 395–396
train_test_split (SciKit-Learn function), 395
trajectory, ellipsis (shape), 334–335
transformation matrices, identification, 113
transformations, application, 378–381
transformed data, usage, *203*
TransformerMixin, usage, 2374
transformers
custom transformers, 372–375
feature transformer, SciKit-Learn offer, 368
operation, verification, 385
OrdinalEncoder transformer, 375–377
trapezoidal composite rule, usage, *286*
trapezoidal method
precision, 265
truncation error, 275
trapezoidal rule, 245–246
application, 254–255, 257
composite trapezoidal rule, *256*, 260–261

open trapezoidal rule, 248
true absolute error, *262*, *263*, *264*
truncation error, 254
usage, 246, 277, 312
weights, computation, 2347
trapezoidal (function), usage, 267
trapezoid IvP solver, 311, 314
order, rank, 316
usage, *313*, *316*, *329*, 330, *336*
trapezoid method, schematic, *310*
tree-based algorithms, 423
values, choices, 428–429
tree-trained random forest classifiers, example, *436*
triangular matrix, 230
trigonometric functions
coefficients, selection, 215
set, consideration, 215
trigonometry, usage, 46
triple root, 87
true absolute error, *75*
display, 44
true errors, 5
calculation/computation, 262, 273, 318
decrease, 65, 71
display, 74–75, *75*
equations, systems (true errors), 129
graphing, 261
increase, 45
log-log plot, display, 319
numerical approximations, *44*
over-estimation, 237
quadrature methods, *274*
ratio, 79
usage, *43*
true negative (TN) value, 410–411
true positive (TP) value, 410–411
true relative forward error, 140–141
true root
calculation, 73
knowledge, 74
knowledge, absence, 95–96
true solution
knowledge, 71
words, usage, 3
true truncation errors
circles, obtaining, *30*
computation, 30
true value, knowledge (absence), 87
truncation error-free algorithm, usage, 144
truncation errors, 15, 22–32
approximation, 25, *30*, 43
behavior, 28–32
behavior, control, 30
composite Newton-Cotes formulas, 259–265, **261**
composite trapezoidal method, 271–274
composite trapezoidal rule, 260
depiction, plots (usage), 30
difference, 23
estimated truncation error, 26
exponential function, series expansion, 23
expression, 252, 266
global truncation errors, 301–303
impact, 269
increase, 28
introduction, 22
limits, 43
local truncation errors, 301–303
maximal values, 255
Newton-Cotes formula, 252–255, **253**
observation, 30
R(x) = |cos x – $T_6(x)$, 31
Richardson's truncation error estimation, 265–266
round-off errors, avoidance, 23
similarity, 30
total error, contrast, 43
trade-off, 42–46, *45*
trade-off, example, 42–43, 44–46, 48–49
trapezoidal rule, 254
true truncation errors, computation, 30
truncation errors, estimation, 25, *30*, 254–255, 267–269
composite quadrature formulas, 265–271
Richardson's error formula, usage, *268*, *271*
Richardson's formula, usage, *286*
Richardson's method, usage, 266
twice continuously differentiable function, 72, 73
two-point Gauss quadrature formula, 283
two-step sizes, Euler's IvP solver (usage), *327*

U

unavoidable output error
Δy, first order estimation, 35
estimation, 35
underfitted model, 403
under-fitting, 359, 393–394
avoidance, 428–429
undistinguishable numbers, 20–22
machine epsilon, value, 21
universal gravitational constant, 329
unknown
coefficients, elimination, 114

grouping, 107
upper bound, display, 305
upper diagonal matrix (computation), naïve Gauss elimination algorithm (usage), 120
upper diagonal system, 108–109
form, understanding, 211
transformation, 119

V

validation error, 398, *402*
calculation, 398
evolution, 401–402
results, 400
values
Octave attributes, 16
prediction, 370
range, plotting, 30
values, algorithm generation, 7
Vandermond interpolating polynomial, evaluation, 230
Vandermond interpolation, 223, 226–228, 241–242
dataset, consideration, 227–228
polynomial, *228*
usage, 236, 238
Vandermond matrix, 147–149, 197
van der Pol, Balthasar, 336
Van der Pol equation
numerical approximation, estimated errors, *337*
numerical oscillation, *336*, 337
Van der Pol oscillator, 336
errors, growth, 338
solution, approximation, 342–343
solution, approximation (global truncation errors), *343*
variables
power-law relation type, 204–205
symmetry, 334
variance
bias variance decomposition, 357
defining, 356
Gaussian white noise, *354*
measurement, 397
variance-bias trade-off, 353–359, 392
vector base, 154, 162
selection, 224
standard representation, usage, 208
usage, 230
vectors
abstract vector spaces (calculations), Octave (usage), 154–155
assignment, one-hot encoder (usage), 377
column vector, 107, 192
column vector, return, 295
construction, *171*
designation, 153
eigenvectors, association, 132–133
indexes, Octave numbering, 89
index numbering, Octave initiation, 65
inner product, 157
inner product, defining, 158
iterations storage, 64
linear independence, 170
Manhattan norm, 130
natural vector base, 154
norms, 130–131, 156–157
notations, 323
notations, usage, 324
Octave numbering, 87
orthogonal vectors, usage, 153
selection, 122
spanning, *193*
vector space. *see* abstract vector spaces
consideration, 168–169
L^p-norm, 157
ordered orthonormal base, 208
spanning, *166*

W

weather data
temperature, *218*
weather data, modeling, 217–219
whitefish
data, **363**
predicted weights, actual weights (contrast), *367*
Wolpert, David, 351
worst-case situation, estimation, 144

X

X_pred (feature matrix), 372
$|x_r - r|$, size, 51

Z

ZΛ decomposition, 179
Z score, *241*